# HANDBOOK OF RESEARCH FOR FLUID AND SOLID MECHANICS

Theory, Simulation, and Experiment

# HANDBOOK OF RESEARCH FOR FLUID AND SOLID MECHANICS

## Theory, Simulation, and Experiment

*Edited by*

**Kaveh Hariri Asli, PhD**

**Soltan Ali Ogli Aliyev, DSc**

**Sabu Thomas, PhD**

**Deepu A. Gopakumar**

AAP | APPLE ACADEMIC PRESS

Apple Academic Press Inc.
3333 Mistwell Crescent
Oakville, ON L6L 0A2 Canada

Apple Academic Press Inc.
9 Spinnaker Way
Waretown, NJ 08758 USA


Printed in the United States of America on acid-free paper
International Standard Book Number-13: 978-1-77188-501-0 (Hardcover)
International Standard Book Number-13: 978-1-315-36570-1 (eBook)



**Library and Archives Canada Cataloguing in Publication**

Handbook of research for fluid and solid mechanics : theory, simulation, and experiment / edited by Kaveh Hariri Asli, PhD, Soltan Ali Ogli Aliyev, DSc, Sabu Thomas, PhD, Deepu A. Gopakumar.
Includes bibliographical references and index.
Issued in print and electronic formats.
ISBN 978-1-77188-501-0 (hardcover).--ISBN 978-1-315-36570-1 (PDF)
1. Fluid mechanics--Mathematical models--Handbooks, manuals, etc. 2. Solids--Mechanical properties--Mathematical models--Handbooks, manuals, etc. 3. Heat--Transmission--Handbooks, manuals, etc. 4. Structural analysis (Engineering)--Handbooks, manuals, etc. I. Hariri Asli, Kaveh, editor II. Aliyev, Soltan Ali Ogli, editor III. Thomas, Sabu, editor IV. Gopakumar, Deepu A., editor
QA901.H36 2017    532    C2017-905340-X    C2017-905341-8

**Library of Congress Cataloging-in-Publication Data**

Names: Hariri Asli, Kaveh, editor.
Title: Handbook of research for fluid and solid mechanics : theory, simulation, and experiment / editors, Kaveh Hariri Asli, PhD [and three others].
Description: Toronto : Apple Academic Press, 2018. | Includes bibliographical references and index.
Identifiers: LCCN 2017035845 (print) | LCCN 2017036737 (ebook) | ISBN 9781315365701 (ebook) | ISBN 9781771885010 (hardcover : alk. paper)
Subjects: LCSH: Fluid mechanics. | Mechanical engineering. | Hydraulic engineering.
Classification: LCC TA357 (ebook) | LCC TA357 .H2935 2018 (print) | DDC 620.1/05--dc23
LC record available at https://lccn.loc.gov/2017035845

Apple Academic Press also publishes its books in a variety of electronic formats. Some content that appears in print may not be available in electronic format. For information about Apple Academic Press products, visit our website at **www.appleacademicpress.com** and the CRC Press website at **www.crcpress.com**

# ABOUT THE EDITORS

**Kaveh Hariri Asli, PhD**

Kaveh Hariri Asli, PhD in Mechanical Engineering and Energy Conversion, has consulted for a number of major corporations. He has over 30 years of experience in practicing mechanical engineering design and teaching. Professor Hariri Asli is the author of many books and over 100 international journals and conference papers in the fields of fluid mechanics, hydraulics, automation and control systems, published by CRC Press, Springer, Nova, etc. He is a member of the editorial boards of many international engineering journals.

**Soltan Ali Ogli Aliyev, DSc**

Soltan Ali Ogli Aliyev, DSc, is Deputy Director of the Department of Mathematics and Mechanics at the National Academy of Science of Azerbaijan (AMEA) in Baku, Azerbaijan. He served as a professor at several universities. He is the author and editor of several books as well as the author of a number of papers published in various journals and conference proceedings.

**Sabu Thomas, PhD**

Sabu Thomas, PhD, is the Director of the International and Inter University Centre for Nanoscience and Nanotechnology and full professor of Polymer Science and Engineering at the School of Chemical Sciences of Mahatma Gandhi University, Kottayam, Kerala, India. Professor Thomas has published over 600 peer-reviewed research papers, reviews, and book chapters. He has co-edited 50 books published by the Royal Society, Wiley, Woodhead, Elsevier, CRC Press, Springer, Nova, etc. He is the inventor of six patents.

**Deepu A. Gopakumar**

Deepu A. Gopakumar is a Research Scholar at the University of South Brittany, Lorient, France. He is engaged in doctoral studies in the area of nanocellulose-based polymer membranes for water treatment. He has also conducted research work at Federal University of Uberlandia, Brazil.

# CONTENTS

# LIST OF CONTRIBUTORS

**Kaveh Hariri Asli**
Department of Mathematics and Mechanics, National Academy of Science of Azerbaijan'AMEA', Baku, Azerbaijan. E-mail: hariri_k@yahoo.com

**Hossein Hariri Asli**
Civil Engineering Department, Faculty of Engineering, University of Guilan, Rasht, Iran. E-mail: hh_asli@yahoo.com

**Soltan Ali Ogli Aliyev**
Department of Mathematics and Mechanics, National Academy of Science of Azerbaijan 'AMEA', Baku, Azerbaijan. E-mail: Soltanaliyev@yahoo.com

**Ramdane Boukellif**
Institute of Mechanics, University of Kassel, Kassel 34125, Germany. E-mail: ramdane.boukellif@uni-kassel.de

**Paul Cunningham**
Department of Aeronautical and Automotive Engineering, Loughborough University, Loughborough, Leicestershire, LE11 3TU, UK

**A. K. Haghi**
University of Guilan, Rasht, P.O. Box 3756, Iran

**Christopher Martin Harvey**
Department of Aeronautical and Automotive Engineering, Loughborough University, Loughborough, Leicestershire, LE11 3TU, UK

**Sh. Maghsoodlou**
University of Guilan, Rasht, P.O. Box 3756, Iran

**S. Muralidhara**
Department of Civil Engineering, BMS College of Engineering, Bangalore, India

**Sriman Narayan, H. N**
Department of Civil Engineering, BMS College of Engineering, Bangalore, India

**S. Poreskandar**
University of Guilan, Rasht, P.O. Box 3756, Iran

**B. K. Raghu Prasad**
Department of Civil Engineering, Indian Institute of Science, Bangalore, India

**Vikram Singh Raghuvanshi**
Department of Materials Science and Metallurgical Engineering, MANIT, Bhopal, Madhya Pradesh, India

**Andreas Ricoeur**
Institute of Mechanics, University of Kassel, Kassel 34125, Germany

**C. Sasikumar**
Department of Materials Science and Metallurgical Engineering, MANIT, Bhopal, Madhya Pradesh, India

**Shahrukh Shamim**
Department of Materials Science and Metallurgical Engineering, MANIT, Bhopal, Madhya Pradesh, India

**Gaurav Sharma**
Department of Materials Science and Metallurgical Engineering, MANIT, Bhopal, Madhya Pradesh, India

**Simon Wang**
Department of Aeronautical and Automotive Engineering, Loughborough University, Loughborough, Leicestershire, LE11 3TU, UK

# LIST OF ABBREVIATIONS

| | |
|---|---|
| 2D | two dimension |
| AZNP | average zone night pressure |
| BL | basic fuzzy logic |
| CCVD | catalytic chemical vapor deposition |
| CFL | compensatory fuzzy logic |
| CMOD | crack mouth opening displacement |
| CNT | carbon nanotubes |
| CV | control volumes |
| CVD | chemical vapor deposition |
| DCB | double cantilever beams |
| DMA | district meter area |
| DNS | direct numerical simulation |
| ELL | economic level of leakage |
| EPS | extended period simulation |
| ERR | energy release rate |
| FD | finite differences |
| FE | finite elements |
| FEM | finite element method |
| FPZ | fracture process zone |
| FRDB | fuzzy relational database |
| FV | finite volume |
| FVM | finite volume method |
| GIS | geographical information system |
| HSC | high strength concrete |
| LEFM | linear elastic fracture mechanics |
| LES | large eddy simulation |
| LOS | level of service |
| MMC | metal-matrix composite |
| MNF | minimum night flow |
| MOC | method of characteristics |
| MWCNT | multi-walled carbon nanotubes |
| NE | new era |
| NRW | non-revenue water |

| | |
|---|---|
| NSD | nanoscale dispersion |
| PBA | polyester binder-assisted |
| PDD | pressure dependent demands |
| PLC | programmable logic control |
| PM | Powder metallurgy |
| PRV | pressure reduction valve |
| PSF | plasma spray forming |
| PSO | particle swarm optimization |
| PVA | polyvinyl alcohol |
| QUAD4 | four-node quadrilateral |
| RANS | Reynolds-averaged Navier–Stokes |
| RTC | real-time control |
| SFRHSC | steel fiber reinforced high strength concrete |
| SIF | stress intensity factors |
| SPS | spark plasma sintering |
| SWCNT | single-walled carbon nanotubes |
| UFW | unaccounted for water |
| WCM | wave characteristic method |

# PREFACE

This book is intended for an undergraduate course with experimental approaches in fluid mechanics, and there is plenty of material for a full year of instruction. The authors cover the first ten chapters as part one and the second four chapters as part two. The more specialized and applied topics, for example, the discussion of numerical methods, or computational fluid dynamics (CFD), inviscid and viscous, steady and unsteady, have been greatly expanded on from Chapters 1 to 10 for students of engineering. The informal, student-oriented style is used and has the flavor of an interactive lecture by the authors.

The focus of the both parts of book is on practical and realistic fluid engineering experiences. A number of photographs and figures are included especially to illustrate new design applications and new instruments. These comprehensive problems grow and recur throughout the book as new concepts arise. The design projects at each chapter allow the students to set sizes and parameters and achieve good design with more than one approach.

New to this book, and to any fluid mechanics book, is a special section on Engineering Equation Solver (EES), which is keyed to many fluid engineering experiences throughout the book. The authors find EES to be an extremely attractive tool for applied engineering problems. Not only does it solve arbitrarily complex systems of equations, written in any order or form, but it also has property evaluations (density, viscosity, enthalpy, entropy, etc.), linear and nonlinear regression, and easily formatted parameter studies and publication-quality plotting.

Part one, which is the discussion of the fluid mechanics and heat transfer has been included to address the uncertainty of engineering. The energy equation discussion and the Bernoulli's equation comes last, after control-volume mass, linear momentum, angular momentum, and energy studies, although some texts begin with an entire chapter on the Bernoulli equation. A few inviscid and viscous flow examples have been added to the basic partial differential equations of fluid mechanics. In both parts of the book more extensive discussion continues to be more successful when one selects scaling variables.

# INTRODUCTION

This book uses many computational methods to address mechanical engineering problems. The proposed methods allowed for any arbitrary combination of devices in the system. Methods are used by scale models and a prototype system. This is not only a platform for solving of problems, but there is also a wealth of information available to help address various technical aspects of troubleshooting of mechanical system failure.

The user will find key websites cited throughout the book, which are useful for equipment selection as well as for troubleshooting mechanical operational problems. Most chapters include a section of recommended resources that the authors have relied upon in their own consulting practice over the years, we believe you will also.

Although our intent was not to create a college textbook, there is value in using this volume with engineering students, either as a supplemental text or as a primary text on fluid and solid mechanics technologies. If used as such, instructors will need to gauge the level of understanding of students before specifying the book for a course, and will need to integrate the sequence and degree of coverage provided in this volume. admittedly, for such a broad and complex subject, it is impossible to provide uniform coverage of all areas in a single volume.

In this book computational and practical methods were used for prediction of system failure.

## CHAPTER 1

# CONTROLLING THE STABILITY OF FLUID JET IN THE ELECTROSPINNING OF FIBERS: MATHEMATICAL MODELING

S. PORESKANDAR, SH. MAGHSOODLOU, and A. K. HAGHI*

*University of Guilan, Rasht, Iran*

*Corresponding author. E-mail: akhaghi@yahoo.com

## CONTENTS

## ABSTRACT

The most vital challenge in the electrospinning process is to achieve uniform nanofibers consistently. In addition, the jet shows some unstable behavior during the process. A better understanding of electrospinning jet movement can be obtained by using modeling and simulation. The main purpose of this chapter is simulating the unstable behavior of the jet and investigating the effects of the most significant parameters (i.e., solution concentration, spinning distance, applied voltage, and flow rate) on the pathway of electrospun polyvinyl alcohol nanofiber jet using a mathematical model. It is observed that, at a constant concentration, increasing applied voltage in the long spinning distance made a shorter pathway of nanofiber jet. Also, the longest pathways were created by decreasing spinning distance and increasing charge accumulation that made the longest pathway.

## 1.1 INTRODUCTION

Nanotechnology is a unique technology that will make most products more effective and less expensive. Research and development in this field guides toward understanding and producing materials with innovative properties.[1] Additionally, the growth in the number of publications and patents in this field has become important in the recent years.[2] Electrospinning, as a significant branch of nanotechnology, must be proved as a cheap and straightforward method to create nanofibers.[3–6] These materials appear in a broad range of potential purposes (i.e., filters, tissue engineering scaffolds, and protective clothing).[7–10] Creating electrospun nanofibers is a process based upon a simple concept that produces nanofibers through an electrically charged polymer solution, explained schematically in Figure 1.1.[11] Thus, this process can be separated into four sections, as shown in Figure 1.2.

The vital challenge in this process is to perform uniform nanofibers consistently.[12,13] On the other hand, the physical characteristics of electrospun nanofibers, such as fiber diameter, depend on numerous parameters which are mainly divided into three categories: solution conditions (i.e., solution concentration), processing conditions (i.e., applied voltage, flow rate, and spinning distance), and ambient conditions (i.e., temperature and humidity).[14–15] Also, controlling the property of the nanofibers is essential for producing well nanofibers.[16] As a concept, successful electrospinning requires an understanding of the complex interaction of electrostatic fields, properties of polymer solutions, and component design and system geometry.[17] In summation, this process deserves special attention and is necessary for predictive tools to better understand the optimization and controlling process.[18] During this process, the jet shows some unstable behavior. Investigating the dynamical behavior of the jet is important for controlling and developing the process. For these reasons, modeling and simulation of electrospinning jet will provide a better understanding of the process.[19,20] In this case, the mathematical models as displayed in Figure 1.3 were utilized.

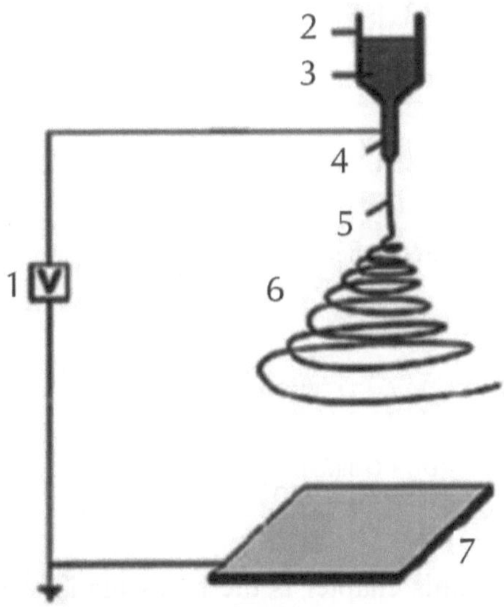


**FIGURE 1.1** Part one of standard electrospinning setup: (1) high voltage, (2) polymer, (3) syringe, (4) needle, (5) straight jet, (6) whipping instability, (7) collector.

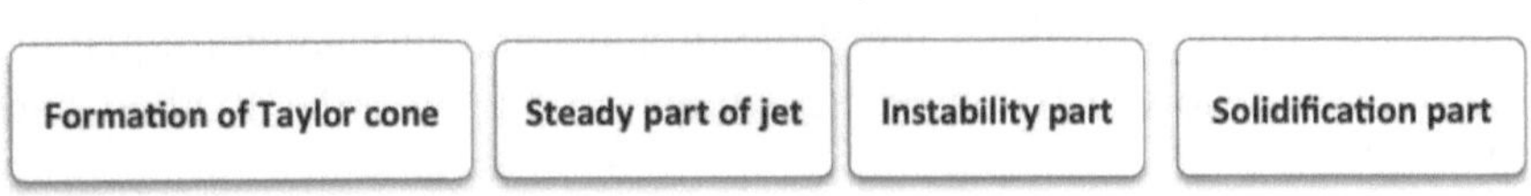


**FIGURE 1.2** Different parts of electrospinning process.

For investigating and modeling the behavior and pathways of electrospun nanofiber, bead-spring model, which describes the fiber as a chain that consisted of beads connected by springs, was proposed. Each bead has a mass $m$ and a charge $e$. Therefore, the momentum equation for the motion of the $i$th bead is shown as eq 1.1.[21]

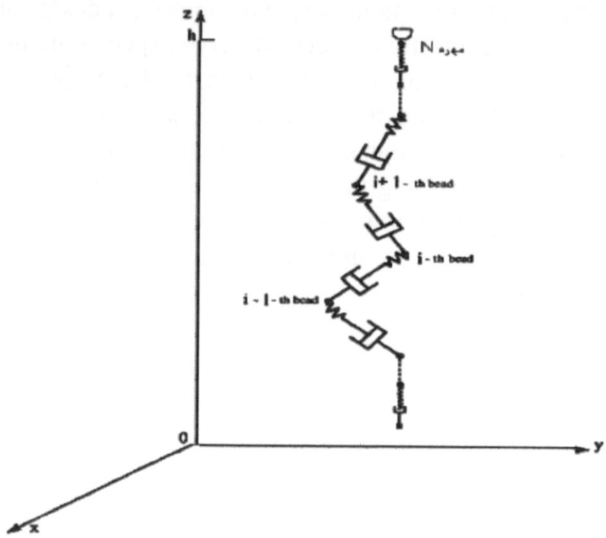


**FIGURE 1.3** Jet path in three-dimensions with beads and springs.

$$m\frac{d^2 r_i}{dt^2} = \sum_{\substack{j=1 \\ j\neq i}}^{N} \frac{e^2}{R_{ij}^2}(r_i - r_j) - e\frac{V_0}{h}\hat{k} + \frac{\pi a_{ui}^2(\bar{\sigma}_{ui} + G\ln(l_{ui}))}{l_{ui}}$$

$$(r_{i+1} - r_i) - \frac{\pi a_{bi}^2(\bar{\sigma}_{bi} + G\ln(l_{bi}))}{l_{bi}}(r_i - r_{i-1}) - \frac{\alpha\pi a_{av}^2 k_i}{\sqrt{(x_i^2 + y_i^2)}}[ix_i + jy_i]. \qquad (1.1)$$

The main aim of this chapter is the investigation of the effects of the most significant parameters (Table 1.1) on the pathway of electrospun polyvinyl alcohol nanofiber. For prediction pathways of electrospun polyvinyl alcohol nanofiber jet, according to the initial conditions obtained using the Taguchi method, the three-dimensional momentum equation (eq 1.1), is resolved (considering Runge–Kutta 4th-order method and MATLAB).

## 1.2 EXPERIMENTAL

In this work, for creating nanofibers with no beads, polyvinyl alcohol powders with high molecular weight (72,000 g/mol) and degree of hydrolysis of >98% from Merck were utilized. The experimental designs using Taguchi method were utilized for investigating the effects of factors on the pathways of creating nanofibers (summarized in Table 1.1).

**TABLE 1.1** Experiment Condition of the Effects of Selected Factors for Pathway Simulation of Polyvinyl Alcohol Nanofibers.

| Experiment Number | Factor | | | |
|---|---|---|---|---|
| | Concentration (%) | Spinning Distance (cm) | Applied Voltage (V) | Flow Rate (mL/h) |
| 1 | 8 | 10 | 15,000 | 0.2 |
| 3 | 8 | 20 | 25,000 | 0.4 |
| 5 | 10 | 10 | 20,000 | 0.4 |
| 6 | 10 | 15 | 25,000 | 0.2 |
| 7 | 12 | 15 | 15,000 | 0.4 |
| 8 | 12 | 20 | 20,000 | 0.2 |

## 1.3 RESULTS AND DISCUSSIONS

As mentioned before, the most vital challenge in this process is to achieve uniform nanofibers consistently. Also, it demonstrates that the jet is stretched as it moves downward from the initial position to the collector. For this reason, simulation electrospun nanofiber jet will provide a better understanding of the process. For prediction pathways of electrospun polyvinyl alcohol nanofiber according to the initial conditions obtained using the Taguchi method, the three-dimensional momentum equation (eq 1.1, is resolved (considering Runge–Kutta 4th-order method and MATLAB). The results of simulating pathways of electrospun polyvinyl alcohol nanofiber are shown in Figures 1.4–1.6. These figures illustrate the consequences of the model data output at various times throughout the calculation of jet pathway. Comparisons between the results of the pathways at different experimental conditions are shown in Figures 1.7 and 1.8. Also, the reasons of difference between pathways are discussed in these figures.

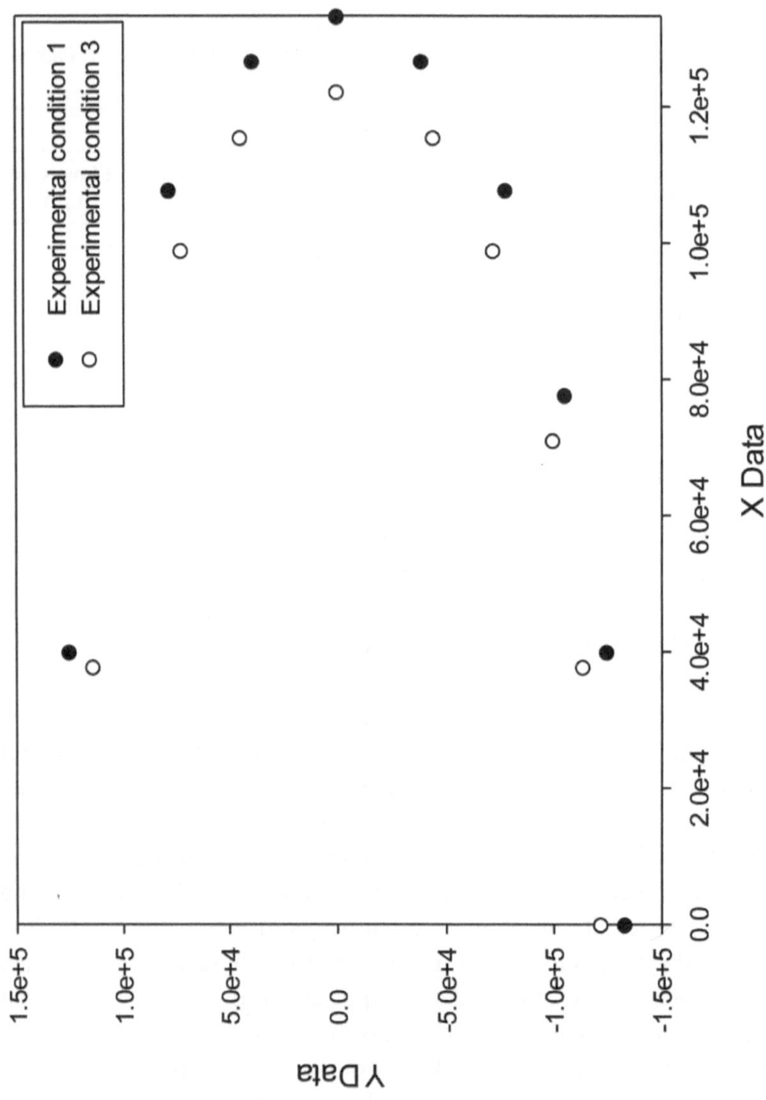


**FIGURE 1.4** Comparison of simulation pathways of polyvinyl alcohol electrospun nanofiber jet for experimental condition 1 and experimental condition 3.

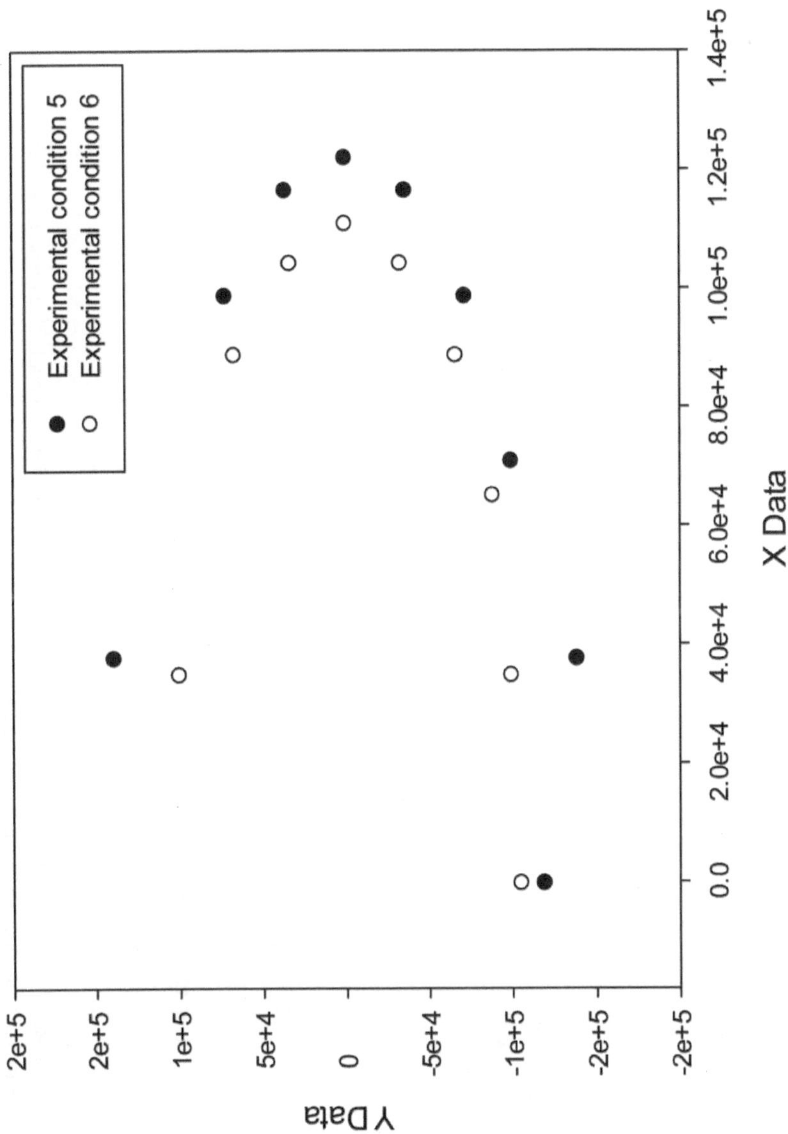


**FIGURE 1.5** Comparing simulation pathways of polyvinyl alcohol electrospun nanofiber jet for experimental condition 5 and experimental condition 6.

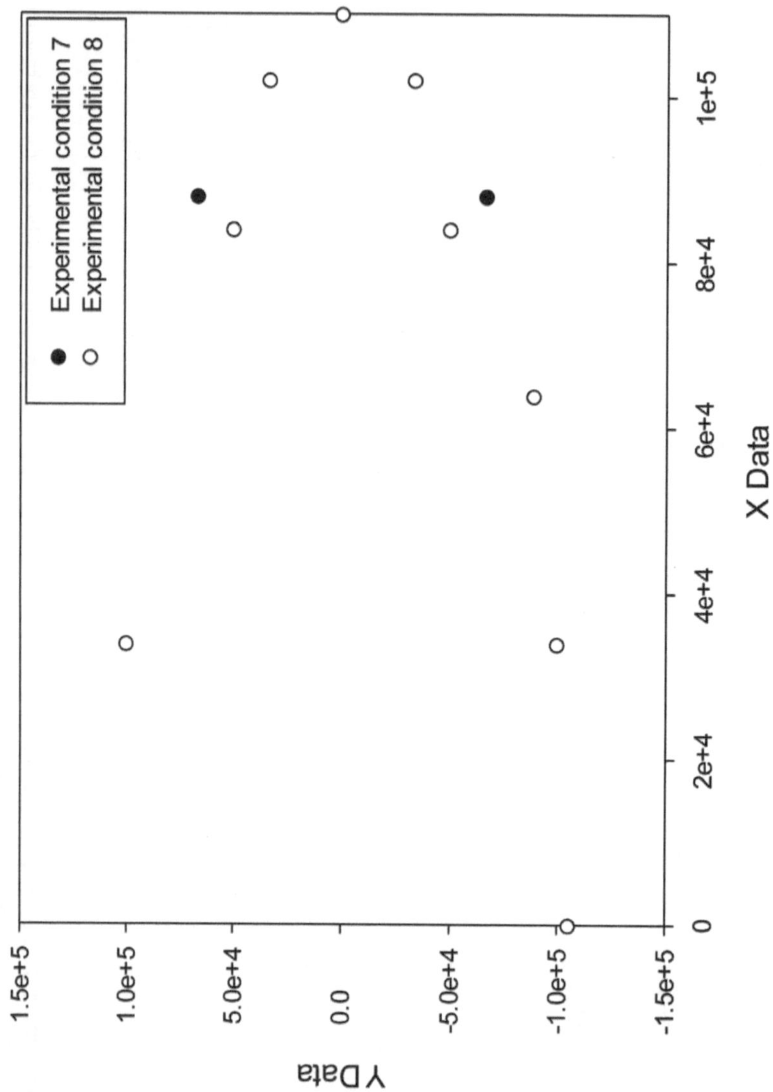


**FIGURE 1.6** Comparison of simulation pathways of polyvinyl alcohol electrospun nanofiber jet for experimental condition 7 and experimental condition 8.

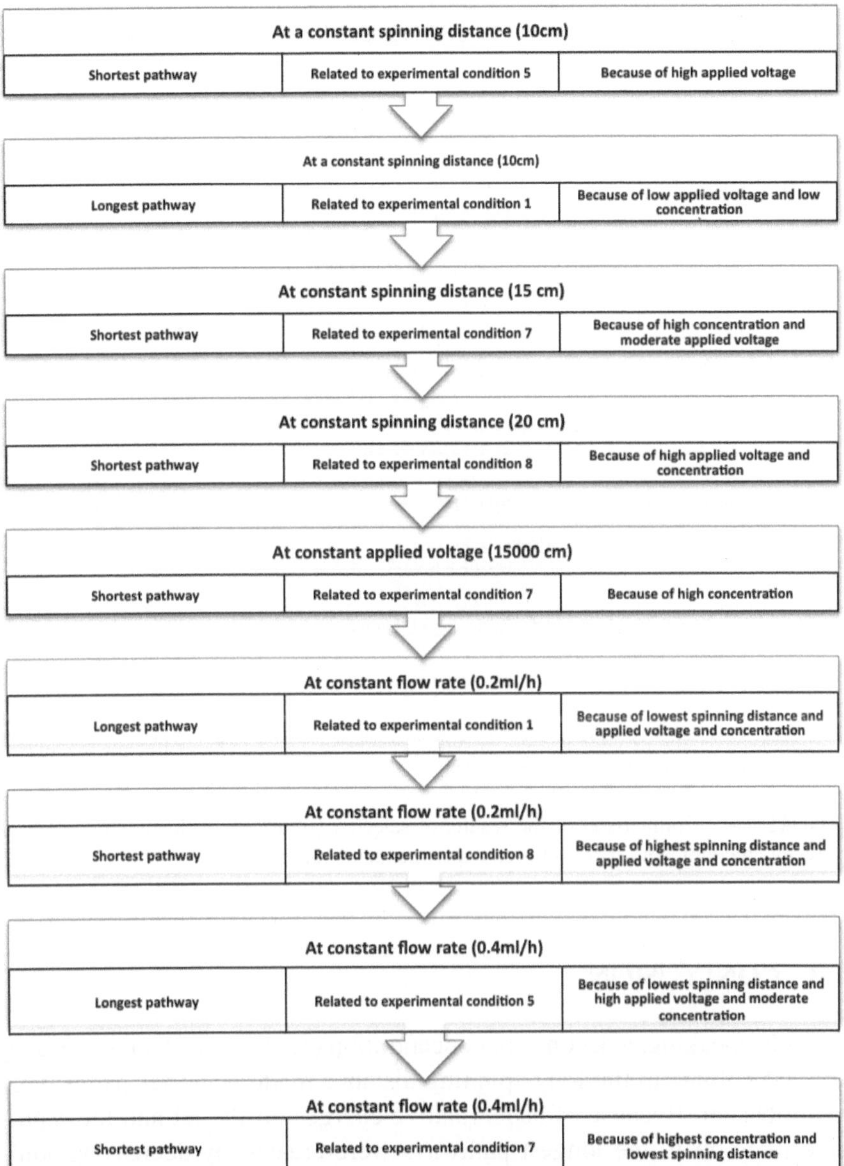


**FIGURE 1.7** Reasons of different effects of selected factors for pathway prediction of polyvinyl alcohol nanofibers between experimental conditions.

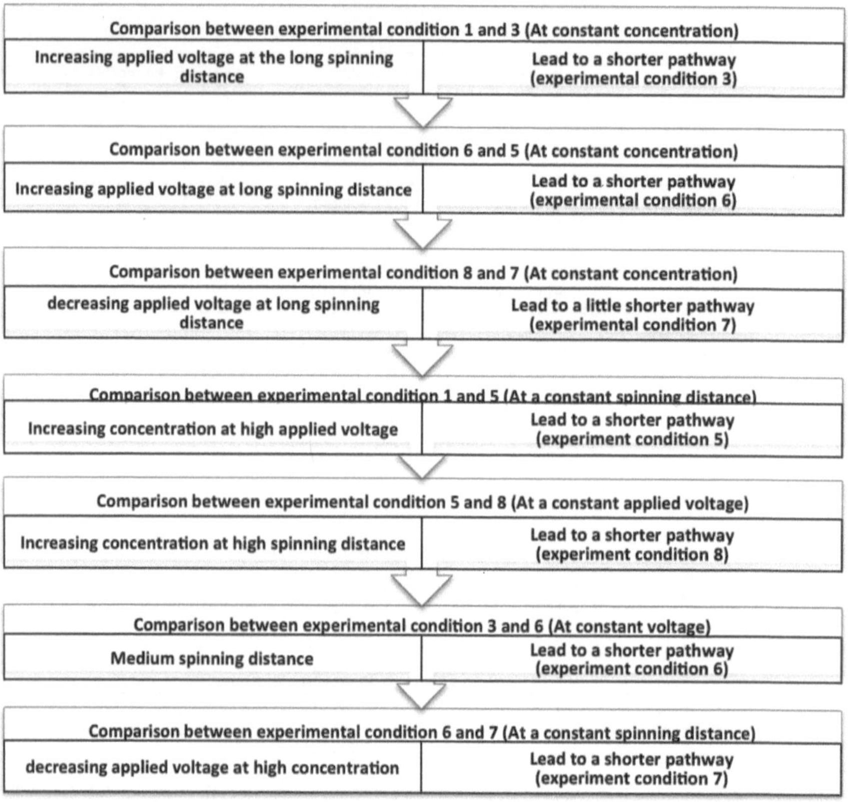


**FIGURE 1.8** Comparison of the results of selected factors for pathway simulation of polyvinyl alcohol nanofibers between experimental conditions.

## 1.4 CONCLUSION

Broadly speaking, at a constant concentration (8, 10, and 12%), increasing applied voltage in the long spinning distance made a shorter pathway of nanofiber jet. Because of accumulative charges on the jet and the repulsive force on it, the longest pathways were created by decreasing spinning distance and increasing charge accumulation that made the longest pathway. At low concentration, high-applied voltages were needed. In addition, by comparison between data and pathways, shortest pathways are led to experimental condition 8. And the longest pathways are led to experimental condition 1. In future works, the effects of these parameters on the pathways of other polymer nanofibers and also on the diameters of polyvinyl alcohol nanofibers will be investigated.

## KEYWORDS


- **electrospun nanofiber**
- **modeling**
- **simulation**
- **pathway**


## REFERENCES


1. Sawhney, A. P. S., et al. Modern Applications of Nanotechnology in Textiles. *Text. Res. J.* **2008,** *78* (8), 731–739.
2. Haghi, A. K. Electrospun Nanofiber Process Control. *Cell. Chem. Technol.* **2010,** *44* (9), 343–352.
3. Keun. S. W., et al. Effect of Ph on Electrospinning of Poly (Vinyl Alcohol). *Mater. Lett.* **2005,** *59* (12), 1571–1575.
4. Chronakis, I. S. Novel Nanocomposites and Nanoceramics Based on Polymer Nanofibers Using Electrospinning Process—A Review. *J. Mater. Process. Technol.* **2005,** *167* (2), 283–293.
5. Wu, Y., et al. Controlling Stability of the Electrospun Fiber by Magnetic Field. *Chaos Solitons Fract.* **2007,** *32* (1), 5–7.
6. Šimko, M.; Erhart, J.; Lukáš, D. A Mathematical Model of External Electrostatic Field of a Special Collector for Electrospinning of Nanofibers. *J. Electrostat.* **2014,** *72* (2), 161–165.
7. Wang, H. S.; Fu, G. D.; Li, X. S. Functional Polymeric Nanofibers from Electrospinning. *Recent Pat. Nanotechnol.* **2009,** *3* (1), 21–31.
8. Reneker, D. H.; Chun, I. Nanometre Diameter Fibres of Polymer, Produced by Electrospinning. *Nanotechnology.* **1996,** *7* (3), 216–223.
9. Ramakrishna, S., et al. Electrospun Nanofibers: Solving Global Issues. *Mater. Today.* **2006,** *9* (3), 40–50.
10. Beachley, V.; Wen, X. Polymer Nanofibrous Structures: Fabrication, Biofunctionalization, and Cell Interactions. *Prog. Polym. Sci.* **2010,** *35* (7), 868–892.
11. Pham, Q. P.; Sharma, U.; Mikos, A. G. Electrospinning of Polymeric Nanofibers for Tissue Engineering Applications: A Review. *Tissue Eng.* **2006,** *12* (5), 1197–1211.
12. De. V, S., et al. The Effect of Temperature and Humidity on Electrospinning. *J. Mater. Sci.* **2009,** *44* (5), 1357–1362.

13. Bognitzki, M., et al. Nanostructured Fibers via Electrospinning. *Adv. Mater.* **2001,** *13* (1), 70–72.
14. Tan, S. H., et al. Systematic Parameter Study for Ultra-Fine Fiber Fabrication via Electrospinning Process. *Polymer.* **2005,** *46* (16), 6128–6134.
15. Bhardwaj, N.; Kundu, S. C. Electrospinning: A Fascinating Fiber Fabrication Technique. *Biotechnol. Adv.* **2010,** *28* (3), 325–347.
16. Huang, Z. M., et al. A Review on Polymer Nanofibers by Electrospinning and Their Applications in Nanocomposites. *Compos. Sci. Technol.* **2003,** *63* (15), 2223–2253.
17. Lukáš, D., et al. Physical Principles of Electrospinning (Electrospinning as a Nanoscale Technology of the Twenty-First Century). *Text. Prog.* **2009,** *41* (2), 59–140.
18. Yarin, A. L.; Koombhongse, S.; Reneker, D. H. Bending Instability in Electrospinning of Nanofibers. *J. Appl. Phys.* **2001,** *89* (5), 3018–3026.
19. Feng, J. J. Stretching of a Straight Electrically Charged Viscoelastic Jet. *J. Non-Newtonian Fluid Mech.* **2003,** *116* (1), 55–70.
20. Liu, L.; Dzenis, Y. Simulation of Electrospun Nanofibre Deposition on Stationary and Moving Substrates. *Micro Nano Lett.* **2011,** *6* (6), 408–411.
21. Dasri, T. Mathematical Models of Bead-Spring Jets during Electrospinning for Fabrication of Nanofibers. *Walailak J. Sci. Technol.* **2012,** *9* (4), 287–296.

## CHAPTER 2

# UPDATES TO CONTROL FLUID JET IN ELECTROSPINNING PROCESS USING TAGUCHI'S EXPERIMENTAL DESIGN

S. PORESKANDAR, SH. MAGHSOODLOU, and A. K. HAGHI*

*University of Guilan, Rasht, Iran*

**Corresponding author. E-mail: akhaghi@yahoo.com*

## CONTENTS

## ABSTRACT

Controlling of fiber diameter during electrospinning is a vital aim for many applications like scaffolds for tissue engineering. A quantitative study on the effects of processing variables enables us to manipulate the properties of electrospun nanofibers. In this article, Taguchi experimental design, as a simple technique, was utilized for investigating the effects of the most important parameters, namely, solution concentration, spinning distance, applied voltage, and volume flow rate on the diameter of electrospun polyvinyl alcohol nanofibers. Various and even contradictory effects, such as increasing or decreasing the diameter of the nanofibers polyvinyl alcohol, are related to different experimental conditions on the diameter of the nanofibers. By comparison, between dates, the longest distribution of diameter electrospun polyvinyl alcohol nanofibers was obtained at the experimental condition with high concentration, moderate spinning distance, low applied voltage, and high flow rate.

## 2.1 INTRODUCTION

With the rapid development and growing role of nanoscience and nanotechnology in recent years, the field of one-dimensional nanostructures has witnessed intensive research due to the unique properties such as a high area to surface and high porosity. The research focused on applications like making scaffolds in tissue engineering.[1–3] Growth and characterization of one-dimensional fibrous materials with a diameter in the range of nanometers have attracted many scientists worldwide.[4–6] Electrospinning, as a simple, inexpensive, and straightforward method for producing fibers in nano-scale, has evinced more interest and attention among the various strategies used so far for producing nanofibers in recent years.[7–10] As indicated in Figure 2.1, the elements required for electrospinning include a polymer source, a high voltage supply, and a collector.[11]

In this process, the polymer solution receives electrical charges from a high voltage supply. By overcoming the repulsive force between the charged ions to fluid surface tension, a fine electrified of the polymer solution jet could be formed.[12–15] Then, this jet stretches and accelerates toward the collector.[1,16] As the process continues, the solvent evaporates during the jet moves, and thereafter, nanofibers are collected on a plate or collector.[14,15,17] A summary of the process of creating electrospun nanofibers is shown in Figure 2.2.

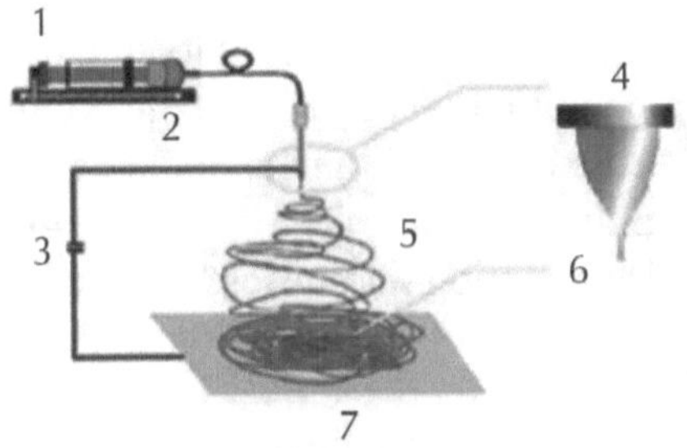


**FIGURE 2.1** Sections of electrospinning process: (1) polymer solution, (2) syringe, (3) high voltage, (4) Taylor cone, (5) whipping instability, (6) nanofibers formation, and (7) collector.

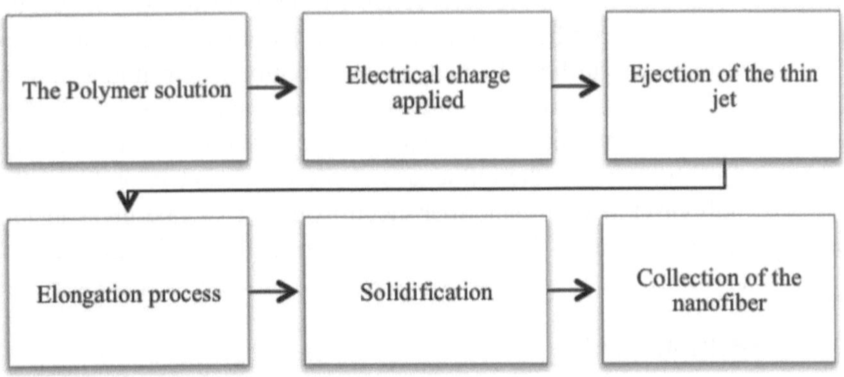


**FIGURE 2.2** Steps of electrospinning process.

Many polymer materials such as polyvinyl alcohol are used for electrospinning.[18,19] Polyvinyl alcohol is a water-soluble and hydrophilic polymer with suitable chemical and thermal stability. It is highly biocompatible and non-toxic. It can be processed easily. These properties lead to the utilization of polyvinyl alcohol in a wide range of applications such as medical and cosmetic uses. Despite the simplicity of the electrospinning technology, industrial applications of this method are still relatively uncommon, mainly due to the notable problems such as very low fiber production rate and difficulties in controlling the process.[20] The most significant challenge in electrospinning process is to achieve uniform nanofibers consistently and reproducibly.[21–24] The key part of electrospinning is how to control the process.[25] Controlling the physical characteristics of electrospun nanofibers such as fiber diameter depends on various parameters, which are principally divided into three categories: solution properties (solution viscosity, solution concentration, polymer molecular weight, and surface tension), processing conditions (applied voltage, volume flow rate, spinning distance, and needle diameter), and ambient conditions (temperature, humidity, and atmospheric pressure).[19,21,26–28] Also, the mechanics of this process require precise attention and necessary predictive tools or direction for better understanding of optimization and controlling process.[29] The design of experiment method seems a vital way to study the effects of changing a wide range of parameters. In this case, Taguchi experimental design is a simple technique which can optimize process parameters by fewer experiments.[30] Thereafter, L9 orthogonal array was chosen according to Taguchi's methodology and the results obtained from the effects of four electrospinning parameters on the diameter of electrospun polyvinyl alcohol nanofiber were investigated. Eventually, scanning electron microscopy was utilized to analyze the nanofibers morphology. Also, Image J software was utilized to measure the average polyvinyl alcohol electrospun nanofibers diameters (measure about 100 nanofibers in each sample). In future works, the influences of these parameters on the pathways of polyvinyl alcohol nanofiber electrospun jet will be investigated with the simulation result from mathematical modeling of this process.

## 2.2 EXPERIMENTAL

### *2.2.1 PREPARATION SOLUTION FOR ELECTROSPINNING*

Polyvinyl alcohol with a molecular weight of 72,000 g/mol and degree of hydrolysis of >98% was obtained from Merck and used as received. Distilled water as a solvent was added to a predetermined substance of polyvinyl alcohol powder to obtain 20 mL of solution with desired concentration. The solution was prepared at 80°C and gently stirred for 30 min to assist the dissolution. After the polyvinyl alcohol had fully dissolved, the solution was transferred to a 5 mL syringe and readied for spinning of nanofibers. The syringe including polyvinyl alcohol solution was placed on a syringe pump (New Era NE-100) used to dispense the solution at a controlled rate. A high voltage DC power supply (Gamma High Voltage

ES-30) was used to create a necessary electric field for electrospinning. The positive electrode of the high voltage supply was attached to the syringe needle and the negative electrode was connected to a flat collector wrapped with aluminum foil for accumulating nanofibers mat. In this work, for creating smooth nanofibers with no beads, polyvinyl alcohol powders with high molecular weight were utilized. In addition, ambient parameters such as temperature and humidity were constant during the electrospinning process.

### 2.2.2 EXPERIMENTAL DESIGN

The aim of the experimental design is to provide reasonable and scientific answers to research questions. In other words, experimental design comprises sequential steps to ensure efficient data gathering process, which can lead to valid statistical inferences.[31] The experimental design using Taguchi method involves arranging an orthogonal array to organize the parameters affecting the process and the levels they should be varied. It determines the factors affecting the product quality with the least number of experiments, thus saving time and resources. The four most significant and influential factors (i.e., solution concentration which changes viscosity and surface tension; spinning distance which determines the space needed for elongation and charge accumulation on the nanofiber; applied voltage which determines the elongation force and the strength of electric field; and volume flow rate which changes of the mass transfer rate) were chosen for creating smooth nanofibers with no beads in this work. Each of these factors is considered to be varied at three levels (summarized in Table 2.1) for investigating similar influences of these parameters in all experimental conditions. The final design experiments were summarized in Table 2.2.

**TABLE 2.1** $L_9$ Orthogonal Array for Selected Factors and Levels.

| Factors | Symbol | Level | | |
|---|---|---|---|---|
| | | 1 | 2 | 3 |
| Concentration (%) | A | 8 | 10 | 12 |
| Voltage (V) | B | 15 | 20 | 25 |
| Flow Rate (mL/h) | C | 0.2 | 0.3 | 0.4 |
| Spinning Distance (cm) | D | 10 | 15 | 20 |

**TABLE 2.2** Summary of Final Design Experiment With $L_9$ Orthogonal Array for Selected Factors and their Corresponding Average Nanofiber Diameter.

| Experiment number | Average Fiber Diameter (nm) | Factor | | | |
|---|---|---|---|---|---|
| | | Concentration (%) | Spinning distance (cm) | Applied voltage (V) | Flow rate (mL/h) |
| 1 | 240.82 | 8 | 10 | 15000 | 0.2 |
| 2 | 206.13 | 8 | 15 | 20000 | 0.3 |
| 3 | 230.76 | 8 | 20 | 25000 | 0.4 |
| 4 | 232.08 | 10 | 20 | 15000 | 0.3 |
| 5 | 299.02 | 10 | 10 | 20000 | 0.4 |
| 6 | 248.85 | 10 | 15 | 25000 | 0.2 |
| 7 | 315.09 | 12 | 15 | 15000 | 0.4 |
| 8 | 297.54 | 12 | 20 | 20000 | 0.2 |
| 9 | 283.64 | 12 | 10 | 25000 | 0.3 |

## 2.3 RESULTS AND DISCUSSION

Controlling the physical characteristics of electrospun nanofibers such as fiber diameter depends on various parameters. Also, the mechanics of this process requires precise attention and necessary predictive tools or direction for better understanding of optimization and controlling process. The design of experiments method seems a vital way to study the effects of changing a wide range of parameters. Thereafter, Taguchi experimental design was chosen in this work and the results obtained from the effects of four electrospinning parameters on the diameter of electrospun polyvinyl alcohol nanofiber were investigated in the next part.

### 2.3.1 INVESTIGATION EFFECTS OF THE CONCENTRATION OF POLYMERIC SOLUTION ON THE DIAMETER OF POLYVINYL ALCOHOL ELECTROSPUN NANOFIBERS JET

The concentrations of polymer solution play a significant role in the fiber formation during the electrospinning technique and have more effect on the morphology of formation nanofibers.[9] In fact, polymer concentration affects both the polymeric solution viscosity and surface tension of the solution.[32] The scanning electron microscopy and distribution of diameter results of the influence of this parameter on the electrospun polyvinyl alcohol nanofiber are shown in Figure 2.3. Comparisons between

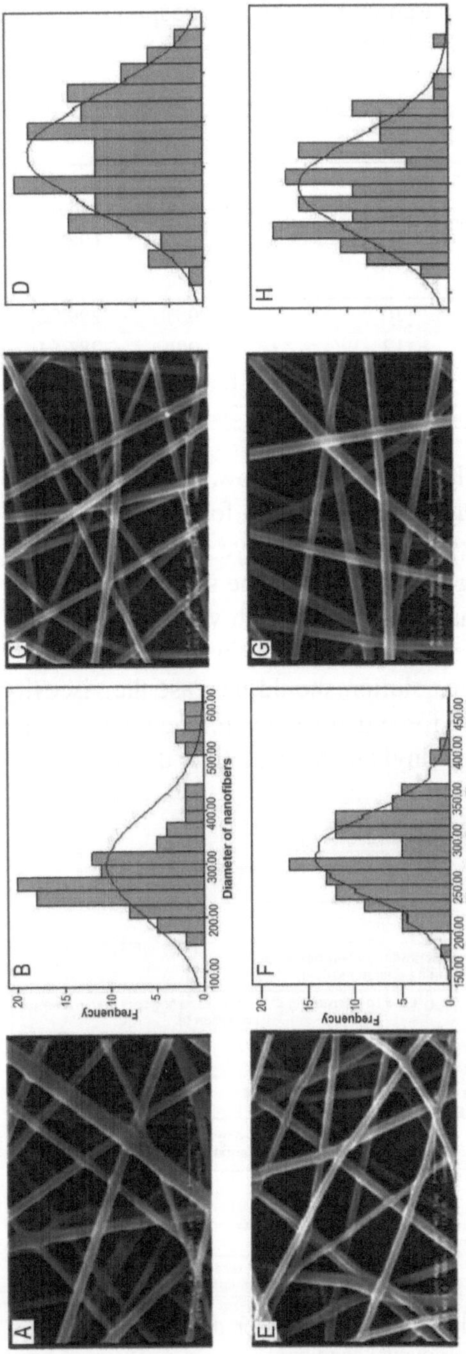


**FIGURE 2.3** Scanning electron microscopy micrographs of electrospun polyvinyl alcohol nanofiber (A: experimental conditions 5, C: experimental conditions 4, E: experimental conditions 9, G: experimental conditions 8) and distribution of electrospun polyvinyl alcohol nanofiber diameter (B: experimental conditions 5, D: experimental conditions 4, F: experimental conditions 9, H: experimental conditions 8).

the results of the influence of this parameter on the electrospun polyvinyl alcohol nanofiber are shown in Table 2.3 and Figure 2.4.

**TABLE 2.3** Investigation Effects of Concentrations on the Average Polyvinyl Alcohol Nanofiber Diameter.

| Factor | | | | | Average Fiber Diameter (nm) |
|---|---|---|---|---|---|
| **Experiment Number** | **Constant Spinning Distance (cm)** | **Flow Rate (mL/h)** | **Concentration (%)** | **Applied Voltage (V)** | |
| 5 | 10 | 0.4 | 10 | 20,000 | 299.02 |
| 9 | | 0.3 | 12 | 25,000 | 283.64 |
| 4 | 20 | 0.3 | 10 | 15,000 | 232.08 |
| 8 | | 0.2 | 20 | 20,000 | 297.54 |

When the concentration of polymer solution is low, viscoelastic force is not enabled enough for overcoming the Coulomb force due to insufficient chain entanglements in the solution and will give a breakdown in an electrical fluid jet before reaching the collector. If the solution had lower concentration, fibers cannot be made due to the high viscosity and difficulties in controlling the flow rate of the solution through capillaries.[33] Increasing the concentration of the solution should increase the viscosity of the solution and the viscoelastic force that can overcome the Coulomb force, resulting in an increased polyvinyl alcohol nanofiber diameter.[22,33–36]

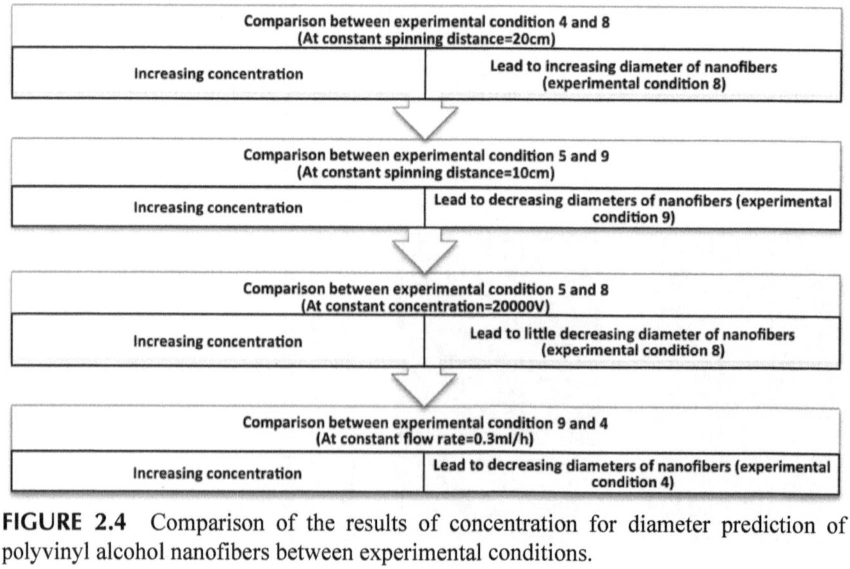


**FIGURE 2.4** Comparison of the results of concentration for diameter prediction of polyvinyl alcohol nanofibers between experimental conditions.

### 2.3.2 INVESTIGATION EFFECTS OF THE APPLIED VOLTAGE ON THE DIAMETER OF POLYVINYL ALCOHOL ELECTROSPUN NANOFIBERS JET

The applied voltage is one of the important and vital parameters which affect the nanofibers diameter.[1,22,37] For a solution with lower viscosity, a high applied voltage is desirable.[38] Lower voltage due to a weaker electrical field and thinner nanofibers was obtained, because of reducing speed of the jet and increase in the flight time of the electrospinning jet.[22,37,38] The scanning electron microscopy and distribution of diameter results by influencing this parameter on the electrospun polyvinyl alcohol nanofiber are shown in Figure 2.5 Comparisons between the results of the influencing this parameter on the electrospun polyvinyl alcohol nanofiber are shown in Table 2.4 and Figure 2.6.

**TABLE 2.4** Investigation Effects of Applied Voltage on the Average Polyvinyl Alcohol Nanofiber Diameter.

| | Factor | | | | Average |
|---|---|---|---|---|---|
| **Experiment Number** | **Constant Concentration (%)** | **Flow Rate (mL/h)** | **Spinning Distance (cm)** | **Applied Voltage (V)** | **Fiber Diameter (nm)** |
| 4 | 10 | 0.3 | 20 | 15000 | 232.08 |
| 5 | | 0.4 | 10 | 20000 | 299.02 |
| 8 | 12 | 0.2 | 20 | 20000 | 297.54 |
| 9 | | 0.3 | 10 | 25000 | 283.64 |

The applied voltage plays twosome behavior on the diameter of the nanofibers.[33] The first is the decrease in the diameter of the nanofibers by increasing the applied voltage[33,36] Because of electric field strength and the influence of fluid under the electrostatic force, the fluid is accelerated, the jet stretches more, and the diameter of nanofibers decreases.[19,33,36] The second type of behavior is increase in the diameter of the nanofibers by increasing the applied voltage. The reason of these various behaviors can be expressed as follows: When using higher voltages, creating nanofibers with larger diameter becomes easier and more fluids are emitted from the tip of the needle, which results in the formation nanofibers with larger diameter.[1,33] As can be seen, the various and even contradictory effects, such as increasing or decreasing the diameter of the nanofibers polyvinyl alcohol under the influence of the applied voltage, can be considered as the reason for the diversity of the influence of this parameter on the diameter of the nanofibers.

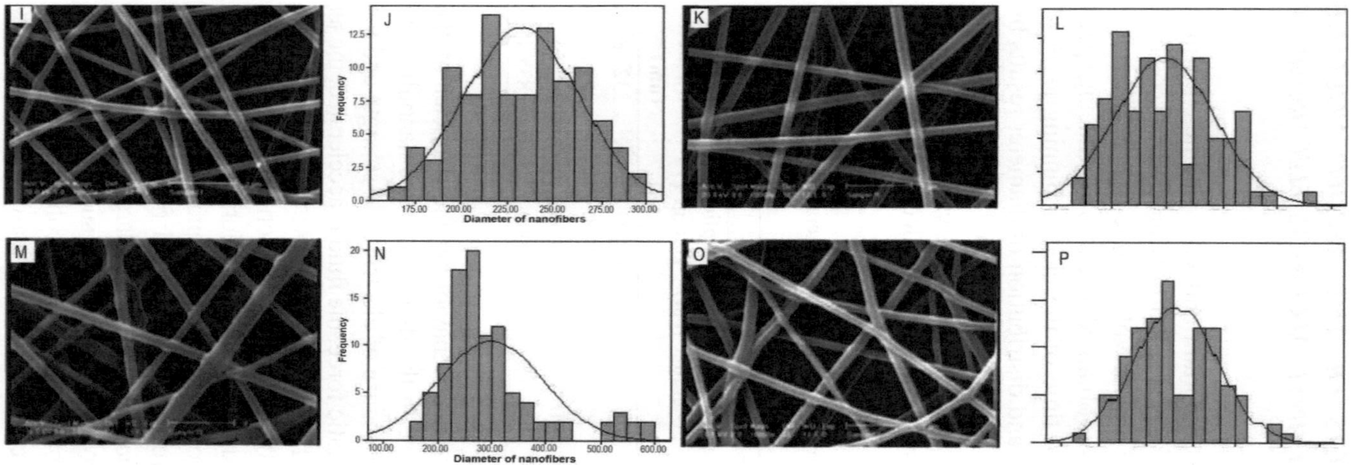


**FIGURE 2.5** Scanning electron microscopy micrographs of electrospun polyvinyl alcohol nanofiber (I: experimental conditions 4, K: experimental conditions 8, M: experimental conditions 5, O: experimental conditions 9) and distribution of electrospun polyvinyl alcohol nanofiber diameter (J: experimental conditions 4, L: experimental conditions 8, N: experimental conditions 5, P: experimental conditions 9).

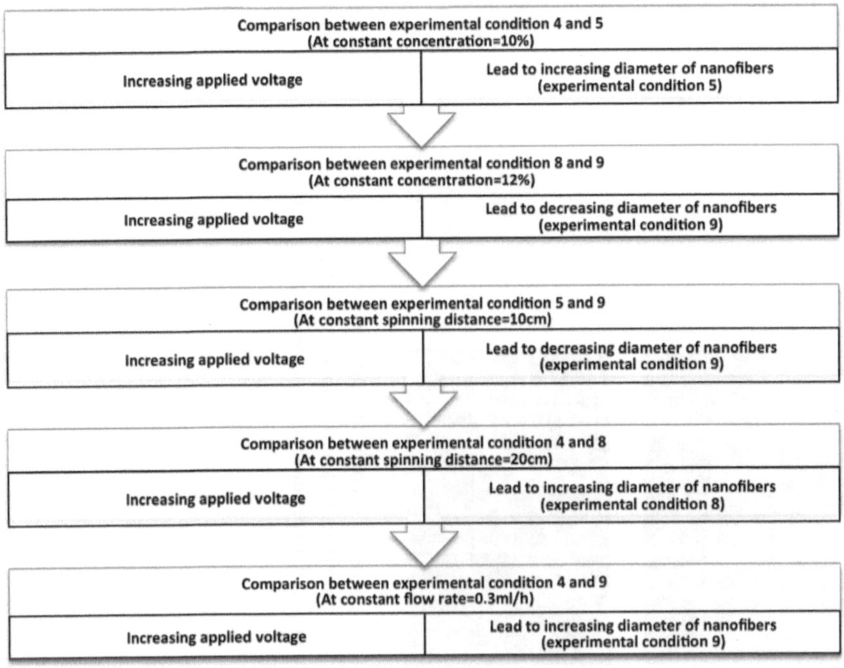


**FIGURE 2.6** Comparison of the result of applied voltage for diameter prediction of polyvinyl alcohol nanofibers between experimental conditions.

### 2.3.3 INVESTIGATION EFFECTS OF THE SPINNING DISTANCE ON THE DIAMETER OF POLYVINYL ALCOHOL ELECTROSPUN NANOFIBERS JET

A suitable spinning distance in electrospinning is one of the important factors which are needed for creating dry nanofibers without solvent.[27,32,39] It has been demonstrated that changing spinning distance will affect increasing or decreasing effects on the nanofibers diameter.[33,40] In addition, this parameter had an efficient effect at the time of formation jet, evaporation of the solvent, the amount of stretching, and electrical field.[33,40] The scanning electron microscopy and distribution of diameter results on the influence of this parameter on the electrospun polyvinyl alcohol nanofiber are shown in Figure 2.7. Comparisons between the results of the influence of this parameter on the electrospun polyvinyl alcohol nanofiber are shown in Table 2.5 and Figure 2.8.

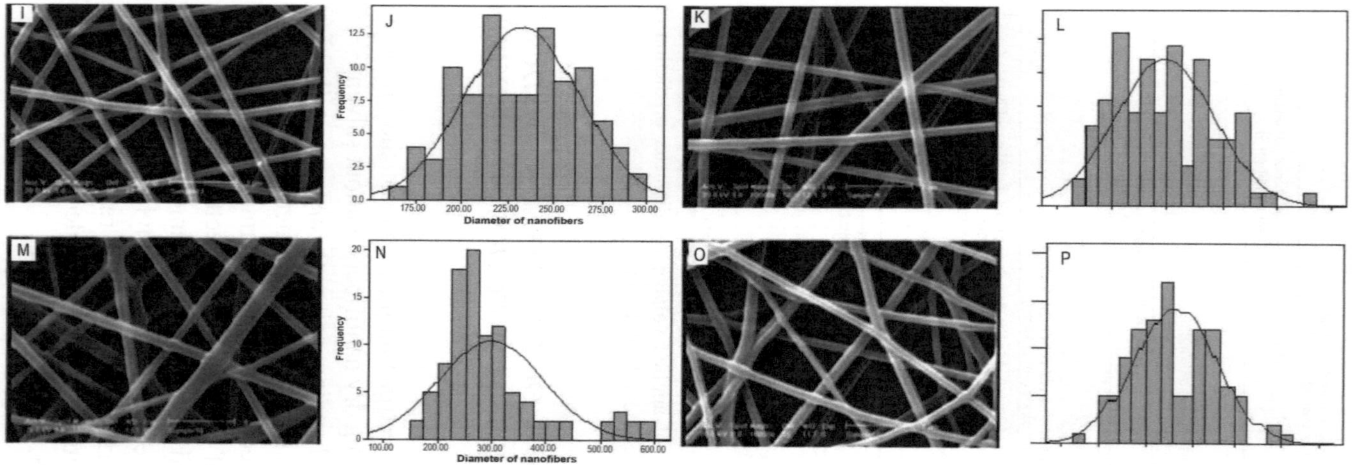


**FIGURE 2.7** Scanning electron microscopy micrographs of electrospun polyvinyl alcohol nanofiber (I: experimental conditions 4, K: experimental conditions 8, M: experimental conditions 5, O: experimental conditions 9) and distribution of electrospun polyvinyl alcohol nanofiber diameter (J: experimental conditions 4, L: experimental conditions 8, N: experimental conditions 5, P: experimental conditions 9).

**TABLE 2.5** Investigation Effects of Spinning Distance on the Average Polyvinyl Alcohol Nanofiber Diameter.

| Factor | | | | | Average Fiber Diameter (nm) |
|---|---|---|---|---|---|
| Experiment Number | Constant Concentration (%) | Flow Rate (mL/h) | Spinning Distance (cm) | Applied Voltage (V) | |
| 4 | 10 | 0.3 | 20 | 15,000 | 232.08 |
| 5 | | 0.4 | 10 | 20,000 | 299.02 |
| 8 | 12 | 0.2 | 20 | 20,000 | 297.54 |
| 9 | | 0.3 | 10 | 25,000 | 283.64 |

At a small spinning distance, nanofibers do not have adequate time for solidification before reaching the collector. So, the formation of instability is observed in nanofiber.[19,27,32,41] With increasing spinning distance, more time is available for stretching in the electrical field and more solvent are evaporating. In this case, thin nanofibers were created.[42] If the spinning distance increases, reduction in the electrical field, decrease in acceleration jet, and less stretching may occur that in turn increase nanofiber diameter.[33,40]

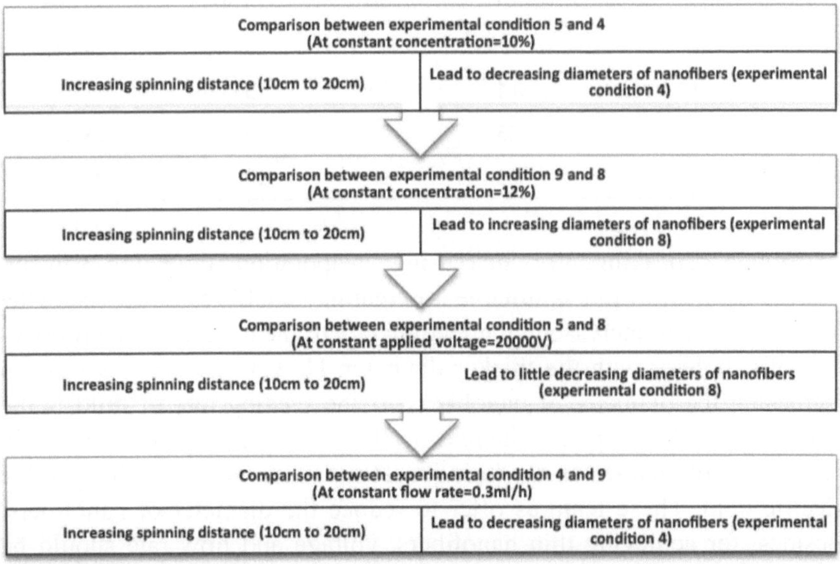


**FIGURE 2.8** Comparison of the results of spinning distance for diameter prediction of polyvinyl alcohol nanofibers between experimental conditions.

### 2.3.4 INVESTIGATION EFFECTS OF THE FLOW RATE ON THE DIAMETER OF POLYVINYL ALCOHOL ELECTROSPUN NANOFIBERS JET

In actuality, controlling flow rate is needed for creating uniform diameter in nanofibers. The influence of flow rate is executed with the changing, transferring rate of solution and speed of the fluid current.[27,32,38] The scanning electron microscopy and distribution of diameter results on the influence of this parameter on the electrospun polyvinyl alcohol nanofiber are shown in Figure 2.9. Comparisons between the results on the influences of this parameter on the electrospun polyvinyl alcohol nanofiber are shown in Table 2.6 and Figure 2.10.

**TABLE 2.6** Investigation Effects of Flow Rate on the Average Polyvinyl Alcohol Nanofiber Diameter.

| | Factor | | | | Average Fiber Diameter (nm) |
|---|---|---|---|---|---|
| **Experiment Number** | **Constant Concentration (%)** | **Flow Rate (mL/h)** | **Spinning Distance (cm)** | **Applied Voltage (V)** | |
| 5 | 10 | 0.4 | 10 | 20,000 | 299.02 |
| 4 | | 0.3 | 20 | 15,000 | 232.08 |
| 6 | | 0.2 | 15 | 25,000 | 248.85 |
| 7 | 12 | 0.4 | 15 | 15,000 | 315.09 |
| 9 | | 0.3 | 10 | 25,000 | 283.64 |
| 8 | | 0.2 | 20 | 20,000 | 297.54 |

Broadly speaking, for increasing evaporation time for solvent, decreasing the flow rate is suitable for creating nanofibers.[38] Increasing the flow rate leads to increase in the feeding rate to the needle tip and production of nanofiber with the thicker diameter. However, there is a limit in increasing the diameter of nanofibers. If the feeding rate of fluid is the same as the taking-up rate, increasing the flow rate leads to increase in the amount of charges on the fluid jet. This causes more stretching in the electric field. These features offer to reduce the diameter of nanofibers. Besides, for achieving thin nanofibers, voltage and flow rate should be changed together. If one of these parameters changes alone, nanofibers with larger diameter are obtained.[36]

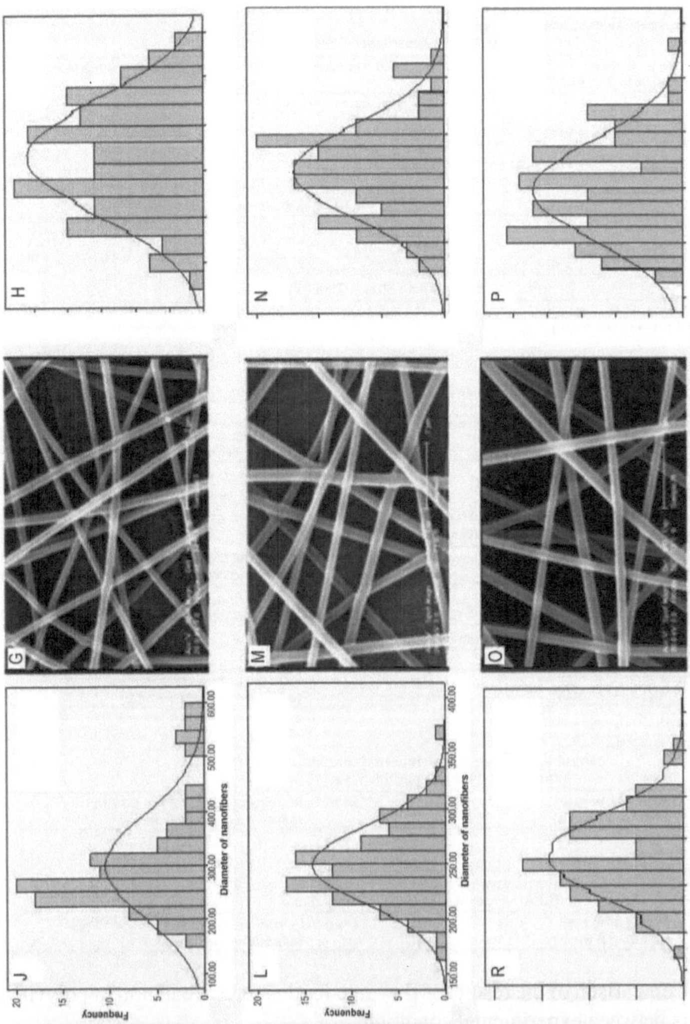


**FIGURE 2.9** Scanning electron microscopy micrographs of electrospun polyvinyl alcohol nanofiber (Q: experimental conditions 5, S: experimental conditions 4, U: experimental conditions 6, W: experimental conditions 7, Y: experimental conditions 9, ZA: experimental conditions 8) and distribution of electrospun polyvinyl alcohol nanofiber diameter (R: experimental conditions 5, T: experimental conditions 4, V: experimental conditions 6, X: experimental conditions 7, Z: experimental conditions 9, ZB: experimental conditions 8).

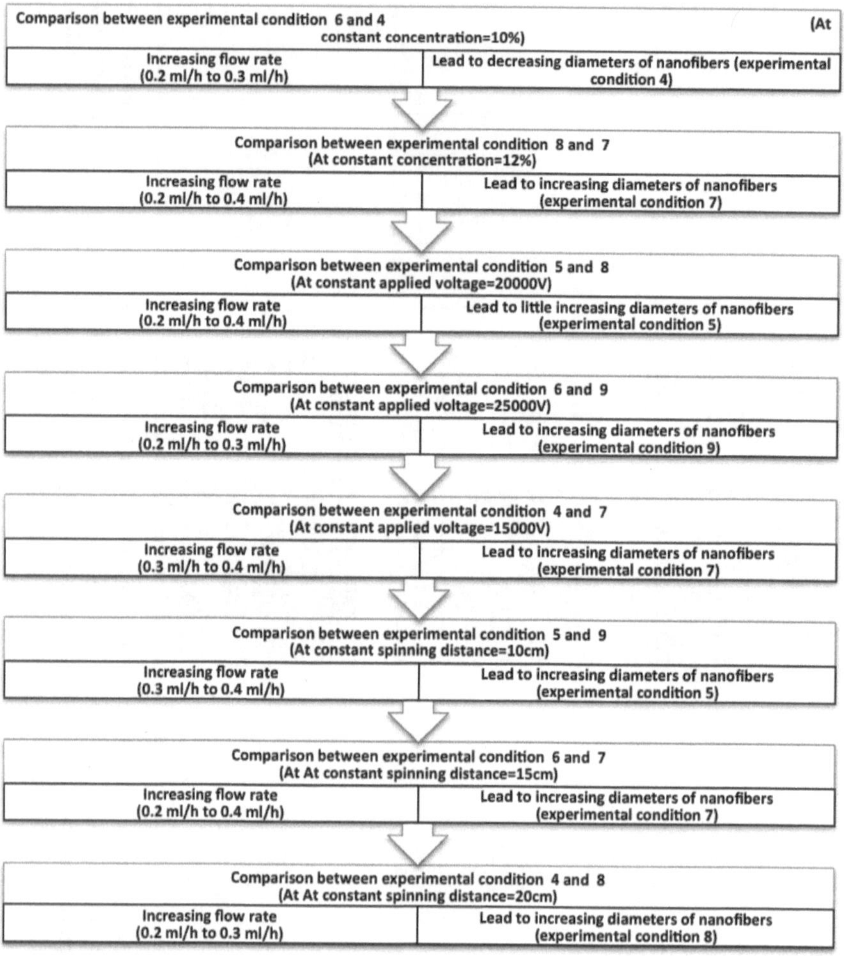


**FIGURE 2.10** Comparison of the results of flow rate for diameter prediction of polyvinyl alcohol nanofibers between experimental conditions.

## 2.4 CONCLUSIONS

Aligned electrospun nanofibers have found importance in many engineering applications. Therefore, the most important part of electrospinning is controlling the process. There are several ways to control aligning the depositing fiber. In this case, understanding the influence of varying parameters is so significant. The effects of four parameters on nanofiber diameter were investigated. Broadly speaking, various and even contradictory effects, such as increase or decrease in the diameter of the nanofibers polyvinyl alcohol, can be the reason for the influence of all parameters related to the conditions on the diameter of the nanofibers. By comparison, between dates, the shortest distribution of diameter electrospun polyvinyl alcohol nanofibers led to experimental condition 2. And the longest distribution of diameter electrospun polyvinyl alcohol nanofibers is led to experimental condition 7. In the near future, the other important tools for a better controlling process, namely modeling and simulating, will be investigated. The effects of these parameters on the pathways of nanofibers will also be investigated.

## KEYWORDS

- **electrospun nanofiber**
- **polyvinyl alcohol**
- **Taguchi experimental design**
- **fiber diameter**

## REFERENCES


1. Huang, Z. M., et al. A Review on Polymer Nanofibers by Electrospinning and Their Applications in Nanocomposites. *Compos. Sci. Technol.* **2003,** *63* (15), 2223–2253.
2. Fang, J., et al. Applications of Electrospun Nanofibers. *Chin. Sci. Bull.* **2008,** *53* (15), 2265–2286.
3. Ramakrishna, S., et al. Electrospun Nanofibers: Solving Global Issues. *Mater. Today.* **2006,** *9* (3), 40–50.
4. Teo, W.; Ramakrishna, S. A Review on Electrospinning Design and Nanofibre Assemblies. *Nanotechnology.* **2006,** *17* (14), R89.
5. Li, D.; Xia, Y. Electrospinning of Nanofibers: Reinventing the Wheel? *Adv. Mater.* **2004,** *16* (14), 1151–1170.
6. Haghi, A.; Akbari, M. Trends in Electrospinning of Natural Nanofibers. *Phys. Status Solidi A.* **2007,** *204* (6), 1830–1834.
7. Baji, A., et al. Electrospinning of Polymer Nanofibers: Effects on Oriented Morphology, Structures and Tensile Properties. *Compos. Sci. Technol.* **2010,** *70* (5), 703–718.
8. Shenoy, S. L., et al. Role of Chain Entanglements on Fiber Formation During Electrospinning of Polymer Solutions: Good Solvent, Non-Specific Polymer–Polymer Interaction Limit. *Polymer.* **2005,** *46* (10), 3372–3384.

9. Rafiei, S., et al. Mathematical Modeling in Electrospinning Process of Nanofibers: A Detailed Review. *Cell. Chem. Technol.* **2013,** *47* (5–6), 323–338.
10. Lannutti, J., et al. Electrospinning for Tissue Engineering Scaffolds. *Mater. Sci. Eng. C.* **2007,** *27* (3), 504–509.
11. Ghochaghi, N. Experimental Development of Advanced Air Filtration Media based on Electrospun Polymer Fibers. In *Mechanical and Nuclear Engineering;* Virginia Commonwealth: Virginia, Richmond, 2014; pp 1–165.
12. Ziabari, M.; Mottaghitalab, V.; Haghi, A. K. Evaluation of Electrospun Nanofiber Pore Structure Parameters. Korean *J. Chem. Eng.* **2008,** *25* (4), 923–932.
13. Sawicka, K. M.; Gouma, P. Electrospun Composite Nanofibers for Functional Applications. *J. Nanopart. Res.* **2006,** *8* (6), 769–781.
14. Li, W. J., et al. Electrospun Nanofibrous Structure: A Novel Scaffold for Tissue Engineering. *J. Biomed. Mater. Res.* **2002,** *60* (4), 613–621.
15. Pham, Q. P.; Sharma, U.; Mikos, A. G. Electrospinning of Polymeric Nanofibers for Tissue Engineering Applications: A Review. *Tissue Eng.* **2006,** *12* (5), 1197–1211.
16. Yousefzadeh, M., et al. A Note on the 3D Structural Design of Electrospun Nanofibers. *JEFF.* **2012,** *7* (2), 17–23.
17. Patan, A. K., et al. Nanofibers – A New Trend in Nano Drug Delivery Systems. *Int. J. Pharm. Res. Anal.* **2013,** *3,* 47–55.
18. Angammana, C. J. *A Study of the Effects of Solution and Process Parameters on the Electrospinning Process and Nanofibre Morphology;* University of Waterloo: Waterloo, Canada, 2011.
19. Rafiei, S., et al. New Horizons in Modeling and Simulation of Electrospun Nanofibers: A Detailed Review. *Cell. Chem. Technol.* **2014,** *48* (5–6), 401–424.
20. Lu, P.; Ding, B. Applications of Electrospun Fibers. *Recent Pat. Nanotechnol.* **2008,** *2* (3), 169–182.
21. Li, Z.; Wang, C. *Effects of Working Parameters on Electrospinning*, in *One-Dimensional nanostructures;* Springer: Heidelberg, 2013; pp 15–28.
22. Bognitzki, M., et al. Nanostructured Fibers via Electrospinning. *Adv. Mater.* **2001,** *13* (1), 70–72.
23. De. V. S., et al., The Effect of Temperature and Humidity on Electrospinning. *J. Mater. Sci.* **2009,** *44* (5), 1357–1362.
24. Zhou, H., *Electrospun Fibers from Both Solution and Melt: Processing, Structure and Property;* Cornell University: Ithaca, NY, 2007.
25. Tan, S. H., et al. Systematic Parameter Study for Ultra-Fine Fiber Fabrication via Electrospinning Process. *Polymer.* **2005,** *46* (16), 6128–6134.
26. Bhardwaj, N.; Kundu, S. C. Electrospinning: A Fascinating Fiber Fabrication Technique. *Biotechnol. Adv.* **2010,** *28* (3), 325–347.
27. Rafiei, S., et al. New Horizons in Modeling and Simulation of Electrospun Nanofibers: A Detailed Review. *Cell. Chem. Technol.* **2014,** *48* (5–6): 401–424.
28. Yarin, A. L.; Koombhongse, S.; Reneker, D. H. Bending Instability in Electrospinning of Nanofibers. *J. Appl. Phys.* **2001,** *89* (5), 3018–3026.
29. Kalita, G., et al. Taguchi Optimization of Device Parameters for Fullerene and Poly (3-Octylthiophene) Based Heterojunction Photovoltaic Devices. *Diamond Relat. Mater.* **2008,** *17* (4), 799–803.

30. Dean, A.; Voss, D. Response Surface Methodology. In *Design and Analysis of Experiments;* Springer-Verlag: New York, 1999; pp 483–529.
31. Sill, T. J.; Recum, H. A. Electrospinning: Applications in Drug Delivery and Tissue Engineering. *Biomaterials. 2008, 29* (13), 1989–2006.
32. Zhang, C., et al. Study on Morphology of Electrospun Poly (Vinyl Alcohol) Mats. *Eur. Polym. J.* **2005,** *41* (3), 423–432.
33. Tao, J.; Shivkumar, S. *M*olecular Weight Dependent Structural Regimes During the Electrospinning of PVA. *Mater. Lett.* **2007,** *61* (11), 2325–2328.
34. Koski, A.; Yim, K.; Shivkumar, S. Effect of Molecular Weight on Fibrous PVA Produced by Electrospinning. *Mater. Lett.* **2004,** *58* (3), 493–497.
35. Rodoplu, D.; Mutlu, M. Effects of Electrospinning Setup and Process Parameters on Nanofiber Morphology Intended for the Modification of Quartz Crystal Microbalance Surfaces. *J. Eng. Fibers Fabrics.* **2012,** *7* (2), 118–123.
36. Zong, X., et al. Structure and Process Relationship of Electrospun Bioabsorbable Nanofiber Membranes. *Polymer.* **2002,** *43* (16), 4403–4412.
37. Ramakrishna, S., et al. Electrospinning Process. In A*n Introduction to Electrospinning and Nanofibers;* World Scientific: Singapore, 2005; pp 135–137.
38. Lee, J. S., et al. Role of Molecular Weight of Atactic Poly (Vinyl Alcohol)(PVA) in the Structure and Properties of PVA Nanofabric Prepared by Electrospinning. *J. Appl. Polym. Sci.* **2004,** *93* (4), 1638–1646.
39. Subbiah, T., et al. Electrospinning of Nanofibers. *J. Appl. Polym. Sci.* **2005,** *96* (2), 557–569.
40. Megelski, S., et al. Micro- and Nanostructured Surface Morphology on Electrospun Polymer Fibers. *Macromolecules.* **2002,** *35* (22), 8456–8466.
41. Yuan, X., et al. Morphology of Ultrafine Polysulfone Fibers Prepared by Electrospinning. *Polym. Int.* **2004,** *53* (11), 1704–1710.
42. Andrady, A. L. *Science and Technology of Polymer Nanofibers;* Wiley: Hoboken, NJ, 2008.

## CHAPTER 3

# MECHANICAL AND PHYSICAL PROPERTIES OF ELECTROSPUN NANOFIBERS: AN ENGINEERING INSIGHT

S. PORESKANDAR, SH. MAGHSOODLOU, and A. K. HAGHI*

*University of Guilan, Rasht, Iran*

*Corresponding author. E-mail: akhaghi@yahoo.com*

## CONTENTS

## ABSTRACT

Electrospinning has emerged as a novel technology to create nanofiber webs. A better understanding of electrospinning jet movement is needed for the investigation of the effects of the important parameters in this process. In addition, the most vital challenge in the electrospinning process is to achieve uniform nanofiber webs. Also, the jet shows different behaviors during the process. Simulation and modeling are the most suitable methods for controlling and predicting pathways of electrospun nanofiber webs. The main purpose of this chapter is to investigate the effects of the most significant parameters (i.e., solution concentration, spinning distance, applied voltage, and flow rate) on the pathway of electrospun polyvinyl alcohol nanofiber jet by using a microscopic model. Therefore, the constitutive equations for the pathways simulation of electrospun nanofibers were used for parameters determination and after that, the process was simulated.


## 3.1 INTRODUCTION

Nanotechnology is known as one of the most promising technologies, attracting the interest of many scientists everywhere in the recent years.[1–3] Research in nanotechnology is directed toward understanding and creating improved materials, devices, and systems that exploit these new properties.[4] Nowadays, with the rapidly developing role of nanoscience, one-dimensional nanostructures has become a field of intensive research.[5–7] By reducing the diameters of polymer fiber from micrometers to nanometers, unique characteristics are changed compared to other known forms of the material in many research studies.[8] These characteristics are summarized in Figure 3.1.

| characteristics of nanofibers | | | |
|---|---|---|---|
| A large surface area to volume ratio | Flexibility in surface functionalities | High porosity | Superior mechanical performance |

**FIGURE 3.1** Unique characteristics of nanofibers.

Because of several desirable properties, these materials have been widely investigated as the best candidates in a broad range of potential applications (i.e., filters, tissue engineering scaffolds, protective clothing).[9–12] Many materials are used for electrospinning (Fig. 3.2).[13,14] A summary of these applications is shown in Figure 3.3.

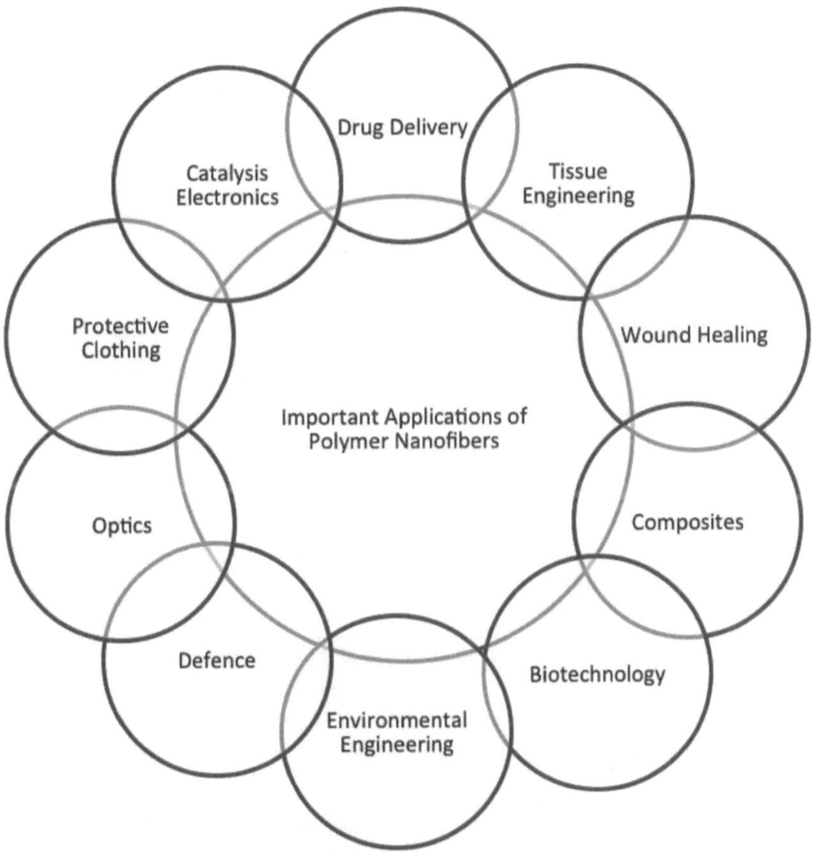


**FIGURE 3.2** Important applications of polymer nanofibers.

Polymer nanofibers can be made by different processing techniques which include the following: (1) drawing,[15] (2) template synthesis,[14] phase separation,[12] (4) self-assembly,[12] (5) electrospinning,[9] (6) gelation,[16] (7) bacterial cellulose,[17] and (8) vapor-phase polymerization.[12] Compared favorably with other methods of nanofiber manufacture, electrospinning stands out as the most versatile and powerful method for forming nanofibers mats.[18] The advantages of electrospinning technology,[13,19,20] are summarized in Figure 3.4.

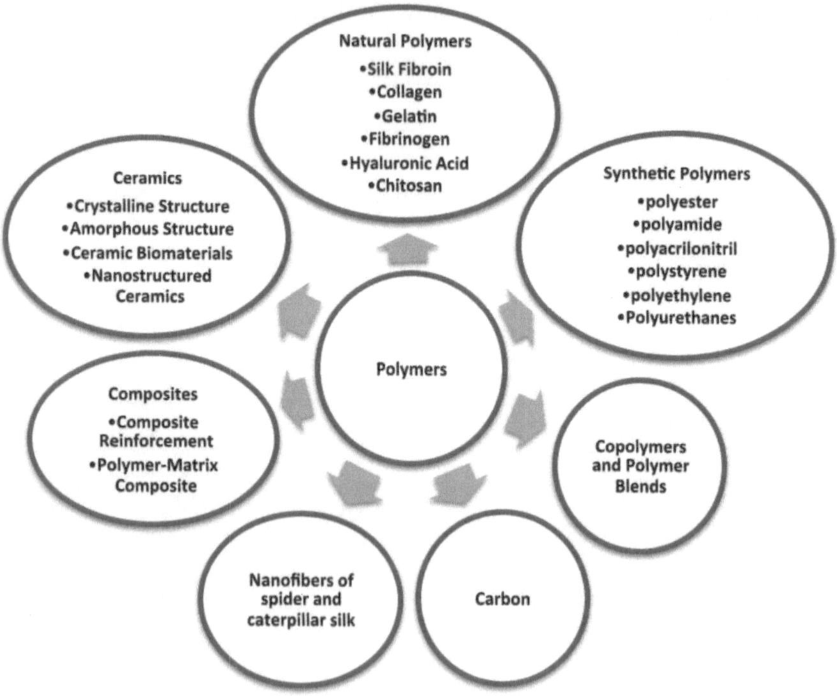


**FIGURE 3.3** Varieties of polymers in electrospinning.

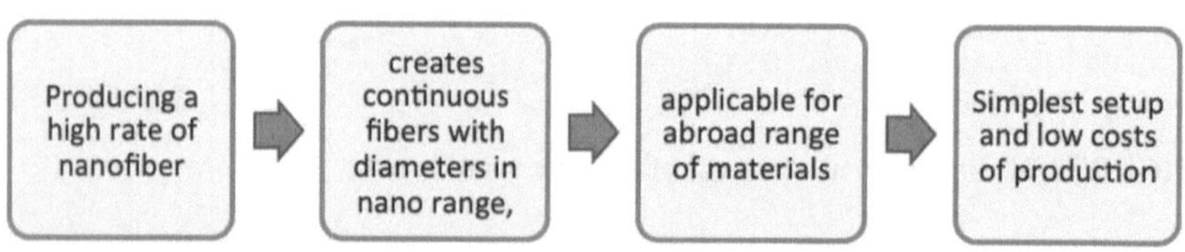


**FIGURE 3.4** The advantages of electrospinning technology.

Employing electrostatic forces to deform materials in the liquid state goes back many centuries,[21,22] but the origin of the electrospinning technique dates back to 100 years ago.[23,24] Many researchers worked on electrospinning to set up an effective factor on this technique.[25] Nowadays, the focus has shifted to developing a better understanding process technique of the electrospinning. Several research groups are interested in the investigation of electrospinning process.[26] In addition, the number of publications in this field has risen significantly in recent years (Fig. 3.5). A brief review of electrospinning process is shown in Table 3.1.

**TABLE 3.1** History of Electrospinning Process.

| Name of Researcher | Year | Subject | References |
|---|---|---|---|
| Lord Rayleigh | 19th century | Understood the technique of electrospinning | [27] |
| William Gilbert | 1600 | Discovered first record of the electrostatic attraction of a liquid | [28] |
| Zeleny | 1914 | Introduced one of the earliest studies of electrified jetting phenomenon | [6] |
| Formhals | 1934 | Invented the experimental setup for the practical production of polymer filaments with an electrostatic force | [21] |
| Vonnegut and Neubauer | 1952 | Produced streams of uniform droplets and invented a simple tool for the electrical atomization | [25] |
| Drozin | 1955 | Examined the dispersion of series of liquids into aerosols under high electric potentials | [29] |
| Simons | 1966 | Patented a tool for producing non-woven fabrics of ultra thin and weightless | [13] |
| Taylor | 1969 | Published his work on the shape of the polymer droplet at the tip of the needle with applying an electric field | [30] |
| Baumgarten | 1971 | Made a tool for electrospinning acrylic fibers with a stainless steel capillary tube and a high-voltage DC current. | [31] |
| | | Estimated the jet speed by using energy balance when a critical voltage was applied | |
| Larrondo and Mandley | 1981 | Produced polyethylene and polypropylene fibers by melting electrospinning successfully | [24] |

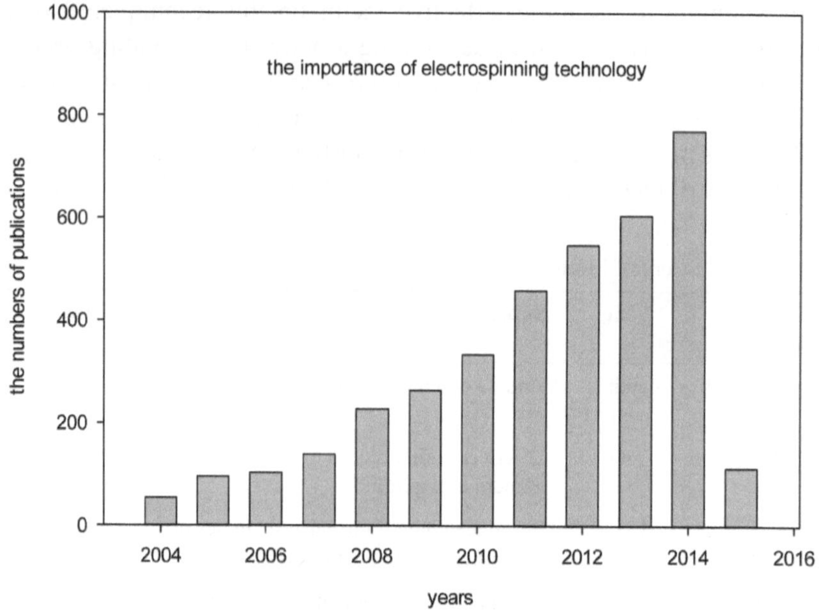


**FIGURE 3.5** The year-wise number of publications about electrospinning.

As remarked before, electrospinning has proven as the most straightforward and simplest process for producing non-woven fibers from a polymer with average diameters in the range of nano- to micrometers.[32,33] A common electrospinning setup is summarized in Figure 3.6.[11]

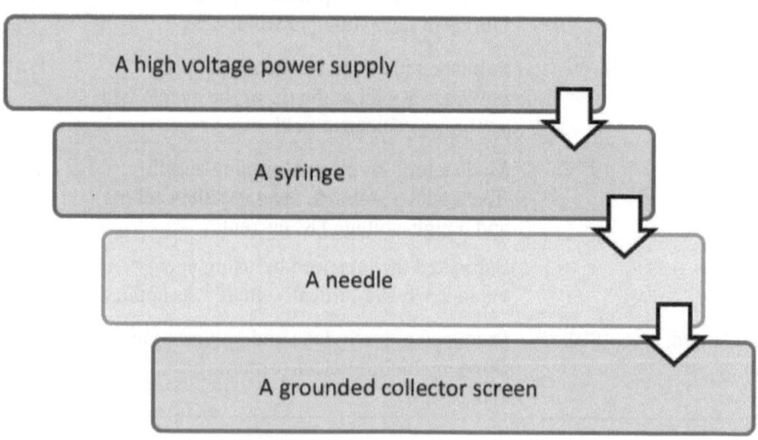


**FIGURE 3.6** A common electrospinning setup.

This process is based upon the simple concept that creates nanofibers through an electrically charged jet of polymer solution, displayed schematically in Figure 3.7.[34] When the voltage is initially applied to the solution fluid, the droplet at the nozzle distorts into the form of a cone. The final conical shape has come to be known as the Taylor cone.[5,34] These changes are due to the rivalry between the increasing solution charge and its surface tension. When the applied voltage is sufficient, the electrostatic force in the polymer and solvent molecules can overcome the surface tension as it has enough charge to overcome surface tension, and a stream is turned out from the tip of the Taylor cone.[35–39] The solution is drawn as a jet toward the collection plate, which will cause the charged solution to speed up toward the collector.[34,40] The solvent gradually evaporates, and a charged, solid polymer fiber is allowed to accumulate on the collection plate.[37,38,41] This process utilizes a high-voltage source to inject the charge of a certain polarity into a polymer solution, which is then accelerated toward a collector of opposite polarity.[42] This process is summarized in Figure 3.8.

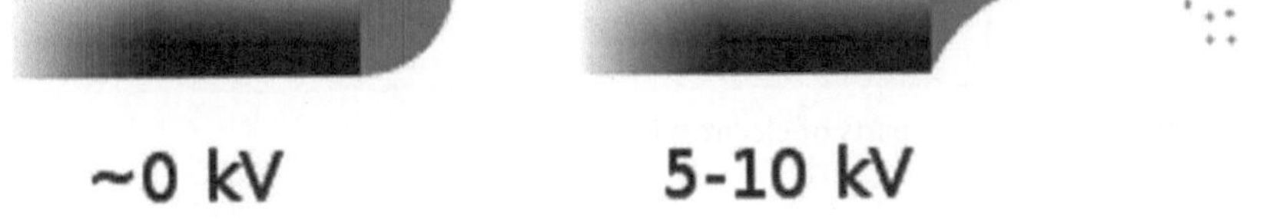


**FIGURE 3.7** Part two of electrospinning setup: Taylor cone. (Reprinted from Keministi. https://commons.wikimedia.org/wiki/File:Taylorcone.png. Published under the https://creativecommons.org/licenses/by-sa/4.0/deed.en)

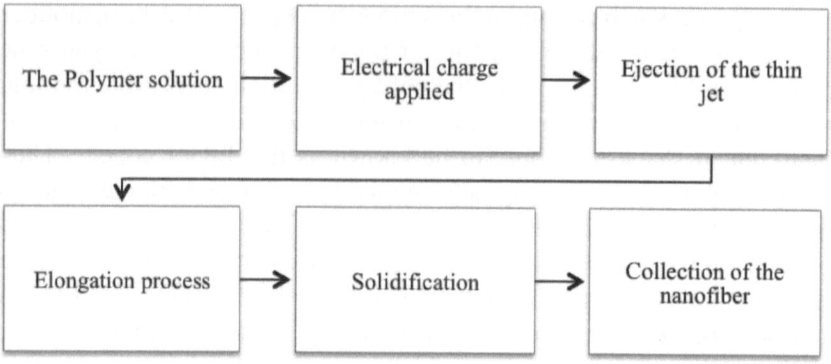


**FIGURE 3.8** Steps of electrospinning process.

In addition, this process is distinguished by four main sections.[43] So, it can be separated into four sections, as shown in Figure 3.9. A description of these stages is discussed briefly in the next section.

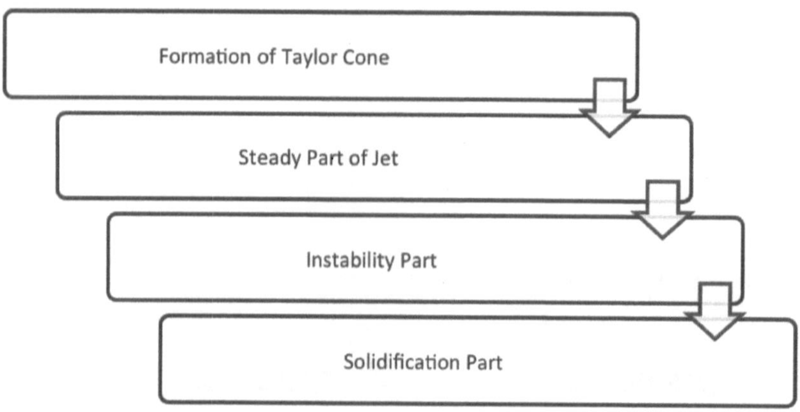


**FIGURE 3.9** Different parts of electrospinning process.

## 3.2 ABRIEFDESCRIPTIONOFSTAGESOFTHEELECTROSPINNING PROCESS

### 3.2.1 *FIRST STAGE: FORMATION OF TAYLOR CONE*

The electrospinning solution is usually an ionic solution that contains charged ions. The amounts of positive and negative charged particles are equal; therefore, the solution is electrically neutral. When an electrical potential difference is given between needle and collector, a hemispherical surface of the polymeric droplet at the orifice of the needle is gradually expanded. When the potential reaches a critical value (eq 3.1), a flow of jet forces the formation to drop and Taylor's cone is formed.[35] A schematic of these steps is shown in Figure 3.10.

$$V_c^2 = 4\frac{H^2}{L^2}\left(\ln\frac{2L}{R} - \frac{3}{2}\right)(0.117\pi\gamma R). \tag{3.1}$$

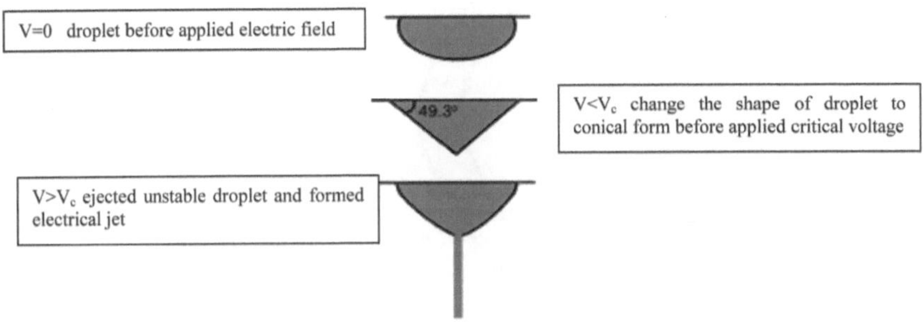


**FIGURE 3.10** Changes in the polymer droplet with applied potential.

### 3.2.2 SECOND STAGE: STEADY PART OF JET

The electrospun nanofiber jet is initiated from the droplet when the repelling forces of the surface charge overcome the surface tension and viscous forces of the droplet;[39] it leads to the elongation of the jet straight in the direction of its axis.[31,35] A stable formation jet, traveling to a collector, is shown in Figure 3.6. During the elongation of the electrified liquid jet, the jet surface area increases dramatically.[44] Therefore, for a thin, stable jet, it is important to look for a balance of viscosity, charge density, and surface tension.[43]

### 3.2.3 THIRD STAGE: INSTABILITY PART

After a small distance of stable traveling, the jet will start to show unstable behavior and eventually separate into many fibers.[31,35,39] As the jet spirals toward the collector, higher order instabilities result in spinning distance.

These instabilities are separated into three sections (which are summarized in Fig. 3.11):[45]

Rayleigh instability (Fig. 3.11a),
bending instability (Fig. 3.11b),
whipping instability (Fig. 3.11c).

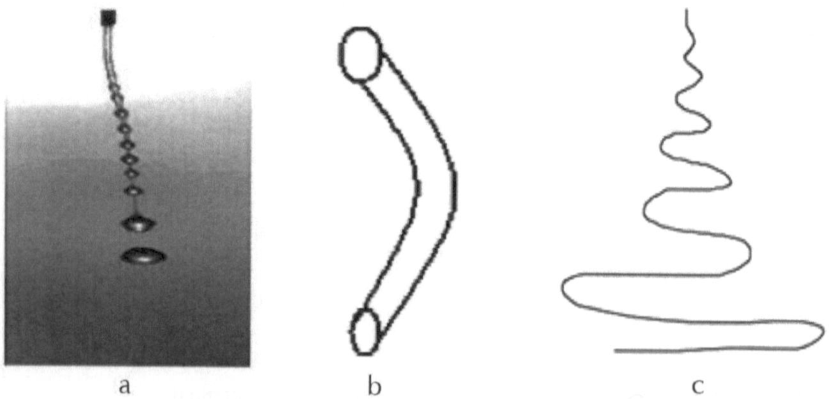


**FIGURE 3.11** (a) Rayleigh instability, (b) bending instability, and (c) whipping instability.[43,46]

The polymer jet is influenced by these instabilities. These instabilities arise owing to the charge–charge repulsion between the excess charges present in the jet, which encourages the thinning and elongation of the jet. At high electric forces, the jet is dominated by bending (axisymmetric) and whipping instabilities (non-axisymmetric), causing the jet to move around, which produces waves in the jet. At higher electric fields and with enough charge density in the jet, the axisymmetric (i.e., Rayleigh and bending) instabilities are suppressed and the non-axisymmetric instability is increased.[24] Recently, instability in electrospinning has received much attention.[35] Each instability grows at different a rate.[43,47] These instabilities vary and increase with distance, electrical field, and fiber diameter at different rates depending on the fluid parameters and performing conditions. Also, they influence the size and geometry of the deposited fibers.[24] Suitable methods for analyzing these behaviors are important. In the later sections, methods for better analyzing of these behaviors will be discussed.

### 3.2.4 FOURTH STAGE: SOLIDIFICATION

For polymers dissolved in non-volatile solvents, water or other suitable liquids can be used to collect the jet, remove the solvent, and develop the polymer fiber.[31] As the jet moves toward the collector, it continues to expand by going past through the loops. Jet solidification is based along the traveling distance of the fibers. The distance between the collector and the capillary tip has a direct effect on the jet solidification and fiber diameter. If the nozzle-to-collector distance is long enough or the whipping instability is high enough, there is more time for fibers to dry before being picked up.[35] The mechanics of this process deserve specific attention and are necessary for predictive tools or way to better understand and optimize the controlling process.[48] For this cause, more details are discussed in the next section.

## 3.3 THE IMPORTANCE OF USAGE MODELING IN ELECTROSPINNING PROCESS

The most significant challenge in this process is to attain uniform nanofibers consistently and reproducibly.[49–51] Depending on several solution parameters, different results can be obtained using the same polymer and electrospinning setup.[40] In addition, controlling the physical characteristics of electrospun nanofibers such as fiber diameter depends on various parameters, which are principally divided into three categories: solution properties (solution viscosity, solution concentration, polymer molecular weight, and surface tension), processing conditions (applied voltage, volume flow rate, spinning distance, and needle diameter), and ambient conditions (temperature, humidity, and atmospheric pressure).[14,52–54] In addition, the type of collectors is vital for formation uniform nanofibers. A summary of the most important parameters and types of the collectors is given in Figures 3.12 and 3.13. Also, the selected parameters with (*) symbol are studied in this work.

Successful electrospinning has involved an understanding of the complex interaction of electrostatic fields, properties of polymer solutions, and component design and system geometry.[20] Studying the dynamical behavior of the jet is of interest for the development of a possible control system.[55] The process is complex as the resulting jet (fiber) diameter is influenced by numerous material, design, and operating parameters. A significant part of the electrospinning process comes from empirical observations, but the complexity of this process makes an empirical

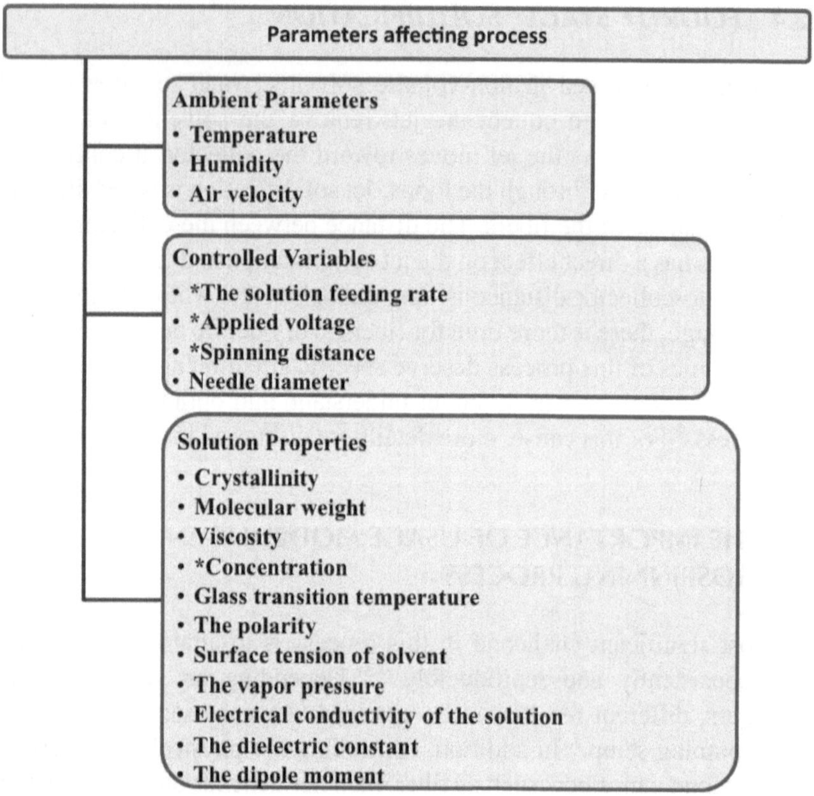


**FIGURE 3.12** Parameters affect the morphology and size of electrospun nanofibers.

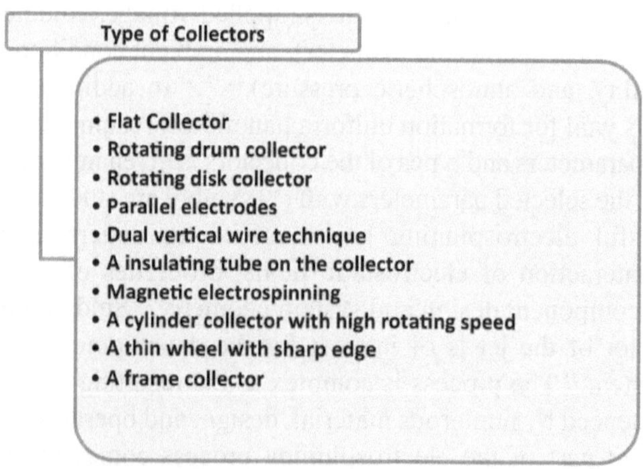


**FIGURE 3.13** Type of collectors.

determination of the parameter effects very difficult, if not impractical.[56] The mechanics of this process deserve specific attention and are necessary for predictive tools or way to better understand and optimize the controlling process.[48] In addition, studying the dynamical behavior of the jet is of interest for the maturation of a possible control system.[55] For predicting electrospun fiber properties and morphology, modeling and simulating are used.[57] Results from modeling also explain how processing parameters and fluid behavior lead to nanofibers with appropriate properties.[58,59] In addition, mathematical and theoretical modeling and simulation procedure will assist in offering an in-depth insight into the physical understanding of complex phenomena during electrospinning and might be very applicable in managing contributing factors to increase the production rate.[52] In addition, they prepared appropriate information about how something will work without actual testing in real.[52,60,61] A brief review of the most important of models is investigated in the next part.

### 3.3.1 MATHEMATICAL MODELING

As mentioned before, for prediction electrospun nanofiber properties and morphology, modeling and simulating are used. An easy and systematic way of controlling the influence of variables was created by utilizing models.[57] It is necessary to develop theoretical/numerical models of electrospinning.[62] As a definition, a model is a schematic description of a system, theory, or phenomenon that accounts for its known or inferred properties and may be used for further study of its characteristics.[8] In addition, they can be useful for investigating the effects of factors perception that cannot be measured experimentally.[63] A real mathematical model might initiate a revolution in the understanding of dynamic and quantum-like phenomena in the electrospinning process.[44,64,65] The analysis and comparison of the model with experiments identify the critical role of the spinning fluid's parameters. In each model, the researcher tried to improve the existing models by changing the tools in electrospinning by using another survey. Therefore, it is attempting to accept a whole view on important models after an investigation about basic objects. Results from modeling also explain how processing parameters and fluid behavior lead to nanofibers with appropriate properties.[58,59] Various developed models can be employed for the analysis of jet deposition and alignment mechanisms on different collecting devices in arbitrary electric fields.[59] A summary of investigated models is shown in Table 3.2.

**TABLE 3.2** A Summary of Investigated Models used in Electrospinning Process.

| Model Name | Equations | Suppositions | Name of Researcher | Year | References |
|---|---|---|---|---|---|
| The electro hydrodynamic of electrospinning<br>(leaky dielectric model) | 1) Navier–Stokes<br>2) Maxwell stress equation<br>3) Coulomb force equation | 1) Dielectric fluid<br>2) Bulk charge = 0<br>3) Axial motion<br>4) Steady part<br>5) Newtonian<br>6) Incompressible | Taylor | 1969 | [66] |
| One-dimensional modes of steady, inviscid, annular liquid jets | 1) Laplace equation<br>2) Bernoulli's equation<br>3) Euler equation | 1) Steady jet<br>2) Inviscid liquid jet<br>3) Annular liquid jet<br>4) Slender jet | Ramos | 1996 | [67] |
| Asymptotic decay of radius of a weakly Conductive viscous jet in an external electric field | 1) Momentum balance<br>2) Mass balance<br>3) Electric charge balance | 1) Newtonian<br>2) Infinite viscose jet<br>3) Weakly conductive jet<br>4) No magnetic effects<br>5) Long slender jet | Spivak, Dzenis | 1998 | [68] |
| Bending instability of electrically charged liquid jets of polymer solutions in electrospinning | 1) Momentum balance<br>2) Viscoelastic force equation | 1) Viscoelastic jet<br>2) Entire jet | Reneker | 2000 | [69] |
| Effects of parameters on nanofiber Diameter determined from electrospinning model | | | Thompson | 2007 | [56] |

**TABLE 3.2** *(Continued)*

| Model Name | Equations | Suppositions | Name of Researcher | Year | References |
|---|---|---|---|---|---|
| Mathematical models of bead-spring jets during electrospinning for fabrication of nanofibers | | | Darsi | 2013 | [70] |
| Experimental characterization of electrospinning: the electrically forced jet and instabilities | 1) Momentum balance<br>2) Mass balance<br>3) Electric charge balance<br>4) Instabilities analysis | 1) Newtonian<br>2) Incompressible<br>3) Long slender jet | Shin, Hohman | 2001 | [71] |
| The stretching of an electrified non-Newtonian jet: a model for electrospinning | 1) Momentum balance<br>2) Mass balance<br>3) Electric charge balance | 1) Non-Newtonian<br>2) Steady jet<br>3) Slender jet | Feng | 2002 | [72] |
| Thermo-electro-hydrodynamic model for electrospinning process | 1) Momentum balance<br>2) Mass balance<br>3) Electric charge balance<br>4) Bratu equation | 1)Thermo-electro-hydrodynamic<br>2) Steady part<br>3) Thermal effects | Wan, Gue, Pan | 2004 | [73] |
| A mathematical model for preparation by AC-electrospinning process | | | He, Wu, Pang | 2005 | [74] |
| A thermo-electro-hydrodynamic model for vibration-electrospinning process | | | Xu, Wang | 2011 | [75] |

All models start with some assumptions. Different stages of electrospun jets have been investigated by different mathematical models during the last decade by one- or three-dimensional techniques.[76] The main focus of this chapter is to investigate the models to give a better understanding of electrospinning process. For this reason, a microscopic model was proposed for investigating pathways of electrospun nanofibers. In addition, polyvinyl alcohol was chosen as a suitable material because of water-soluble and hydrophilic polymer, suitable chemical and thermal stability, biocompatible and non-toxic.[52] Also, the Taguchi experimental design was chosen, and the results obtained from the effects of four electrospinning parameters on the pathways of electrospun polyvinyl alcohol nanofiber were utilized in the simulation program. In this work, the initial values of these parameters were determined by governing equations obtained from investigation resources or experimentally. In the next part, governing relations in the microscopic model was investigated. Also, the governing relations for initial parameters of simulation electrospun nanofiber jet were discussed.

### 3.3.2 GOVERNING RELATIONS IN THE MICROSCOPIC MODEL

For investigating the fiber motion and investigating the whipping instability of jet, we proposed a microscopic model. As mentioned in Figure 3.14, we consider the jet as a chain that consisted of beads joined by springs with a viscoelastic model in spinning distance (between nozzle and collector).

Mass losses due to evaporation in this model were neglected in the main equation.[16,22,70] The total number of beads, $N$, increases over time as new electrically charged beads are introduced to represent the flow of solution into the jet. The first formation drop in nozzle part as $i = N$ and the last formation drop as $i = 1$ were considered. For investigating all of the forces act along the jet, three beads as $i$, $i + 1$, $i - 1$ were determined. The distance between $i$, $i + 1$ with $u$ index and $i - 1$, $i$ with the $d$ index is shown. Because all beads are motions in different directions and axis, then in general the momentum equations include many forces (eqs 3.2 and 3.4) like Coulomb force, electric force, viscoelastic force, and surface tension force in three dimensions. $R_{ij}$ is the distance between bead $i$ and bead $j$ and is expressed as eq 3.3. All the equations are presented in Table 3.3.[22,70]

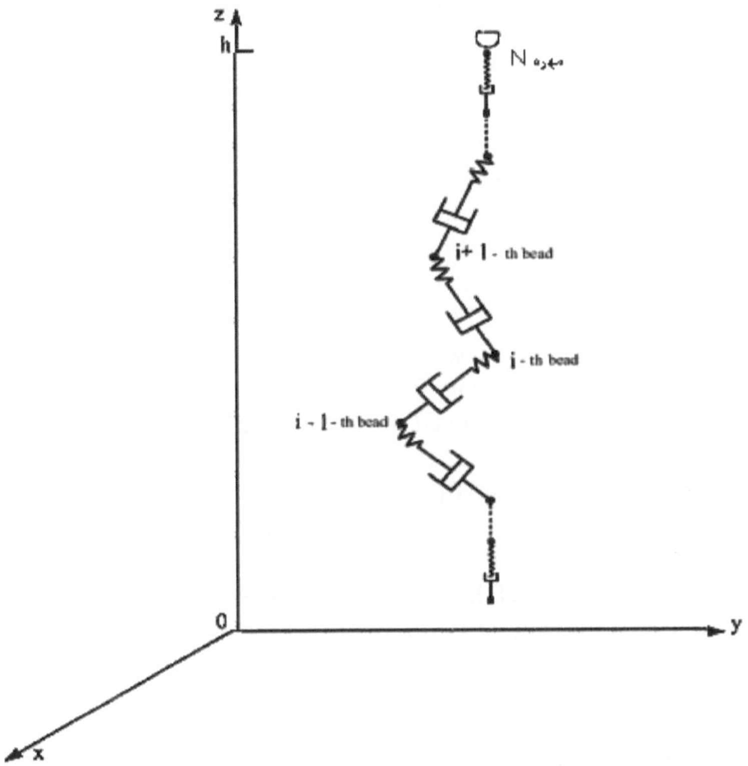


**FIGURE 3.14** Jet path in three dimensions with beads and springs.

**TABLE 3.3** Equations **of Modeling of Electrospinning Process.**

| | |
|---|---|
| $\Sigma F_i = \Sigma F_{\text{total}}$ | (3.2) |
| $R_i = iX_i + jY_i + kZ_i$ | (3.3) |
| $m\dfrac{d^2R_i}{dt^2} = \sum\limits_{\substack{j=1 \\ j \neq i}}^{N} \dfrac{e^2}{R_{ij}^3}\left[R_i - R_j\right] - e\dfrac{V_0}{h}K + \dfrac{\pi a_{ui}^2 \sigma_{ui}}{L_{ui}}\left[R_{i+1} - R_i\right]$ $- \dfrac{\pi a_{di}^2 \sigma_{di}}{L_{di}}\left[R_i - R_{i-1}\right] - \dfrac{\alpha\pi(a)_{av}^2 K_i}{\left(x_i^2 + y_i^2\right)^{\frac{1}{2}}}\left[i\left\|X_i\right\|\text{sign}\left(X_i\right) + j\left\|Y_i\right\|\text{sign}\left(Y_i\right)\right]$ | (3.4) |

Since all the differential equations are defined, the MATLAB script 2013a can be planned for simulation. In this case, new beads are introduced into the system and removed if they progress to the collector. The fluid jet is simulated for polyvinyl alcohol with MATLAB software. Initial conditions were resolved to run MATLAB program and also simulation pathways of the electrospun jet. In this case, obtaining the correct values of them is important. In addition, some of these parameters are related to each other (i.e., viscosity and elastic modulus, viscosity and concentration, and the charge of the solution and elastic modulus). For this reason, a correct equation should be demanded for each parameter. For these causes, surface tension, viscosity, the mass of consumable polymer, elastic modulus, specific density, length and radius of the needle, and charge of the solution should be calculated in run simulation program in this work. In the next part, the constitute relations for initial parameters of simulation electrospun nanofiber jet are discussed.

## 3.4 THE CONSTITUTE RELATIONS FOR INITIAL PARAMETER OF SIMULATION ELECTROSPUN NANOFIBER JET

### *3.4.1 DESIGN OF EXPERIMENT*

For creating nanofibers with no beads, polyvinyl alcohol powders with high molecular weight were utilized in this work. The design of experiment method seems a vital path to study the effects of changing a wide range of parameters. For this purpose, Taguchi experimental design is a simple technique which can optimize process parameters by fewer experiments.[77] Thereafter, $L_9$ orthogonal array was selected according to Taguchi's methodology. The four most significant, influential factors (i.e., solution concentration which changes viscosity and surface tension, spinning distance which determined the enough space for elongation and charge accumulation on the nanofiber, applied voltage which determines the elongation force and the strength of electric field, and volume flow rate, which changes of the mass transfer rate) were chosen for creating smooth nanofibers with no beads in this work. Each of these factors is considered to be varied at three levels for investigating the similar influences of these parameters in all experimental conditions. The final design experiments are summarized in Table 3.4. Also, the scanning electron microscopy and distribution of electrospun polyvinyl alcohol nanofiber diameter are shown in Figure 3.15.

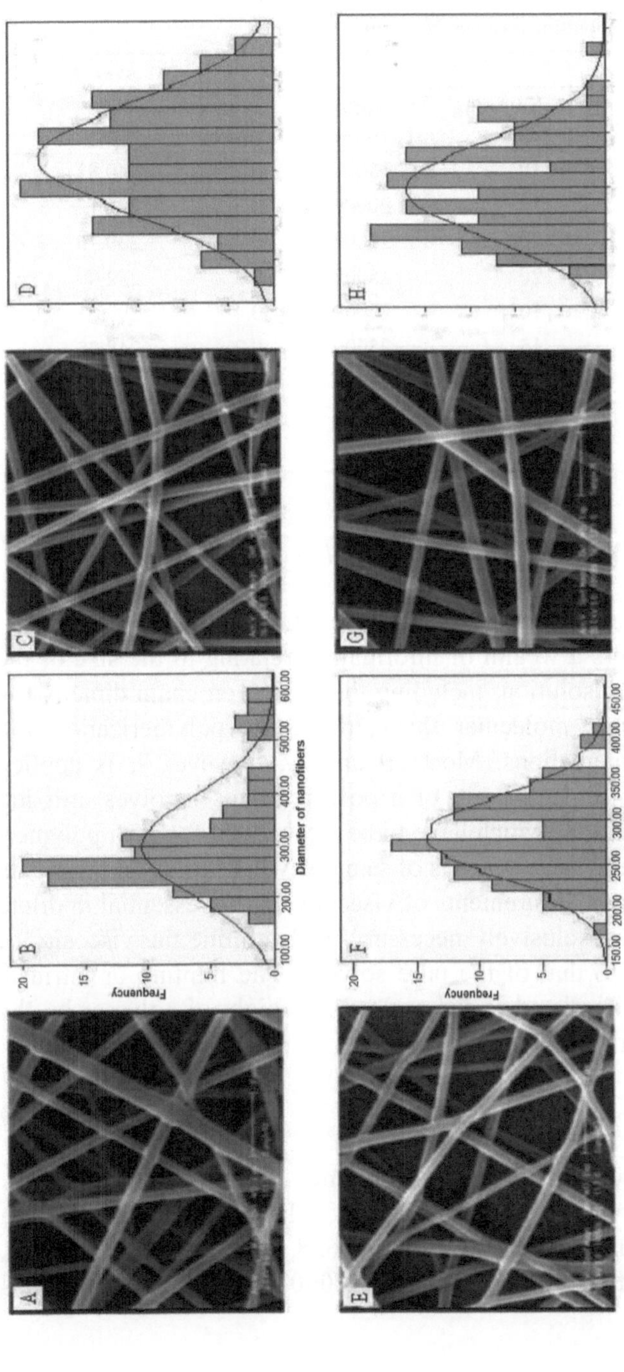


**FIGURE 3.15** Scanning electron microscopy micrographs of electrospun polyvinyl alcohol nanofiber (A: experimental conditions 5, C: experimental conditions 4, E: experimental conditions 9, G: experimental conditions 8) and distribution of electrospun polyvinyl alcohol nanofiber diameter (B: experimental conditions 5, D: experimental conditions 4, F: experimental conditions 9, H: experimental conditions 8).

**TABLE 3.4** Summary of Final Design Experiment with $L_9$ Orthogonal Array for Selected Factors and Their Corresponding Average Nanofiber Diameter.

| Experiment number | Factor | | | | Average Fiber Diameter (nm) |
|---|---|---|---|---|---|
| | Concentration (%) | Spinning distance (cm) | Applied voltage (V) | Flow rate (ml/h) | |
| 1 | 8 | 10 | 15000 | 0.2 | 240.82 |
| 2 | 8 | 15 | 20000 | 0.3 | 206.13 |
| 3 | 8 | 20 | 25000 | 0.4 | 230.76 |
| 4 | 10 | 20 | 15000 | 0.3 | 232.08 |
| 5 | 10 | 10 | 20000 | 0.4 | 299.02 |
| 6 | 10 | 15 | 25000 | 0.2 | 248.85 |
| 7 | 12 | 15 | 15000 | 0.4 | 315.09 |
| 8 | 12 | 20 | 20000 | 0.2 | 297.54 |
| 9 | 12 | 10 | 25000 | 0.3 | 283.64 |

### 3.4.2 VISCOSITY OF THE SOLUTION

One of the important parameters in a simulation program is viscosity. This parameter offers a wealth of information relating to the size of the polymer molecule in solution, including the effects on chain dimensions of polymer, structure, molecular shape, degree of polymerization and polymer–solvent interactions. Most commonly, however, it is applied to estimate the molecular weight of a polymer. This involves utilizing semi-empirical equations which have to be established for each polymer/solvent/temperature system analysis of samples whose molecular weights are known. Absolute measurements of viscosity are not essential in dilute solutions since it is exclusively necessary to determine the viscosity of a polymer relative to that of the pure solvent. The limiting or intrinsic viscosity, quantity is related to the molecular weight of polymer by the semi-empirical Mark–Houwin equation:[78]

$$[\eta] = K.\bar{M}_{v}^{a}, \tag{3.5}$$

where $K$ and $\alpha$ are constants for a given polymer, solvent, and temperature. Generally, for polyvinyl alcohol in the molecular weight range between 69,000 and 690,000, $K$ and $\alpha$ are 6.51 and 0.628, respectively. In addition, the intrinsic viscosity of this polymer with 72000 g/l molecular weight was

0.73. Also, the relation between relative viscosity and specific viscosity is calculated from the following equation.[78]

$$\eta_{sp} = \eta_r - 1 = \frac{\eta_s - \eta_0}{\eta_0}. \tag{3.6}$$

In this work, Ram Mohan Rao and Yaseen equation were utilized for specific viscosity:[78]

$$[\eta] = \left[\ln \eta_r + \eta_{sp}\right](2c)^{-1}. \tag{3.7}$$

Equation 3.7 was solved by MATLAB program and the values of specific viscosity (Table 3.5) were obtained as initial parameters for viscosity in MATLAB script.

### 3.4.3 SURFACE TENSION OF THE SOLUTION

The other important parameter in the simulation of the electrospun jet is surface tension. It is a property of fluids that makes the outer layer act as an elastic sheet. In other words, the surface tension is equal to the force of section per unit length. In addition, a governing equation was obtained between surface tension and voltage. As mentioned before, electrospinning solution is usually an ionic solution that contains charged ions. The amounts of positive and negative charged particles are equal; therefore, the solution is electrically neutral. When an electrical potential difference is given between needle and collector, a hemispherical surface of the polymeric droplet at the orifice of the needle is gradually expanded. When the potential reaches a critical value, a flow of jet starts forces the formation to drop.[35] As mentioned in eq 3.8, the relation between surface tension and critical voltage was observed. This behavior may arise as a result of surface tension differences between several solutions. It has been demonstrated that the surface tension of polyvinyl alcohol solutions is proportional to the degree of polymerization and to the extent of hydrolysis. The surface tension decreases with increasing concentration of the polymer in the solution. A low surface tension is desirable in electrospinning, as it thins out the critical voltage $V_c$ needed for the ejection of the jet from the Taylor's cone as shown below:[29,79]

$$V_c^2 = 4\frac{H^2}{L^2}\left(\ln\frac{2L}{R} - \frac{3}{2}\right)(0.117\pi\gamma R). \tag{3.8}$$

In summation, the amount of $V_c$ and $H$ were utilized from an initial condition of the design of experiment in Table 3.4. Also, the amount of $\rho$,[80,81] $L$, and $R$ were defined:

$$\begin{aligned} &\rho = 1.25\text{g/cm}^3, \\ &L = 34\,\text{mm}, \\ &D = 0.7\,\text{mm} \rightarrow R = 0.35\,\text{mm}. \end{aligned} \tag{3.9}$$

### 3.4.4 CHARGE OF THE SOLUTION

The amount of force between two charges that has a certain distance from each other is called the electrical force. For determining the amount of the charge of solving in simulation program, the dimensionless amount of the charge was used:

$$e = \sqrt{\frac{mL_{\text{el}}^3 G^2}{\eta_{sp}^2}}. \tag{3.10}$$

The amount of mass of polymer powder for preparing $v$ = 20 ml of polyvinyl alcohol solution was determined by the following equation:

$$\%c = \frac{m}{v}. \tag{3.11}$$

In addition, the amount of $L_{\text{el}}$, elastic modulus, and specific viscosity were determined by previous equations.

### 3.4.5 ELASTIC MODULUS

Elastic modulus is defined as the ratio of stress (force per unit area) along an axis to strain (ratio of deformation over initial length) along that axis. During electrospinning, the stable jet ejected from Taylor's cone is subjected to tensile stresses and may undergo significant elongational flow. The characteristics of this elongation flow can be determined by examining the elasticity of the solution. The longest relaxation time ($\lambda$) of the molecules in solution can be estimated from the Rouse model:[79]

$$\lambda = \frac{6\eta_s[\eta]M_w}{\pi^2 RT}. \tag{3.12}$$

In summation, a relation exists between the viscosity $\eta$ and the longest relaxation time $\lambda$:[82]

$$\eta = G\lambda. \tag{3.13}$$

By combining eqs 3.12 and 3.13, the new longest relation time can be utilized for determining the elastic modulus:

$$\eta_{sp} = G\frac{6\eta_s[\eta]M_w}{R.T\pi^2}. \tag{3.14}$$

At last, initial parameters were calculated from experimental design or constitute eqs 3.5–3.14 for running simulation program. A summary of all dates is shown in Table 3.5.

## 3.5 RESULTS AND DISCUSSION

As mentioned before, controlling the unstable behavior of electrospun nanofibers depends on several parameters. Also, the jet is stretched as it travels downward from the initial position to the collector. For this reason, the modeling and simulation method is a vital path to study the effects of changing a wide range of parameters and jet behavior. Initial parameters were needed for running the simulation program. In addition, some of these parameters are related to each other (i.e., viscosity and elastic modulus, viscosity and concentration, and the charge of the solution and elastic modulus). For this reason, a correct equation should be needed for each parameter. In this work, surface tension, viscosity, the mass of consumable polymer, elastic modulus, specific density, length and radius of the needle, and charge of the solution should be calculated for running simulation. For predicting the behavior and pathways of electrospun polyvinyl alcohol nanofiber according to the initial conditions obtained using the Taguchi method, the three-dimensional momentum equation (eq 3.4) is resolved (considering Runge–Kutta 4th-order method and MATLAB). The results of simulating electrospinning process are shown in Figures 3.16–3.25.

**TABLE 3.5** Calculation Parameters for Electrospinning of Polyvinyl Alcohol Using in this Model.

| Number of experiment | 1 | 2 | 3 | 4 | 5 | 6 | 7 | 8 | 9 |
|---|---|---|---|---|---|---|---|---|---|
| Concentration | 8 | 8 | 8 | 10 | 10 | 10 | 12 | 12 | 12 |
| Spinning distance | 10 | 15 | 20 | 20 | 10 | 15 | 15 | 20 | 10 |
| Applied voltage | 15000 | 20000 | 25000 | 15000 | 20000 | 25000 | 15000 | 20000 | 25000 |
| Flow rate | 0.2 | 0.3 | 0.4 | 0.3 | 0.4 | 0.2 | 0.4 | 0.2 | 0.3 |
| Number of beads | 10 | 10 | 10 | 10 | 10 | 10 | 10 | 10 | 10 |
| Specific density | 1.25 | 1.25 | 1.25 | 1.25 | 1.25 | 1.25 | 1.25 | 1.25 | 1.25 |
| Viscosity | 0.00106 | 0.00106 | 0.00106 | 0.00107 | 0.00107 | 0.00107 | 0.00109 | 0.00109 | 0.00109 |
| Surface tension | 134.1 | 105.9 | 93.1 | 33.5 | 238.4 | 165.6 | 59.6 | 59.6 | 372.5 |
| Mass of polymer | 1.6 | 1.6 | 1.6 | 2 | 2 | 2 | 2.4 | 2.4 | 2.4 |
| Elastic modulus | $8.223 \times 10^5$ | $8.223 \times 10^5$ | $8.223 \times 10^5$ | $8.301 \times 10^5$ | $8.301 \times 10^5$ | $8.301 \times 10^5$ | $8.456 \times 10^5$ | $8.456 \times 10^5$ | $8.456 \times 10^5$ |
| Charge of solution | $6.152 \times 10^9$ | $6.152 \times 10^9$ | $6.152 \times 10^9$ | $6.878 \times 10^9$ | $6.878 \times 10^9$ | $6.878 \times 10^9$ | $7.535 \times 10^9$ | $7.535 \times 10^9$ | $7.535 \times 10^9$ |
| Initial radius of syringe | 0.035 | 0.035 | 0.035 | 0.035 | 0.035 | 0.035 | 0.035 | 0.035 | 0.035 |
| Time | $0–10^{-7}$ | $0–10^{-7}$ | $0–10^{-7}$ | $0–10^{-7}$ | $0–10^{-7}$ | $0–10^{-7}$ | $0–10^{-7}$ | $0–10^{-7}$ | $0–10^{-7}$ |
| Step time | $10^{-12}$ | $10^{-12}$ | $10^{-12}$ | $10^{-12}$ | $10^{-12}$ | $10^{-12}$ | $10^{-12}$ | $10^{-12}$ | $10^{-12}$ |

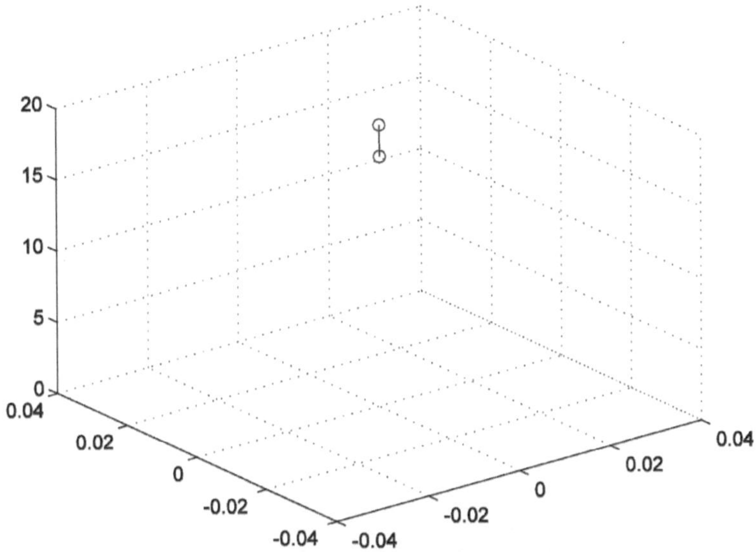


**FIGURE 3.16** Jet path calculated for $N = 10$ at time $= 0–10^{-7}$ s, with step time period $10^{-12}$.

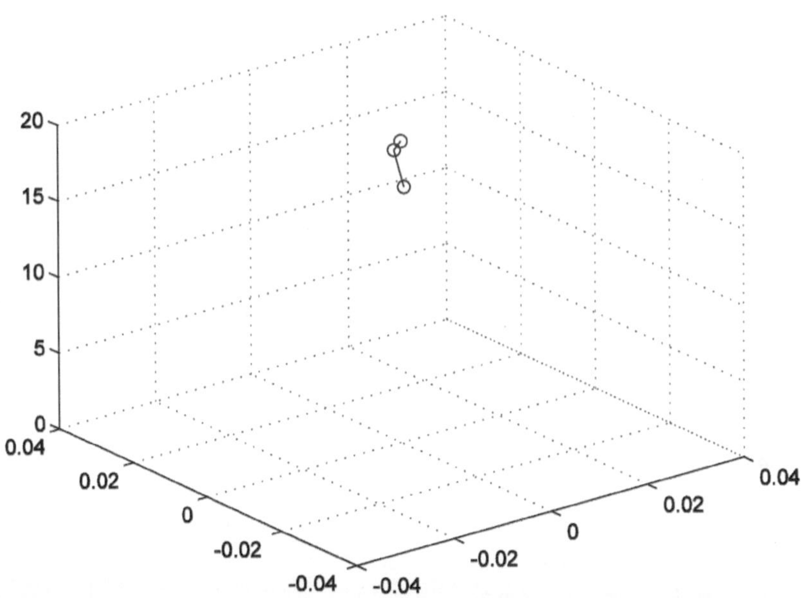


**FIGURE 3.17** Jet path calculated for $N = 10$ at time $= 0–10^{-7}$ s, with step time period $10^{-12}$.

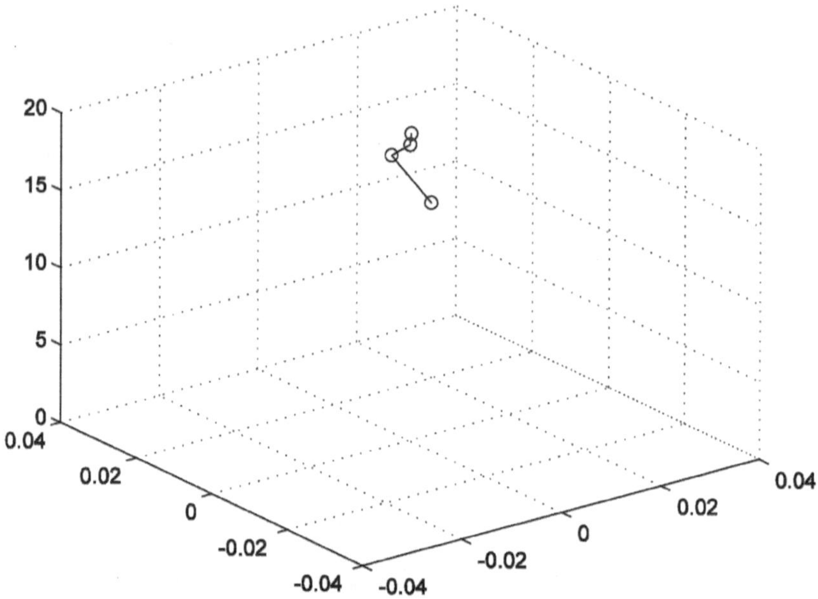


**FIGURE 3.18** Jet path calculated for $N = 10$ at time = $0$–$10^{-7}$ s, with step time period $10^{-12}$.

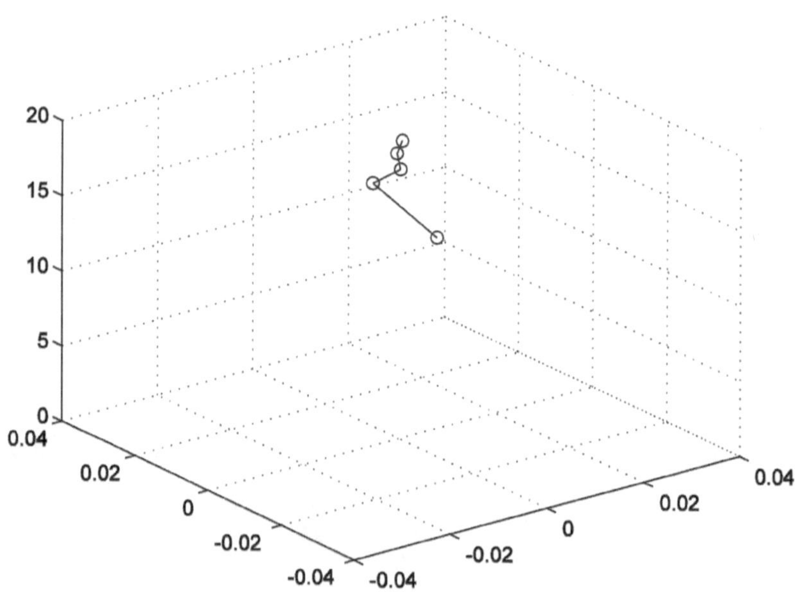


**FIGURE 3.19** Jet path calculated for $N = 10$ at time = $0$–$10^{-7}$ s, with step time period $10^{-12}$.

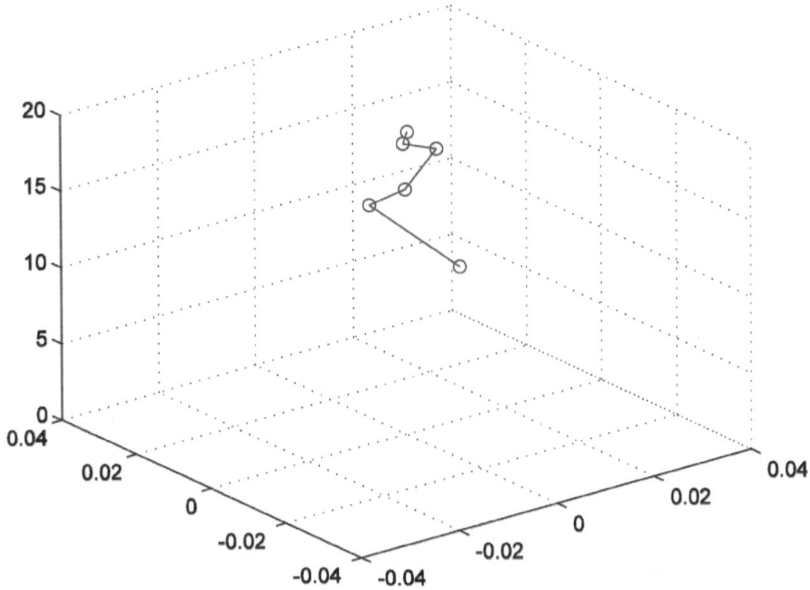


**FIGURE 3.20** Jet path calculated for $N = 10$ at time $= 0\text{–}10^{-7}$ s, with step time period $10^{-12}$.

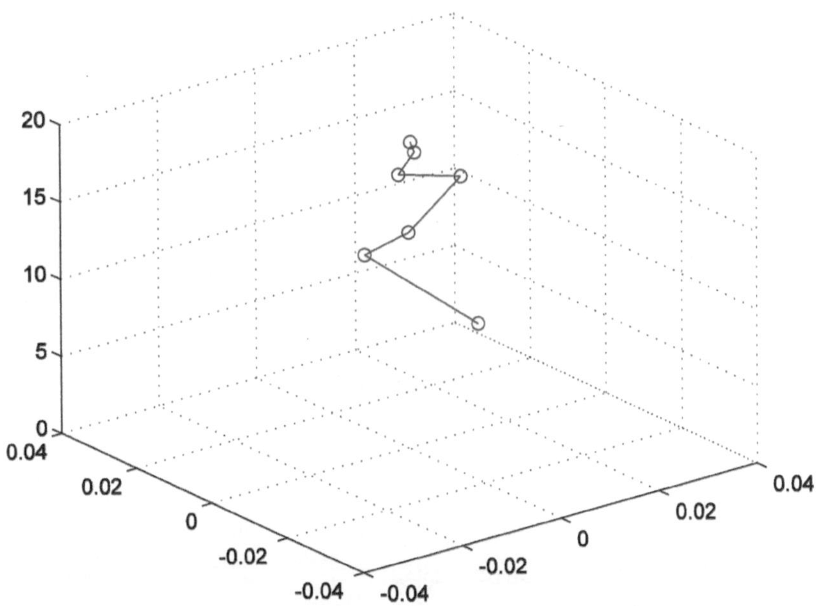


**FIGURE 3.21** Jet path calculated for $N = 10$ at time $= 0\text{–}10^{-7}$ s, with step time period $10^{-12}$.

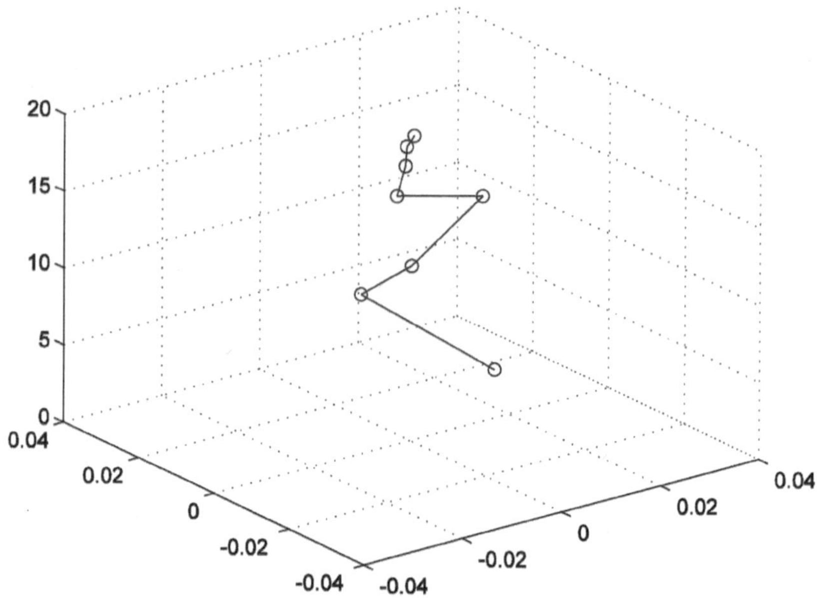


**FIGURE 3.22** Jet path calculated for $N = 10$ at time = 0–$10^{-7}$ s, with step time period $10^{-12}$.

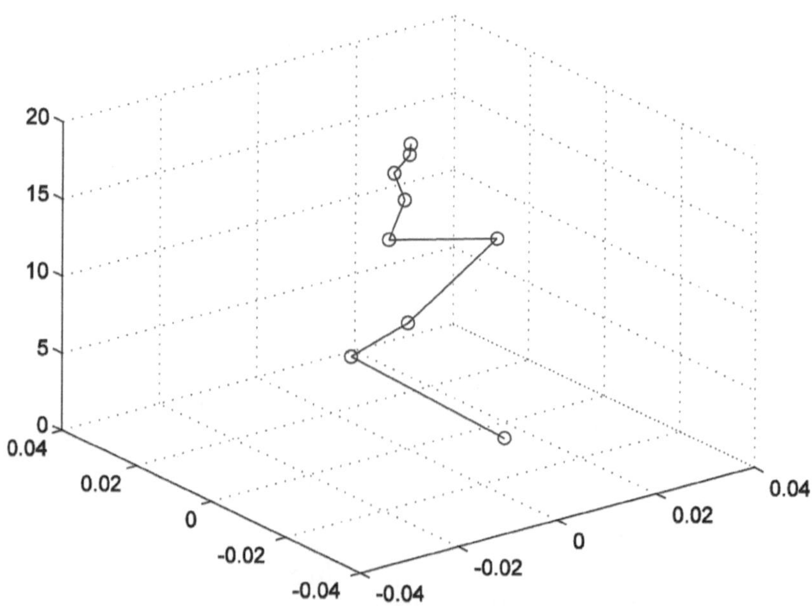


**FIGURE 3.23** Jet path calculated for $N = 10$ at time = 0–$10^{-7}$ s, with step time period $10^{-12}$.

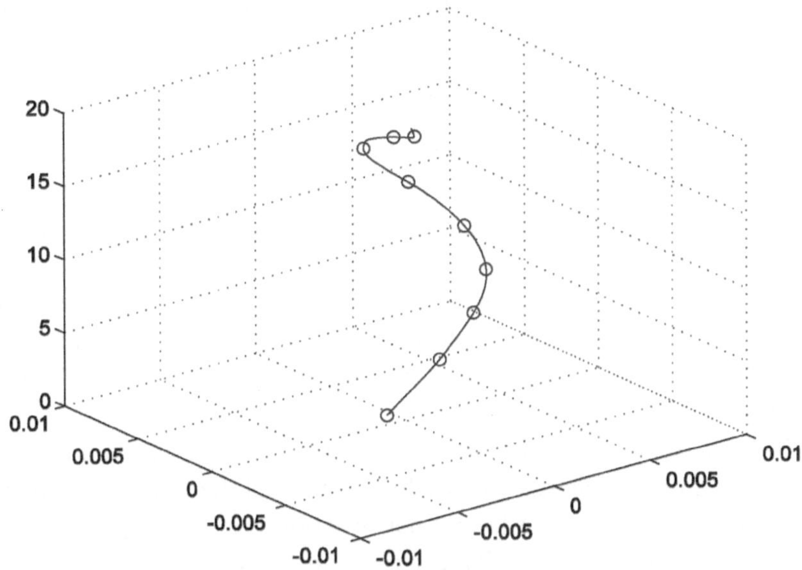


**FIGURE 3.24** Path of a single bead calculated for $N = 10$ at times ranging from 0 to $10^{-7}$ s, with step time period $10^{-12}$.

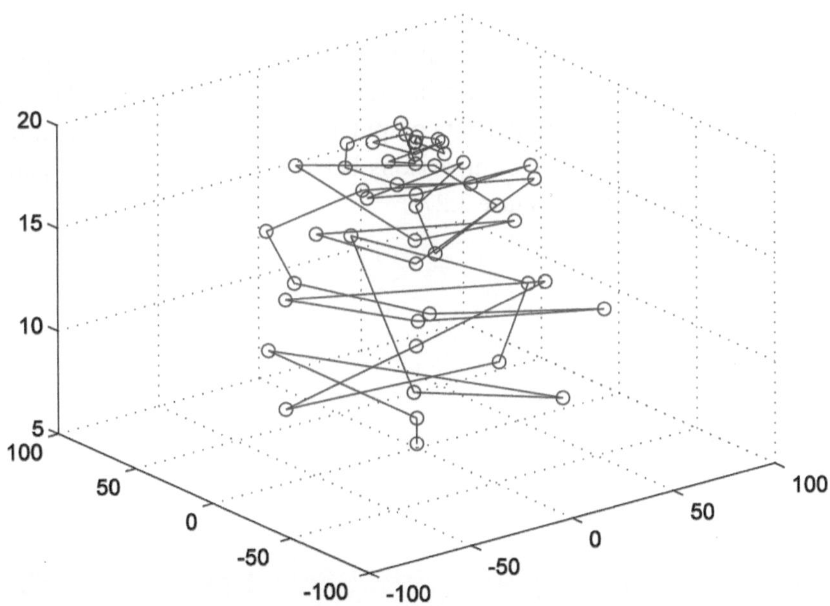


**FIGURE 3.25** Path of a single bead calculated for $N = 50$ at times ranging from 0 to $10^{-7}$ s, with step time period $10^{-12}$.

These figures illustrate the increases in jet segment length as time develops. It demonstrates that the jet is stretched as it moves downward from the initial position to the collector. In addition, new beads are inserted into the calculation.

As the beads progress downwards, the perturbation added to the $x$ and $y$ coordinates begins to grow as it fully develops into the bending instability, at which point the loops continue to grow outward as the jet moves down.

Figure 3.24 shows the path of a single bead at period times (in Table 3.5). The bead does not follow a spiral path in its motion. This type of behavior corresponds to the observed path of a jet during electrospinning.

It shows that the longitudinal stress caused by the external electric field acting on the charge carried by the jet stabilized the straight jet for some distance. Then a lateral perturbation grew in response to the repulsive forces between adjacent elements of the charge carried by the jet.

In Figure 3.25, simulation running can be seen for 50 bead-viscoelastic elements. By considering more beads in lower time, the accuracy of simulation can be increased.

In addition, the results of simulating pathways of electrospun polyvinyl alcohol nanofiber are shown in Figure 3.26. These figures illustrate the consequences of the model data output at several times throughout the calculation of jet pathway. Comparisons between the results of the pathways are summarized in Figure 3.27.

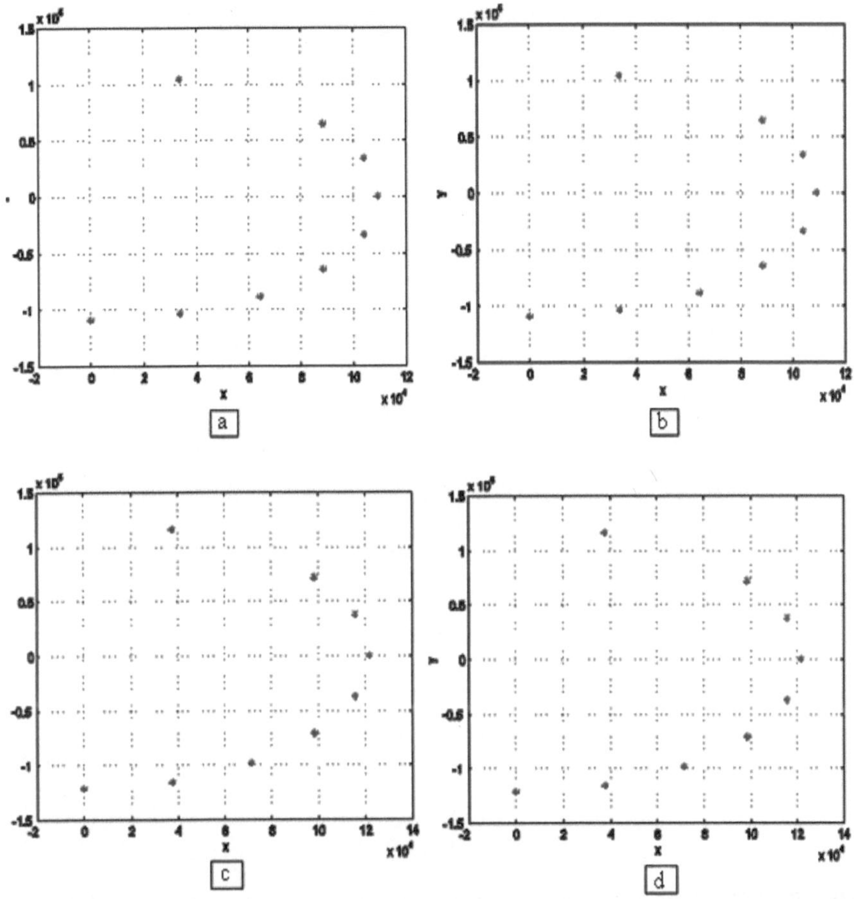


**FIGURE 3.26** Comparison of simulation pathways of polyvinyl alcohol electrospun nanofiber jet for experimental condition 1(a), experimental condition 3(b), experimental condition 6(c), and experimental condition 5(d).

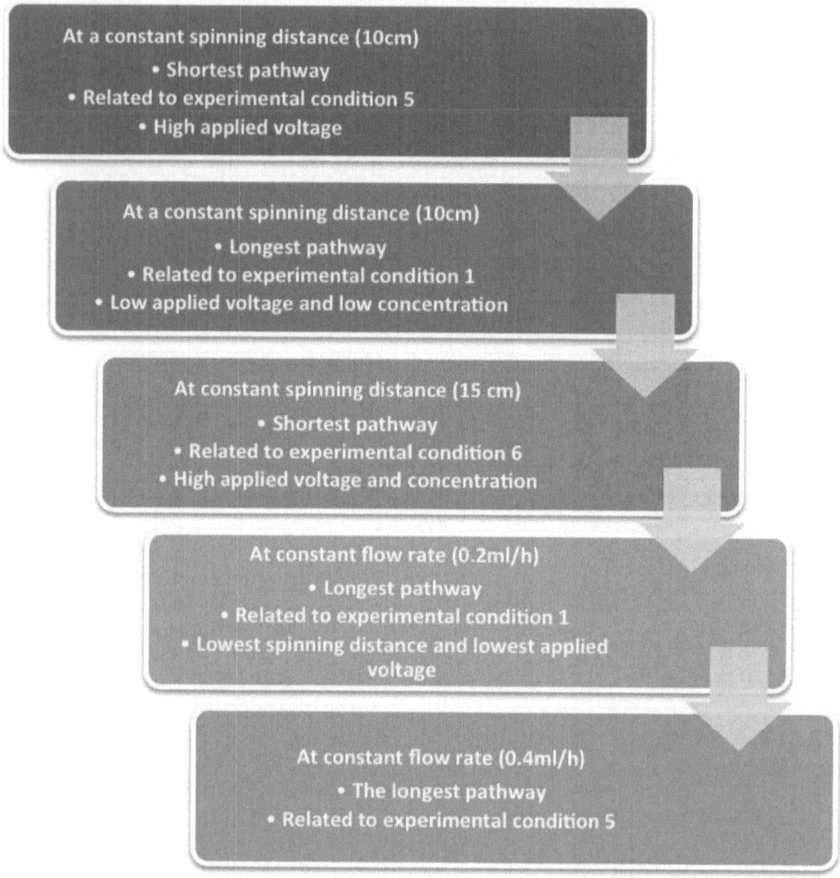


**FIGURE 3.27** Reasons of different effects of selected factors for pathway prediction of polyvinyl alcohol nanofibers between experimental conditions.

## 3.6 CONCLUSIONS

Producing nanofibers by electrospinning is a simple technique and is widely utilized for a variety of applications. During the recent years, research attention has focused more on the optimization of this method to solve the problems which make electrospinning uncontrollable. In addition, the electrospinning process is a fluid dynamics-related problem. For achieving this goal, using simple models can be useful. The analysis and comparison of the model with experiments identify the critical role of the spinning fluid's parameters. Different models were utilized for predicting ideal electrospinning process and electrospun fiber properties. In each model, the researchers tried to improve the existing models or changing the tools in electrospinning by using another view. In this chapter electrospun nanofiber process, and also the importance of utilizing mathematical models for developing the electrospinning process, is investigated and briefly reviewed. In addition, various and even contradictory effects can be found to explain that the influence of all parameters is related to the conditions on the electrospun nanofibers. For achieving a suitable condition, the initial values of these parameters were defined by governing equations obtained from investigation resources or experimentally obtained by Taguchi experimental design. These initial values were utilized for resolving the three-dimensional momentum equation (eq 3.4) by considering Runge–Kutta 4th-order method and MATLAB. All mathematical models and governing equations are investigated in future chapters.

## KEYWORDS


- **electrospun nanofiber**
- **controlling process**
- **microscopic model**
- **constitutive equations**
- **simulation pathway**


## REFERENCES


1. Lu, P.; Ding, B. Applications of Electrospun Fibers. *Recent Pat. Nanotechnol.* **2008,** *2* (3), 169–182.
2. Reneker, D. H.; Yarin, A. L.; Zussman, E.; Xu, H. Electrospinning of Nanofibers from Polymer Solutions and Melts. *Adv. Appl. Mech.* **2007,** *41,* 343–346.
3. Vonch, J.; Yarin, A.; Megaridis, C. M. Electrospinning: A Study in the Formation of Nanofibers. *J. Undergrad. Chem. Res.* **2007,** *1,* 1–6.
4. Sawhney, A. P. S.; Condon, B.; Singh, K. V.; Pang, S. S.; Li, G.; Hui, D. Modern Applications of Nanotechnology in Textiles. *Text. Res. J.* **2008,** *78* (8), 731–739.

5. Huang, Z. M.; Zhang, Y. Z.; Kotaki, M.; Ramakrishna, S. A Review on Polymer Nanofibers by Electrospinning and Their Applications in Nanocomposites. *Compos. Sci. Technol.* **2003,** *63* (15), 2223–2253.
6. Fang, J.; Niu, H. T.; Lin, T.; Wang, X. G. Applications of Electrospun Nanofibers. *Chin. Sci. Bull.* **2008,** *15,* 2265–2286.
7. Ramakrishna, S.; Fujihara, K.; Teo, W. E.; Yong, T.; Ma, Z.; Ramaseshan, R. Electrospun Nanofibers: Solving Global Issues. *Mater. Today.* **2006,** *3,* 40–50.
8. Rafiei, S.; Maghsoodloo, S.; Noroozi, B.; Mottaghitalab, V.; Haghi, A. K.; Mathematical Modeling in Electrospinning Process of Nanofibers: A Detailed Review. *Cellul. Chem. Technol.* **2013,** *47* (5), 323–338.
9. Mazoochi, T.; Hamadanian, M.; Ahmadi, M.; Jabbari, V. Investigation on the Morphological Characteristics of Nanofiberous Membrane as Electrospun in the Different Processing Parameters. *Int. J. Ind. Chem.* **2012,** *1,* 1–8.
10. Wang, H. S.; Fu, G. D.; Li, X. S. Functional Polymeric Nanofibers from Electrospinning. *Recent Pat. Nanotechnol.* **2009,** *3* (1), 21–31.
11. Agarwal, S.; Wendorff, J. H.; Greiner, A. Use of Electrospinning Technique for Biomedical Applications. *Polymer.* **2008,** *26,* 5603–5621.
12. Beachley, V.; Wen, X. Polymer Nanofibrous Structures: Fabrication, Biofunctionalization, and Cell Interactions. *Prog. Polym. Sci.* **2010,** *35* (7), 868–892.
13. Patan, A. K.; Sasikanth, K.; Sreekanth, N.; Suresh, P.; Brahmaiah, B. Nanofibers–A New Trend in Nano Drug Delivery Systems. *Int. J. Pharm. Res. Biomed. Anal.* **2013,** *3,* 47–55.
14. Angammana, C. J. A Study of the Effects of Solution and Process Parameters on the Electrospinning Process and Nanofibre Morphology. Ph.D. Thesis, The University of Waterloo, 2011.
15. Ramakrishna, S.; Fujihara, K., et al. *An Introduction to Electrospinning and Nanofibers;* World Scientific: Singapore, 2005; pp 1–383.
16. Karra, S. Modeling Electrospinning Process and a Numerical Scheme Using Lattice Boltzmann Method to Simulate Viscoelastic Fluid Flows. MS. Thesis, Texas A&M University, May 2007.
17. Brown, E. E.; Laborie, M. P. G. Bioengineering Bacterial Cellulose/Poly (Ethylene Oxide) Nanocomposites, *Biomacromolecules.* **2007,** *8* (10), 3074–3081.
18. Ciechańska, D. Multifunctional Bacterial Cellulose/Chitosan Composite Materials for Medical Applications. *Fibres Text. East. Eur.* **2004,** *12* (4), 69–72.
19. Wang, C.; Cheng Y. W.; Hsu C. H.; Chien, H. S.; Tsou, S. Y. How to Manipulate the Electrospinning Jet with Controlled Properties to Obtain Uniform Fibers with the Smallest Diameter?—A Brief Discussion of Solution Electrospinning Process. *J. Polym. Res.* **2011,** *18* (1), 111–123.
20. Lukáš, D.; Sarkar, A.; Martinová, L.; Vodsed'álková, K.; Lubasova, D.; Chaloupek, J.; Pokorný, P.; Mikeš, P.; Chvojka, J.; Komarek, M. Physical Principles of Electrospinning (Electrospinning as a Nano-Scale Technology of the Twenty-First Century). *Text. Prog.* **2009,** *2,* 59–140.
21. Tao, J.; Shivkumar, S. Molecular Weight Dependent Structural Regimes during the Electrospinning of PVA, *Mater. Lett.* **2007,** *61* (11–12), 2325–2328.
22. Zeng, Y.; Pei, Z.; Wang, X.; Chen, S. In *Numerical Simulation of Whipping Process in Electrospinning,* Proceedings of the Eighth WSEAS International Conference on

Applied Computer and Applied Computational Science, Shanghai, China, 309–317, 2009.
23. Kowalewski, T. A.; Blonski, S.; Barral, S. Experiments and Modelling of Electrospinning Process. *Bull. Pol. Acad. Sci. Tech. Sci.* **2005,** *53* (4), 385–394.
24. Baji, A.; Mai, Y. W.; Wong, S. C.; Abtahi, M.; Chen, P. Electrospinning of Polymer Nanofibers: Effects on Oriented Morphology, Structures and Tensile Properties. *Compos. Sci. Technol.* **2010,** *70* (5), 703–718.
25. Zong, X.; Kim, K.; Fang, D.; Ran, S.; Hsiao, B. S.; Chu, B. Structure and Process Relationship of Electrospun Bioabsorbable Nanofiber Membranes. *Polymer.* **2002,** *43* (16), 4403–4412.
26. Haghi, A. K. Electrospun Nanofiber Process Control. *Cellul. Chem. Technol.* **2010,** *44* (9), 343–352.
27. Keun, S. W.; Ho, Y. J.; Seung, L. T.; Park, W. H. Effect of pH on Electrospinning of Poly (Vinyl Alcohol). *Mater. Lett.* **2005,** *12,* 1571–1575.
28. Gilbert, W. *De Magnete;* Courier Dover Publications: New York, 1958; p 366.
29. Garg, K.; Bowlin, G. L. Electrospinning Jets and Nanofibrous Structures. *Biomicrofluidics.* **2011,** *5* (1), 13403.
30. Taylor, G. I. The Scientific Papers of Sir Geoffrey Ingram Taylor. In *Mechanics of Fluids: Miscellaneous Papers;* Batchelor, G. K., Eds.; Cambridge University Press: New York, 1971; Vol. 4, pp 590.
31. Reneker, D. H.; Chun, I. Nanometre Diameter Fibres of Polymer, Produced by Electrospinning, *Nanotechnology.* **1996,** *7* (3), 216–223.
32. Stanger, J. J.; Tucker, N.; Kirwan, K.; Coles, S.; Jacobs, D.; Staiger, M. P. Effect of Charge Density on the Taylor Cone in Electrospinning. *Int. J. Mod. Phys. B.* **2009,** *23* (6), 54–59.
33. Chronakis, I. S. Novel Nanocomposites and Nanoceramics Based on Polymer Nanofibers Using Electrospinning Process—A Review. *J. Mater. Process. Technol.* **2005,** *167* (2–3), 283–293.
34. Pham, Q. P.; Sharma, U.; Mikos, A. G. Electrospinning of Polymeric Nanofibers for Tissue Engineering Applications: A Review. *Tissue Eng.* **2006,** *12* (5), 1197–1211.
35. Ghochaghi, N. Experimental Development of Advanced Air Filtration Media Based on Electrospun Polymer Fibers. Ph.D. Thesis, Virginia Commonwealth University, December 2014.
36. Ziabari, M.; Mottaghitalab, V.; Haghi, A. K. Evaluation of Electrospun Nanofiber Pore Structure Parameters. *Korean J. Chem. Eng.* **2008,** *25* (4), 923–932.
37. Sawicka, K. M.; Gouma, P. Electrospun Composite Nanofibers for Functional Applications. *J. Nanopart. Res.* **2006,** *8* (6), 769–781.
38. Li, W. J.; Laurencin, C. T.; Caterson, E. J.; Tuan, R. S.; Ko, F. K. Electrospun Nanofibrous Structure: A Novel Scaffold for Tissue Engineering. *J. Biomed. Mater. Res.* **2002,** *4,* 613–621.
39. Brooks, H.; Tucker, N. Electrospinning Predictions Using Artificial Neural Networks. *Polymer.* **2015,** *58,* 22–29.
40. Sill, T. J.; Recum, H. A. Electrospinning: Applications in Drug Delivery and Tissue Engineering. *Biomaterials.* **2008,** *29* (13), 1989–2006.
41. Yousefzadeh, M.; Latifi, M.; Amani, T. M.; Teo, W. E.; Ramakrishna, S. A Note on the 3D Structural Design of Electrospun Nanofibers, *J. Eng. Fabr. Fiber.* **2012,** *7* (2), 17–23.

42. Machmudah, S.; Murakami, K.; Okubayashi, S.; Goto, M. Generation of PVP Fibers by Electrospinning in One-Step Process under High-Pressure $CO_2$. *Int. J. Ind. Chem.* **2013,** *4* (1), 1–6.
43. Zhang, S. Mechanical and Physical Properties of Electrospun Nanofibers. MS Thesis. NC State University, August 2009.
44. Šimko, M.; Erhart, J.; Lukáš, D. A Mathematical Model of External Electrostatic Field of a Special Collector for Electrospinning of Nanofibers. *J. Electrost.* **2014,** *72* (2), 161–165.
45. He, J. H.; Wu, Y.; Zuo, W. W. Critical Length of Straight Jet in Electrospinning. *Polymer.* **2005,** *46* (26), 12637–12640.
46. Hohman, M. M.; Shin, M.; Rutledge, G.; Brenner, M. P. Electrospinning and Electrically Forced Jets. II. Applications. *Phys. Fluids.* **2001,** *13* (8), 2221–2236.
47. Wu, Y.; Yu, J. Y.; He, J. H.; Wan, Y. Q. Controlling Stability of the Electrospun Fiber by Magnetic Field. *Chaos Solitons Fractals.* **2007,** *32* (1), 5–7.
48. Yarin, A. L.; Koombhongse, S.; Reneker, D. H. Bending Instability in Electrospinning of Nanofibers. *J. Appl. Phys.* **2001,** *89* (5), 3018–3026.
49. Li, Z.; Wang, C. *Effects of Working Parameters on Electrospinning;* Springer: Berlin, 2013; pp 15–28.
50. Bognitzki, M.; Czado, W.; Frese, T.; Schaper, A.; Hellwig, M.; Steinhart, M.; Greiner, A.; Wendorff, J. Nanostructured Fibers via Electrospinning. *Adv. Mater.* **2001,** *13* (1), 70–72.
51. De, V. S.; Van, C. T.; Nelvig, A.; Hagström, B.; Westbroek, P.; De, C. K. The Effect of Temperature and Humidity on Electrospinning. *J. Mater. Sci.* **2009,** *44* (5), 1357–1362.
52. Rafiei, S.; Maghsoodloo, S.; Saberi, M.; Lotfi, S.; Motaghitalab, V.; Noroozi, B.; Haghi, A. K. New Horizons in Modeling and Simulation of Electrospun Nanofibers: A Detailed Review. *Cellul. Chem. Technol.* **2014,** *48* (5), 401–424.
53. Bhardwaj, N.; Kundu, S. C. Electrospinning: A Fascinating Fiber Fabrication Technique. *Biotechnol. Adv.* **2010,** *28* (3), 325–347.
54. Tan, S. H.; Inai, R.; Kotaki, M.; Ramakrishna, S. Systematic Parameter Study for Ultra-Fine Fiber Fabrication via Electrospinning Process. *Polymer.* **2005,** *46* (16), 6128–6134.
55. Feng, J. J. Stretching of a Straight Electrically Charged Viscoelastic Jet. *J. Nonnewton. Fluid Mech.* **2003,** *116* (1), 55–70.
56. Thompson, C. J.; Chase, G. G.; Yarin, A. L.; Reneker, D. H. Effects of Parameters on Nanofiber Diameter Determined from Electrospinning Model. *Polymer.* **2007,** *48* (23), 6913–6922.
57. Thompson, C. J. An Analysis of Variable Effects on a Theoretical Model of the Electrospin Process for Making Nanofibers. MS Thesis, The Graduate Faculty of the University of Akron, May 2007.
58. Liu, L.; Dzenis, Y. Simulation of Electrospun Nanofibre Deposition on Stationary and Moving Substrates. *Micro Nano Lett.* **2011,** *6* (6), 408–411.
59. Greenfeld, I.; Arinstein, A.; Fezzaa, K.; Rafailovich, M. H.; Zussman, E. Polymer Dynamics in Semidilute Solution during Electrospinning: A Simple Model and Experimental Observations. *Phys. Rev. E.* **2011,** *84* (4), 041806.

60. Collins, A. J.; Meyr, D.; Sherfey, S.; Tolk, A.; Petty, M. *The Value of Modeling and Simulation Standards;* Virginia Modeling, Analysis and Simulation Center, Old Dominion University: Suffolk, VA, 2011; pp 1–8.
61. Banks, J. *Handbook of Simulation;* John Wiley: Canada, 1998; p 837.
62. Kowalewski, T. A.; Barral, S.; Kowalczyk, T. Modeling Electrospinning of Nanofibers. In: Pyrz, R., Rauhe, J. C., Eds.; IUTAM Symposium on Modelling Nanomaterials and Nanosystems; IUTAM Bookseries: Dordrecht, 2009; pp 279–292.
63. Patanaik, A.; Jacobs, V.; Anandjiwala, R. D. *Experimental Study and Modeling of the Electrospinning Process*, The 86th Textile Institute World Conference, Hong Kong, *2008*; pp 1160–1168.
64. Reneker, D. H.; Yarin, A. L. Electrospinning Jets and Polymer Nanofibers. *Polymer.* **2008,** *49* (10), 2387–2425.
65. Lyons, J.; Li, C.; Ko, F. Melt-Electrospinning Part I: Processing Parameters and Geometric Properties. *Polymer.* **2004,** *45* (22), 7597–7603.
66. Melcher, J. R.; Taylor, G. I. Electrohydrodynamics: A Review of the Role of Interfacial Shear Stresses. *Annu. Rev. Fluid Mech.* **1969,** *1* (1), 111–146.
67. Ramos, J. I. One-Dimensional Models of Steady, Inviscid, Annular Liquid Jets. *Appl. Math. Model.* **1996,** *20* (8), 593–607.
68. Spivak, A. F.; Dzenis, Y. A. Asymptotic Decay of Radius of a Weakly Conductive Viscous Jet in an External Electric Field. *Appl. Phys. Lett.* **1998,** *73* (21), 3067–3069.
69. Reneker, D. H.; Yarin, A. L.; Fong, H.; Koombhongse, S. Bending Instability of Electrically Charged Liquid Jets of Polymer Solutions in Electrospinning. J. *Appl. Phys.* **2000,** *87* (9), 4531–4547.
70. Dasri, T. Mathematical Models of Bead-Spring Jets during Electrospinning for Fabrication of Nanofibers. *Walailak J. Sci. Technol.* **2012,** *9* (4), 287–296.
71. Shin, Y. M.; Hohman, M. M.; Brenner, M. P.; Rutledge, G. C. Experimental Characterization of Electrospinning: The Electrically Forced Jet and Instabilities. *Polymer.* **2001,** *42* (25), 09955–09967.
72. Feng, J. J. The Stretching of an Electrified Non-Newtonian Jet: A Model for Electrospinning. *Phys. Fluids.* **2002,** *14* (11), 3912–3926.
73. Wan, Y. Q.; Guo, Q.; Pan, N. Thermo-Electro-Hydrodynamic Model for Electrospinning Process. *Int. J. Nonlinear Sci. Numer. Simul.* **2004,** *5* (1), 5–8.
74. He, J. H.; Wu, Y.; Pang, N. A Mathematical Model for Preparation by AC-Electrospinning Process. *Int. J. Nonlinear Sci. Numer. Simul.* **2005,** *6* (3), 243–248.
75. Xu, L.; Wang, L.; Faraz, N. A Thermo-Electro-Hydrodynamic MODEL for Vibration-Electrospinning Process. *Therm. Sci.* **2011,** *15* (Suppl. 1), S131–S135.
76. Haghi, A. K.; Zalkov, G. E. Mathematical Models on the Transport Properties of Electrospun Nanofibers, CRC Press: Toronto, 2013; pp 195–218.
77. Kalita, G.; Adhikari, S.; Aryal, H. R.; Somani, P. R.; Somani S. P.; Sharon, M.; Umeno, M. Taguchi Optimization of Device Parameters for Fullerene and Poly (3-Octylthiophene) based Heterojunction Photovoltaic Devices. *Diam. Relat. Mater.* **2008,** *17* (4), 799–803.
78. Kwaambwa, H. M.; Goodwin, J. W.; Hughes, R. W.; Reynolds, P. A. Viscosity, Molecular Weight and Concentration Relationships at 298K of Low Molecular Weight Cis-Polyisoprene in a Good Solvent. *Colloids Surf. A Physicochem. Eng. Asp.* **2007,** *294* (1–3), 14–19.

79. Koski, A.; Yim, K.; Shivkumar, S. Effect of Molecular Weight on Fibrous PVA Produced by Electrospinning. *Mater. Lett.* **2004,** *58* (3–4), 493–497.
80. Supaphol, P.; Chuangchote, S. On the Electrospinning of Poly (Vinyl Alcohol) Nanofiber Mats: A Revisit. *J. Appl. Polym. Sci.* **2008,** *108* (2), 969–978.
81. Chuangchote, S.; Sirivat, A.; Supaphol, P. Mechanical and Electro-Rheological Properties of Electrospun Poly (Vinyl Alcohol) Nanofibre Mats Filled with Carbon Black Nanoparticles. *Nanotechnology*. **2007,** *18* (14), 145705–145713.
82. Adam, M.; Delsanti, M. Viscosity and Longest Relaxation Time of Semi-Dilute Polymer Solutions. I. Good Solvent. *J. Physique*. **1983,** *44* (10), 1185–1193.

CHAPTER 4

# NUMERICAL MODELING FOR HOMOGENEOUS AND STRATIFIED FLOWS: FROM THEORY TO PRACTICE

KAVEH HARIRI ASLI[1*], SOLTAN ALI OGLI ALIYEV[1], and HOSSEIN HARIRI ASLI[2]

[1]*Department of Mathematics and Mechanics, National Academy of Science of Azerbaijan 'AMEA', Baku, Azerbaijan*

[2]*Civil Engineering Department, Faculty of Engineering, University of Guilan, Rasht, Iran*

[*]*Corresponding author. E-mail: hariri_k@yahoo.com*

## CONTENTS

## ABSTRACT

One of the problems in the study of fluid flow in plumbing systems is the behavior of stratified fluid in the channels. Mostly, steady flows initially are ideal, then the viscous and turbulent fluid in the pipes. If you look at the deep pool filled with water, and on its surface to create a disturbance, then the surface of the water will begin to propagate. Their origin is explained by the fact that the fluid particles are located near the cavity. They create disturbances which will seek to fill the cavity under the influence of gravity. This work showed that the development of this phenomenon leads to the spread of waves on the water.


## 4.1 INTRODUCTION

The fluid particles in a wave do not move up and down around in circles. The waves of water are neither longitudinal nor transverse. They seem to be a mixture of both. The radius of the circles varies with depth of moving fluid particles. They keep reducing as long as the radius does not become equal to zero.

If we analyze the propagation velocity of waves on water, it will be revealed that the velocity of waves depends on length of waves. The speed of long waves is proportional to the square root of the acceleration of gravity multiplied by the wave length:

$$v_{\Phi} = \sqrt{g\lambda}.$$

The cause of these waves is the force of gravity.

For short waves the restoring force is due to surface tension force, and therefore the speed of these waves is proportional to the square root of the surface tension and wavelength private. The numerator of which is the surface tension, and in the denominator, the product of the wavelength to the density of water:

$$v_{\Phi} = \sqrt{\frac{\sigma}{\lambda \rho}}.$$

Suppose there is a channel with a constant slope bottom, extending to infinity along the axis $Ox$. And let the feed in a field of gravity flows, incompressible fluid. It is assumed that the fluid is devoid of internal friction. Friction is neglected on the sides and bottom of the channel. The liquid level at bottom of the channel is $h$. It is a small quantity compared with the characteristic dimensions of the flow, the size of the bottom roughness, and so on.

$$\text{Let } h = \xi + h_0,$$

where $h_0$ denotes the ordinate of the free liquid surface (Fig. 4.1). Free liquid surface $h_0$ which is in equilibrium in the gravity field is flat. As a result of any external influence, liquid surface in a location is removed from its equilibrium position. There is a movement spreading across the entire surface of the liquid in the form of waves, called gravity.

They are caused by the action of gravity field. This type of waves occurs mainly on the liquid surface. They capture the inner layers and are deeper for smaller liquid surfaces.[1–15]

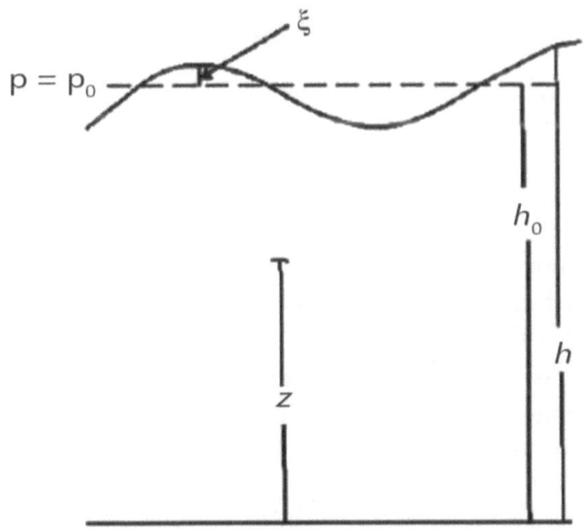


**FIGURE 4.1** Schematic showing the layer of fluid of variable depth.

where $h_0$ is the level of the free surface; $\xi$ a deviation from the level of the liquid free surface; $h$ the depth of the fluid; and $z$ is the vertical coordination of any point in the water column. We assume that the fluid flow is characterized by a spatial variable $x$ and time dependent $t$.

Thus, it is believed that the fluid velocity $u$ has a nonzero component $u_x$ which will be denoted by $u$ (other components can be neglected in addition, the level of $h$ depends only on $x$ and $t$).

Let us consider such gravitational waves, in which the speed of moving particles is so small that for the Euler equation, one can neglect $(u\nabla)u$ compared with $\partial u / \partial t$ During the time period $\tau$, committed by the fluid particles in the wave, these particles pass the distance of the order of the amplitude $\alpha$.

Therefore, the speed of their movement will be:

$$u \sim \alpha / \tau.$$

Rate $u$ varies considerably over time intervals of the order $\tau$ and for distances of the order of wavelength $\lambda$ along the direction of wave propagation.

Therefore, the derivative of the velocity time order $u / \tau$ and the coordinates order $u / \lambda$ lead to the condition:

$$(u\nabla)u << \partial u / \partial t,$$

which is equivalent to the requirement

$$\frac{1}{\lambda}\left(\frac{a}{\lambda}\right)^2 << \frac{a}{\tau}\frac{1}{\tau} \qquad a << \lambda \tag{4.1}$$

That is, amplitude of the wave must be small compared with the wavelength. Consider the propagation of waves in the channel $Ox$ directed along the axis for fluid flow along the channel.

Channel cross-section can be of any shape and it changes along its length with changes in liquid level. The cross-sectional area of the liquid in the channel is denoted by

$$h = h\,(x, t).$$

The depth of the channel and basin is assumed to be small compared with the wavelength.

We write the Euler equation in the form of:

$$\frac{\partial u}{\partial t} = -\frac{1}{\rho}\frac{\partial p}{\partial x}, \tag{4.2}$$

$$\frac{1}{\rho}\frac{\partial p}{\partial z} = -g, \tag{4.3}$$

where $\rho$ is density, $p$ is pressure, and $g$ is acceleration of free fall.

Quadratic in velocity members is omitted, since the amplitude of the waves is still considered low.

From the second equation we have that at the free surface:

$$z = h\,(x,\, t),$$

where $p = p_0$ should be satisfied:

$$p = p_0 + \rho g\,(h - z),$$

$$\frac{\partial u}{\partial t} = -g\frac{\partial h}{\partial x}. \tag{4.4}$$

To determine $u$ and $h$ we use the continuity equation for the case under consideration.

Consider the volume of fluid contained between two planes of the cross-section of the canal at a distance $dx$ from each other. Let $(hu)_x$ be the amount of fluid passing through a cross-section $x$ per unit time. At the same time through another section $x + dx$, the forthcoming amount of fluid $(hu)_{x+dx}$.

Therefore, the volume of fluid between the planes is

$$(hu)_{x+dx} - (hu)_x = \frac{\partial(hu)}{\partial x}dx.$$

By virtue of incompressibility of the liquid, a change could occur only due to changes in its level.Changing the volume of fluid between these planes in a unit time is equal to $\frac{\partial h}{\partial t}dx$. Consequently, we can write:

$$\frac{\partial(hu)}{\partial x}dx = -\frac{\partial h}{\partial t}dx \text{ and } \frac{\partial(hu)}{\partial x} + \frac{\partial h}{\partial t} = 0,\ t > 0,\ -\infty < x < \infty. \tag{4.5}$$

Since, $h = h_0 + \xi$ where $h_0$ denotes the ordinate of the free liquid surface, a state of relative equilibrium evolving under the influence of gravity is:

$$\frac{\partial \xi}{\partial t} + h_0\frac{\partial u}{\partial x} = 0. \tag{4.6}$$

Thus, we obtain the following system of equations describing the fluid flow in the channel:

$$\frac{\partial \xi}{\partial t}+h_0\frac{\partial u}{\partial x}=0,\quad \frac{\partial u}{\partial t}+g\frac{\partial \xi}{\partial x}=0,\quad t>0,\quad -\infty<x<\infty. \tag{4.7}$$

## 4.2 MATERIALS AND METHODS

The phase velocity $h_0$ is expressed in terms of frequency $v_\Phi$ and wavelength $\lambda$ (or the angular frequency $f$) and wave number by formulas $\omega = 2\pi f$ and $k = 2\pi / \lambda$.

The concept of phase velocity can be used if the harmonic wave propagates without changing shape.

This condition is always performed in linear environments. When the phase velocity depends on the frequency, it is equivalent to talk about the velocity dispersion. In the absence of any dispersion the waves assumed with a rate equal to the phase velocity.

Experimentally, the phase velocity at a given frequency can be obtained by determining the wavelength of the interference experiments. The ratio of phase velocities in the two media can be found on the refraction of a plane wave at the plane boundary of these environments. This is because the refractive index is the ratio of phase velocities.

It is known that the wave number $k$ satisfies the wave equation are not any values $\omega$. To establish this connection, it is sufficient to substitute the solution of the form $\exp[i(\omega t - kx)]$ in the wave equation.

The complex form is the most convenient and compact. We can show that any other representation of harmonic solutions, including the form of a standing wave, leads to the same connection between $\omega$ and $k$.

By substituting the wave solution into the equation for a string, we can see that the equation becomes an identity for:

$$\omega^2 = k^2 v_\phi^2.$$

Exactly the same relation follows from the equations for waves in the gas, the equations for elastic waves in solids, and the equations for electromagnetic waves in vacuum. The presence of energy dissipation[16–18] leads to the appearance of the first derivatives (forces of friction) in the wave equation. The relationship between frequency and wave number becomes the domain of complex numbers. For example, the telegraph equation (for electric waves in a conductive line) yields:

$$\omega^2 = k^2 v_\phi^2 + \frac{i \cdot \omega R}{L}.$$

The relation connecting frequency and wave number (wave vector), in which the wave equation has a wave solution, is called a dispersion relation, the dispersion equation, or dispersion. This type of dispersion relation determines the nature of the wave. Since the wave equations are partial differential equations of second order in time and coordinates, the dispersion is usually a quadratic equation in the frequency or wave number.

The simplest dispersion equations presented above for the canonical wave equation are also two very simple solutions:

$$\omega = -kv_\phi. \text{ and } \omega = -kv_\phi.$$

We know that these two solutions represent two waves traveling in opposite directions. By its physical meaning the frequency is a positive value so that the two solutions must define two values of the wave number, which differ in sign. The act permits the dispersion and, generally speaking, the existence of waves with all wave numbers that is of any length, and, consequently, any frequencies. The phase velocity of these waves is

$$v_\Phi = \frac{\omega}{k},$$

which coincides with the most velocity, appears in the wave equation, and is a constant which depends only on the properties of the medium.

The phase velocity depends on the wave number and, consequently, on the frequency. The dispersion equation for the telegraph equation is an algebraic quadratic equation that has complex roots. By analogy with the theory of oscillations, the presence of imaginary part of the frequency means the damping or growth of waves. It can be noted that the form of the dispersion law determines the presence of damping or growth.

In general terms, the dispersion can be represented by the equation:

$$\Phi\left(\omega, \vec{k}\right) = 0,$$

where $\Phi$ is a function of frequency and wave vector.

By solving this equation for $\omega$ you can obtain an expression for the phase velocity:

$$v_{\Phi} = \frac{\omega}{k} = f\left(\omega, \vec{k}\right).$$

By definition, the phase velocity is a vector directed normal to phase surface.

Then, more correctly, one can write the last expression in the following form:

$$\vec{v}_{\Phi} = \frac{\lambda}{T} = \frac{\omega}{k^2} \cdot \vec{k} = f\left(\omega, \vec{k}\right).$$

## 4.3 RESULTS AND DISCUSSION

The most important subject of this research is wave physics, which has the primary practical significance.

If we refer to dimensionless parameters and variables:

$$\tau = t\sqrt{\frac{g}{h_0}}, \quad X = \frac{x}{h_0}, \quad U = u\frac{1}{\sqrt{gh_0}}, \quad \delta = \frac{\xi}{h_0},$$

The system of equations becomes:

$$\frac{\partial \delta}{\partial \tau} + \frac{\partial U}{\partial X} = 0, \qquad \frac{\partial U}{\partial \tau} + \frac{\partial \delta}{\partial X} = 0, \quad t = 0, \qquad -\infty < X < \infty, \tag{4.8}$$

Consider plane harmonic longitudinal waves, that is, we seek the solution of eq 4.9 as the real part of the following complex expressions:

$$\Psi = \Psi^0 \exp[i(k_* X + \omega_* \tau)], \qquad \Psi^0 = \Psi^0_* + i\Psi^0_{**} \qquad |\Psi^0| << 1$$

$$k_* = k + k_{**}, \qquad \omega_* = \omega + i\omega_{**}, \tag{4.9}$$

which determines the amplitude of the perturbations of displacement and velocity.

$$\Psi = \delta, U, \ \text{a} \ \ \Psi^0 = \delta^0, U^0.$$

There are two types of solutions:

*Type I*: Solution or wave of the first type, when:

$k_* = k =$ A real positive number ($k > 0$, $k_{**} = 0$).

In this case, we have:

$$\Psi = \left(\Psi_*^0 + i\Psi_{**}^0\right)\exp\left[i\left(kX + \omega\tau + i\omega_{**}\tau\right)\right] = \left(\Psi_*^0 + i\Psi_{**}^0\right)\exp\left(-\omega_{**}\tau\right)\times$$
$$\left[\cos\left(kX + \omega\tau\right) + i\sin\left(kX + \omega\tau\right)\right]$$

$$\operatorname{Re}\{\Psi\} = \exp\left(-\omega_{**}\tau\right)\left|\Psi^0\right|\sin\left[\phi + \left(kX + \omega\tau\right)\right],$$

$$\left|\Psi^0\right| = \sqrt{\Psi_*^{02} + \Psi_{**}^{0\,2}}, \qquad \phi = arctg\left(-\Psi_*^0 / \Psi_{**}^0\right)$$

Thus, the decision of the first type is a sinusoidal coordinate and $\omega_{**} > 0$, decaying exponentially in time perturbation, which is called $k$ wave:

$$\Psi(k) = \left|\Psi^0\right|\exp\left[-\omega_{**}(k)\,\tau\right]\sin\left\{\phi + \frac{2\pi\left[X + v_{\Phi}(k)\,\tau\right]}{\lambda(k)}\right\} \tag{4.10}$$

where

$v_{\Phi}(k) = \omega(k) / k$, $\lambda(k) = 2\pi / k$), $\varphi =$ initial phase.

Here, $v_{\Phi}(k)$ is the phase velocity or the velocity of phase fluctuations, $\lambda(k)$ the wavelength, and $\omega_{**}(k)$ denotes the damping oscillations in time.

In other words, $k$ waves have uniform length, but time-varying amplitude.

These waves are analog of free oscillations.

*Type II*: Decisions, or wave, the second type, when:

$\omega_* = \omega =$a.

Real positive number ($\omega > 0, \omega_{**} = 0$).

In this case, we have;

$$\psi = \left(\psi_*^0 + i\,\psi_{**}^0\right)\exp\left[i\left(kX + \omega\tau + ik_{**}z\right)\right] = \left(\Psi_*^0 + i\Psi_{**}^0\right)\exp\left(-k_{**}X\right)\times$$
$$\left[\cos\left(kX + \omega\tau\right) + i\sin\left(kX + \omega\tau\right)\right]$$
$$\operatorname{Re}\{\Psi\} = \exp\left(-k_{**}X\right)\left|\Psi^0\right|\sin\left[\phi + \left(kX + \omega\tau\right)\right].$$

Thus, the solution of the second type is a sinusoidal oscillation in time (excited, for example, any stationary source of external monochromatic vibrations at) $X = 0$, decaying exponentially along the length of the amplitude.

Such disturbances, which are analogous to a wave of forced oscillations, are called $\omega$ waves:

$$\Psi(\omega) = \left|\Psi^0(\omega)\right| \exp\left(-k_{**}(\omega)X\right) \sin\left\{\phi + \frac{2\pi\left[X + v_{\delta}(\omega)\tau\right]}{\lambda(\omega)}\right\}, \tag{4.11}$$

$$v_{\delta}(\omega) = \omega / k(\omega),$$
$$\lambda(\omega) = 2\pi / k(\omega).$$

Here, $k_{**}(\omega)$ = damping vibrations in length.

In other words, $\omega$ waves are stationary in time but varying in length amplitudes.

Cases $k < 0$, $k_{**} < 0$ and $k > 0$, $k_{**} > 0$ are consistent with attenuation of amplitude, the disturbance regime in the direction of phase fluctuations, or phase velocity.

Let us obtain the characteristic equation, linking $k_*$ and $\omega_*$.

After substituting eq 4.9 in the system of equation 4.10, we obtain:

$$\delta^0 \frac{\omega_*}{k_*} + U^0 = 0, \qquad U^0 \frac{\omega_*}{k_*} + \delta^0 = 0. \tag{4.12}$$

The condition of the existence of a system of linear homogeneous algebraic equation 4.9 with respect to perturbations of a nontrivial solution implies the desired characteristic, or dispersion, which has one solution:

$$v_{\Phi} = \sqrt{gh_0}. \tag{4.13}$$

Thus, we obtain a solution representing a sinusoidal in time and coordinate free undammed oscillations. Such behaviors of the waves are due to the absence of any dissipation in the fluid. The fluid is incompressible and ideal. There is no heat mass transfer.

Equation 4.9 with respect to perturbations takes the form of wave equations:

$$\frac{\partial^2 \xi}{\partial t^2} = gh_0 \frac{\partial^2 \xi}{\partial x^2} \quad \text{and} \quad \frac{\partial^2 u}{\partial t^2} = gh_0 \frac{\partial^2 u}{\partial x^2}. \tag{4.14}$$

Note that in gas dynamics $v_\Phi = \sqrt{gh_0}$ is equivalent to the speed of sound.

The dynamics and heat and mass transfer of vapor bubble in a binary solution of liquids, as in Ref. 8 was studied for significant thermal, diffusion and inertial effect. It was assumed that binary mixture with a density $\rho_l$, consisting of components 1 and 2, respectively, have the density $\rho_1$ and $\rho_2$.

Moreover:

$$\rho_1 + \rho_2 = \rho_1,$$

where it is the mass concentration of component 1 of the mixture[19,20] This model assumes homogeneity of the temperature in phases.[21,22]

The intensity of heat transfer for one of the dispersed particles with an endless stream of carrier phase will be set by the dimensionless parameter of Nusselt $Nu_l$.

Bubble dynamics is described by the Rayleigh equation:

$$R\dot{w}_l + \frac{3}{2}w_l^2 = \frac{p_1 + p_2 - p_\infty - 2\sigma/R}{\rho_l} - 4v_1\frac{w_l}{R} \tag{4.15}$$

where $p_1$ and $p_2$ are the pressure components of vapor in the bubble, $p_\infty$ is the pressure of the liquid away from the bubble, σ and $v_1$ are the surface tension coefficients of kinematic viscosity for the liquid. Consider the condition of mass conservation at the interface.[23]

Mass flow $j_i^{\text{th}}$ component ($i = 1,2$) of the interface $r = R(t)$ in $j_i^{\text{th}}$ phase per unit area and per unit of time characterizes the intensity of the phase transition is given by:[24–30]

$$j_i = \rho_i\left(\dot{R} - w_l - w_i\right), \quad (i = 1,2), \tag{4.16}$$

where $w_i$ is the diffusion velocity component on the surface of the bubble. The relative motion of the components of the solution near the interface is determined by Fick's law.[25,26]

$$\rho_1 w_1 = -\rho_2 w_2 = -\rho_l D \left.\frac{\partial k}{\partial r}\right|_R \tag{4.17}$$

If we add eq 4.16, while considering that $\rho_1 + \rho_2 = \rho_l$ and draw eq 4.17, we obtain.[27,28]

$$\dot{R} = w_l + \frac{j_1 + j_2}{\rho_l} \tag{4.18}$$

Multiplying the first equation 4.16 with $\rho_2$ the second in $\rho_1$ and subtract the second equation from the first in view of eq 4.17, we obtain

$$k_R j_2 - (1 - k_R) j_1 = -\rho_l D \frac{\partial k}{\partial r}\bigg|_R .$$

Here $k_R$ is the concentration of the first component at the interface.

With the assumption of homogeneity of parameters inside the bubble changes in the mass of each component due to phase transformations can be written as:

$$\frac{d}{dt}\left(\frac{4}{3}\pi R^3 \rho_i'\right) = 4\pi R^2 j_i \ \text{ or } \ \frac{R}{3}\dot{\rho}_i' + \dot{R}\rho_i' = j_i, \ (i = 1,2). \tag{4.19}$$

Express the composition of a binary mixture in mole fractions of the component relative to the total amount of substance in liquid phase:

$$N = \frac{n_1}{n_1 + n_2}. \tag{4.20}$$

The number of moles of $i^{\text{th}}$ component $n_i$, which occupies the volume $V$, expressed in terms of its density:

$$n_i = \frac{\rho_i V}{\mu_i}. \tag{4.21}$$

Substituting eq 20 in eq 21, we obtain:

$$N_1(k) = \frac{\mu_2 k}{\mu_2 k + \mu_1 (1 - k)}, \tag{4.22}$$

By law, Raul partial pressure of the component above the solution is proportional to its molar fraction in the liquid phase, that is:

$$p_1 = p_{S1}(T_v)\, N_1(k_R) \text{ and } p_2 = p_{S2}(T_v)\,[1 - N_1(k_R)]. \tag{4.23}$$

Equations of state phases have the form:

$$p_i = BT_v \rho_i' / \mu_i, \; (i = 1,2), \tag{4.24}$$

where, $B$ is Gas constant, $T_v$ is the temperature of steam, $\rho_i'$ is the density of the mixture components in the vapor bubble, $\mu_i$ is molecular weight, and $p_{Si}$ is saturation pressure.

The boundary conditions $r = \infty$ and on a moving boundary can be written as:

$$k|_{r=\infty} = k_0, \; k|_{r=R}, = k_R, \; T_l|_{r=\infty} = T_{0,} \; T_l|_{r=R} = T_v, \tag{4.25}$$

$$j_1 l_1 + j_2 l_2 = \lambda_l D \left.\frac{\partial T_l}{\partial r}\right|_{r=R}, \tag{4.26}$$

where $l_i$ is specific heat of vaporization.[29,30]

By the definition of Nusselt parameter the dimensionless parameter characterizing the ratio of particle size and the thickness of thermal boundary layer in the phase around the phase boundary are determined from additional considerations or experience.[31,32]

The heat of the bubble's intensity with the flow of the carrier phase will be further specified as:

$$\left(\lambda_l \frac{\partial T_l}{\partial r}\right)_{r=R} = Nu_l \cdot \frac{\lambda_l (T_0 - T_v)}{2R}. \tag{4.27}$$

Nigmatulin et al.[33] obtained an analytical expression for the Nusselt parameter:

$$Nu_l = 2\sqrt{\frac{\omega R_0^2}{a_l}} = 2\sqrt{\frac{R_0}{a_l}\sqrt{\frac{3\gamma p_0}{\rho_l}}} = 2\sqrt{\sqrt{3\gamma} \cdot Pe_l}, \tag{4.28}$$

where $a_l = \lambda_l / \rho_l c_l$ is the thermal diffusivity of fluid,

$$Pe_l = \frac{R_0}{a_l}\sqrt{\frac{p_0}{\rho_l}} = \text{Peclet number.}$$

The intensity of mass transfer of the bubble with the flow of the carrier phase will continue to ask by using the dimensionless parameter Sherwood $Sh$:

$$\left( D\frac{\partial k}{\partial r} \right)_{r=R} = Sh \cdot \frac{D(k_0 - k_R)}{2R},$$

here, $D$ is Diffusion coefficient, $k$ is the concentration of dissolved gas in liquid, and the subscripts 0 and $R$ refer to the parameters in an undisturbed state and at the interface.

We define a parameter in the form of Sherwood:[33]

$$Sh = 2\sqrt{\frac{\omega R_0^2}{D}} = 2\sqrt{\frac{R_0}{D}\sqrt{\frac{3\gamma p_0}{\rho_l}}} = 2\sqrt{\sqrt{3\gamma} \cdot Pe_D}, \tag{4.29}$$

where

$$Pe_D = \frac{R_0}{D}\sqrt{\frac{p_0}{\rho_l}} = \text{diffusion Peclet number.}$$

The system of equations 4.15–4.29 is a closed system of equations describing the dynamics and heat transfer of insoluble gas bubbles with liquid.

If we use eqs 4.15–4.29, we obtain relations for the initial concentration of component 1:

$$k_0 = \frac{1 - \chi_2^0}{1 - \chi_2^0 + \mu(\chi_1^0 - 1)}, \quad \mu = \frac{\mu_2}{\mu_1}, \quad \chi_i^0 = \frac{p_{si0}}{p_0}, \quad i = 1,2, \tag{4.30}$$

where, $\mu_2$, $\mu_1$ are molecular weights of the liquid components of the mixture and $psi_0$ is saturated vapor pressure of the components of the mixture at an initial temperature of the mixture $T_0$, which are determined by integrating the Clausius-Clapeyron relation. The parameter $\chi_i^0$ is equal to:

$$\chi_i^0 = \exp\left[ \frac{l_i \mu_i}{B} \left( \frac{1}{T_{ki}} - \frac{1}{T_0} \right) \right]. \tag{4.31}$$

Gas phase liquid components in the derivation of eq 4.31 seemed perfect gas equations of state:

$$p_i = \rho_i B T_i / \mu_i,$$

where, $p_i$ is universal gas constant, $p_i$ is the vapor pressure inside the bubble $T_i$ to the temperature in the ratio of eq 4.31, $T_{ki}$ is the temperature evaporating the liquid components of binary solution at an initial pressure $p_0$, and $l_i$ is the specific heat of vaporization.

The initial concentration of the vapor pressure of component $p_0$ is determined from the relation:

$$c_0 = \frac{k_0 \chi_1^0}{k_0 \chi_1^0 + (1 - k_0) \chi_2^0}. \tag{4.32}$$

The problem of radial motions of a vapor bubble in binary solution[34] was solved. It was investigated at various pressure drops in the liquid for different initial radii $R_0$ for a bubble. It is of great practical interest of aqueous solutions of ethanol and ethylene glycol.

It revealed an interesting effect. The parameters characterized the dynamics of bubbles in aqueous ethyl alcohol. The field of variable pressure lies between the limiting values of the parameter $p_0$ for pure components.

The pressure drops and, consequently, the role of diffusion are assumed unimportant. The pressure drop, along with the heat dissipation, is included in diffusion dissipation. The rate of growth and collapse of the bubble is much higher than in the corresponding pure components of the solution under the same conditions. A completely different situation exists during the growth and collapse of vapor bubble in aqueous solutions of ethylene glycol.

In this case, the effect of diffusion resistance led to the inhibition of the rate of phase transformations. The rate of growth and collapse of the bubble is much smaller than the corresponding values for the pure components of the solution. Further research and calculations have to give a physical explanation for the observed effect. Nagiyev[35] studied the influence of heat transfer and diffusion on damping of free oscillations of a vapor bubble binary solution.

The research found the dependence of the damping rate of oscillations of a bubble of water solutions of ethanol, methanol, and toluene monotonic on $k_0$.

A similar dependence was mentioned for the aqueous solution of ethylene glycol with a characteristic minimum at:

$$k_0 \approx 0.02$$

Moreover, for $0.01 \leq k_0 \leq 0.3$ decrement, binary solution has less damping rates for pulsations of a bubble in pure (one-component) water and ethylene glycol.

This means that in the range of concentrations of water:

$$0.01 \leq k_0 \leq 0.3.$$

Pulsations of the bubble (for water solution of ethylene glycol) decay much more slowly and there is inhibition of the process of phase transformations. A similar process was revealed for forced oscillations of bubbles in an acoustic field.[36–38]

The influence of non-stationary heat and mass transfer[39] processes was investigated in the propagation of waves in a binary solution of liquids with bubbles. The influence of component composition and concentration of binary solution was investigated on the dispersion, dissipation, and attenuation of monochromatic waves in two-phase, two-component media.

The aqueous solution of ethyl alcohol in aqueous ethylene glycol decrements showed less relevant characteristics of pure components in the solution.

The unsteady interphase heat transfer[40] revealed a calculation. It refers to the structure of stationary shock waves in bubbly binary solutions. The problem signifies in effect a violation of monotonicity behavior of the calculated curves for concentration, indicating the presence of diffusion resistance.

In some of binary mixtures, it is seen as the effect of diffusion resistance. It leads to inhibition of the intensity of phase transformations.

The physical explanation revealed the reason for an aqueous solution of ethylene glycol. The pronounced effect of diffusion resistance is related to the solution with limited ability. It diffuses through the components of $D = 10^{-9}$ (m$^2$/scc).

$D$, the diffusion coefficient volatility of the components, is very different and thus leads to greatly different concentrations of the components in the solution and vapor phase.

In the case of aqueous solution of ethanol, volatility component is roughly the same $\chi_1^0 \approx \chi_2^0$.

In accordance with eq 4.3, $c_0 \approx k_0$; so the finiteness of the diffusion coefficient does not lead to significant effects in violation of the thermal and mechanical equilibrium phases.

Figures 4.2 and 4.3 show the dependence $k_0$ ($c_0$)of ethyl alcohol and ethylene glycol's aqueous solutions. From Figure 4.4 it is clear that for almost the entire range of $k_0$, $k_0 \approx c_0$.

At the same time for an aqueous solution of ethylene glycol, by the calculations and Figure 4.4, $0.01 \leq k_0 \leq 0.3$, $k_0 \leq c_0$, and when $k_0 > 0.3$, $k_0 \sim c_0$.

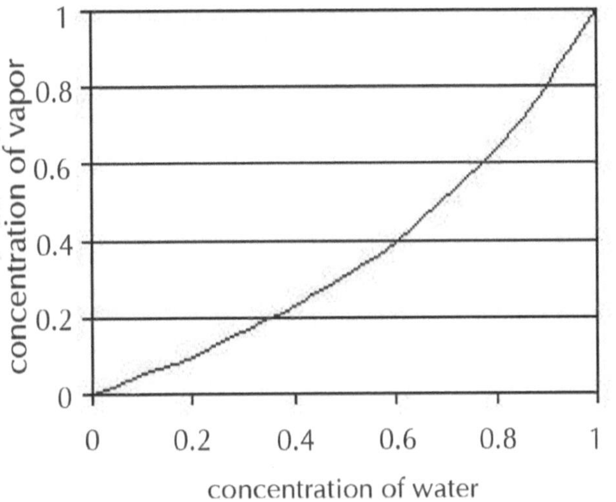


**FIGURE 4.2** The dependence ($k_0$ ($c_0$)) for an aqueous solution of ethanol.

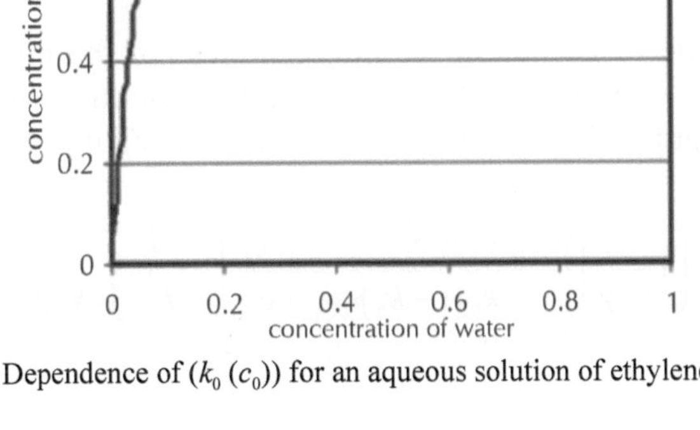


**FIGURE 4.3** Dependence of ($k_0$ ($c_0$)) for an aqueous solution of ethylene glycol.

Figures 4.4 and 4.5 show the boiling point of the concentration for the solution of two systems.

when $k_0 = 1$, $c_0 = 1$ and get clean water to steam bubbles. It is for boiling of a liquid at

$$T_0 = 373°\text{K}.$$

If $k_0 = 0$, $c_0 = 0$ and has correspondingly pure bubble ethanol ($T_0 = 373°$K) and ethylene glycol ($T_0 = 470°$K).

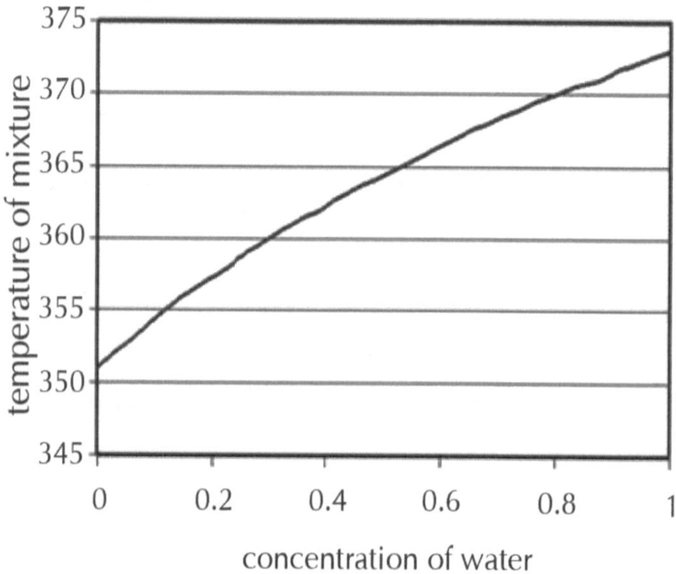


**FIGURE 4.4** The dependence of the boiling temperature of the concentration of the solution to an aqueous solution of ethanol.

It should be noted that all works regardless of the problems in the mathematical description of the cardinal effects of component composition of the solution show the value of the parameter $\beta$ equal:

$$\beta = \left(1 - \frac{1}{\gamma}\right) \frac{(c_0 - k_0)(N_{c_0} - N_{k_0})}{k_0(1 - k_0)} \frac{c_l}{c_{pv}} \left(\frac{c_{pv} T_0}{L}\right)^2 \sqrt{\frac{a_l}{D}}, \tag{4.33}$$

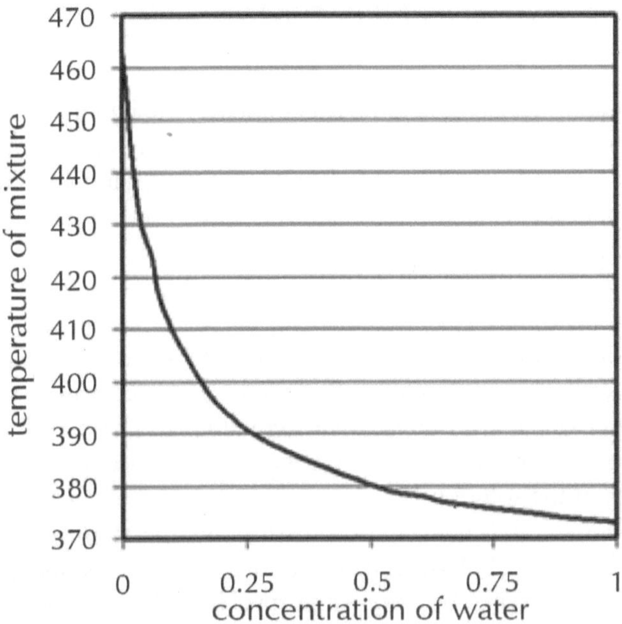


**FIGURE 4.5** The dependence of the boiling point of the concentration of the solution to an aqueous solution of ethylene glycol.

where $N_{k_0}$ and $N_{c_0}$ are molar concentrations of component 1 in the liquid and steam, respectively.

$$N_{k_0} = \frac{\mu k_0}{\mu k_0 + 1 - k_0},$$

$$N_{c_0} = \frac{\mu c_0}{\mu c_0 + 1 - c_0} \quad \gamma = \text{Adiabatic index},$$

$c_l$ and $c_{pv}$, respectively, are the specific heats of liquid and vapor at constant pressure and $a_l$ is thermal diffusivity.

$$L = l_1 c_0 + l_2(1 - c_0).$$

We also note that option (4) is a self-similar solution describing the growth of a bubble in a superheated solution. This solution has the form:[41]

$$R = 2\sqrt{\frac{3}{\pi}}\frac{\lambda_l \Delta T \sqrt{t}}{L\rho_v \sqrt{a_l}(1+\beta)}, \tag{4.34}$$

here, $\rho_v$ is vapor density, $t$ is time, $R$ is radius of the bubble, $\lambda_l$ is the coefficient of thermal conductivity, and $\Delta T$ is overheating of the liquid.

Figures 4.6 and 4.7 show the dependence $\beta$ ($k_0$) for the binary solutions. For aqueous ethanol, $\beta$ is negative for any value of concentration and dependence on $k_0$ is monotonic.

For an aqueous solution of ethylene glycol, $\beta$ is positive and has a pronounced maximum at $k_0 = 0.02$.

As a result of this work at low pressure drops (respectively superheating and super cooling of the liquid), diffusion does not occur in aqueous solutions of ethyl alcohol. By approximate equality of $k_0$ and $c_0$ all calculated dependence lie between the limiting curves for the case of one component constituents of the solution.

They include dependence of pressure, temperature, vapor bubble radius, the intensity of phase transformations, and so on from time to time. The pressure difference becomes important in diffusion processes. Mass transfer between bubble and liquid is a more intensive mode than single component constituents of the solution. In particular, the growth rate of the bubble in a superheated solution is higher than in pure water and ethyl alcohol. It is because of the negative $\beta$.

In an aqueous solution of ethylene glycol, there is the same perturbations due to significant differences between $k_0$ and $c_0$. It is especially the case when $0.01 \le k_0 \le 0.3$, the effect of diffusion inhibition contributes to a significant intensity of mass transfer. In particular, during the growth of the bubble, the rate of growth in solution is much lower than in pure water and ethylene glycol. It is because of the positive $\beta$ by (3.3.5).

Moreover, the maximum braking effect is achieved at the maximum value of $\beta$, when $k_0 = 0.02$. A similar pattern is observed at the pulsations and the collapse of the bubble. The dependence of the damping rate of fluctuations in an aqueous solution of ethyl alcohol from the water concentration is monotonic as shown elsewhere.[42–48] Aqueous solution of ethylene glycol dependence of the damping rate has a minimum at $k_0 = 0.02$, $0.01 \le k_0 \le 0.3$.

The function decrement is small with respect to the large difference between $k_0$ and $c_0$, and $\beta$ takes a large value. These ranges of concentrations in the solution have significant effect of diffusion inhibition. For aqueous solutions of glycerin, methanol, toluene, and so on, calculations are performed. The comparison with experimental data confirms the possibility of theoretical prediction of the braking of heat and mass transfer.

The dependence of the parameter $\beta$ and decrement of oscillations of a bubble from the equilibrium concentration of the mixture components were analyzed. Therefore, in every solution, the concentration of the components of a binary mixture was determined.

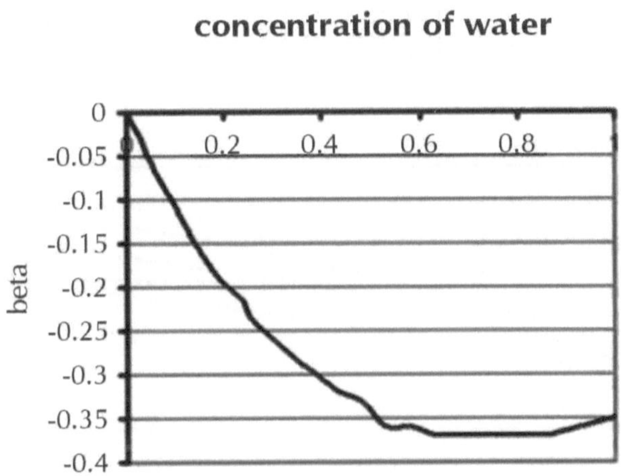


**FIGURE 4.6** The dependence $\beta(k_0)$ for an aqueous solution of ethanol.

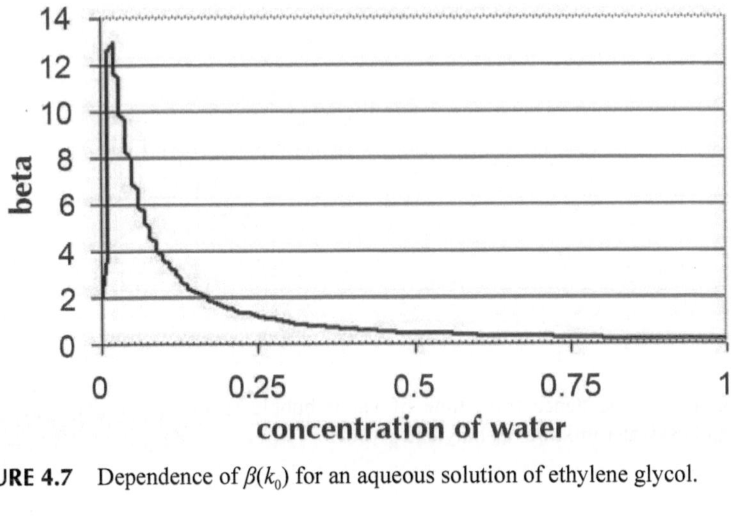


**FIGURE 4.7** Dependence of $\beta(k_0)$ for an aqueous solution of ethylene glycol.

Figures 4.8 and 4.9 are illustrated by theoretical calculations. These figures are defined on the example of aqueous solutions of ethyl alcohol and ethylene glycol (antifreeze used in car radiators). It is evident that the first solution is not suitable for the task.

The aqueous solution of ethylene glycol with a certain concentration, theoretically, boils much more slowly than with clean water and ethylene glycol. This confirms the reliability of the method.

The calculation shows that solution never freezes. The same method can offer concrete solutions for cooling the hot parts and components of various machines and mechanisms.

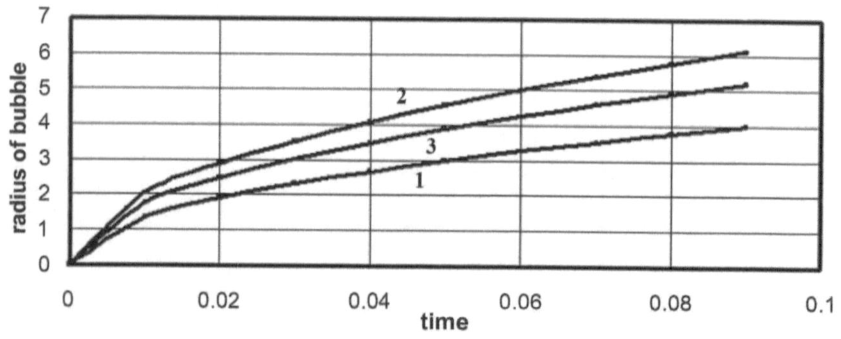


**FIGURE 4.8** Dependence from time of vapor bubble radius: (1) water, (2) ethyl spirit, and (3) water mixtures of ethyl spirit.

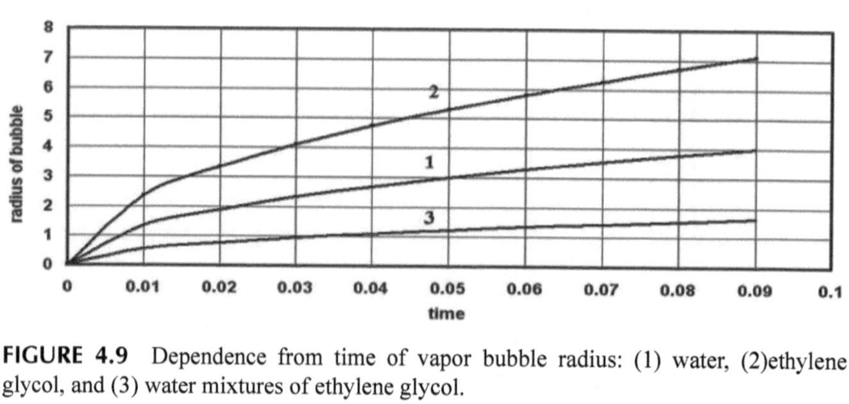


**FIGURE 4.9** Dependence from time of vapor bubble radius: (1) water, (2)ethylene glycol, and (3) water mixtures of ethylene glycol.

## 4.4 CONCLUSION

The solution of the reduced system of equations revealed an interesting effect. The parameters were characterized by the dynamics of bubbles in aqueous ethyl alcohol in the field of variable pressure. They lay between the limiting values of relevant parameters for the pure components. It was for this case that pressure drops and consequently the role of diffusion were unimportant.

A completely different situation is observed during the growth and collapse of vapor bubble in aqueous solutions of ethylene glycol. The effect of diffusion resistance leads to inhibition of the rate of phase transformations. For pure components of the solution, the rate of growth and collapse of the bubble is much smaller than the corresponding values.

## KEYWORDS

- **stratified fluid**
- **wavelength**
- **wave equation**
- **wave propagation**

## REFERENCES

1. Kraichnan, R. H.; Montgomery, D. "Two-Dimensional Turbulence." *Rep. Prog. Phys.* **1967,** *43* (547), 1417–1423.
2. Tullis, J. P. *"Control of Flow in Closed Conduits";* Colorado State University: Fort Collins, Colorado, 1971; pp 315–340.
3. Wood, D. J.; Dorsch, R. G.; Lightener, C. "Wave-Plan Analysis of Unsteady Flow in Closed Conduits". *J. Hyd. Div. ASCE.* **1966,** *92,* pp 83–110.
4. Wood, D. J.; Jones, S. E. "Water Hammer Charts for Various Types of Valves". *Proc. ASCE.* **1973,** *99,*167–178.
5. Leon Arturo, S. "An Efficient Second-Order Accurate Shock-Capturing Scheme for Modeling One and Two-Phase Water Hammer Flows". Ph.D. Thesis, March 29, 2007; pp 4–44.
6. Vallentine, H. R. "Rigid Water Column Theory for Uniform Gate Closure", *J. Hyd. Div. ASCE.* **1965,** 55–243.

7. Pickford, J. *"Analysis of Surge";* Macmillan: London, 1969; pp153–156.
8. Committee on Pipeline Location of the Pipeline Division, ASCE; *Pipeline Design for Water and Wastewater,* American Society of Civil Engineers: New York, 1975; p 54.
9. Xu, B.; Ooi, K. T.; Mavriplis, C.; Zaghloul, M. E. *Viscous Dissipation Effects for Liquid Flow in Microchannels;* Micorsystems: 2002; pp 53–57.
10. Fedorov, A. G.; Viskanta, R. Three-Dimensional Conjugate Heat Transfer into Micro-channel Heat Sink for Electronic Packaging. *Int. J. Heat Mass Transfer.* **2000,** *43,* 399–415.
11. Tuckerman, D. B. Heat Transfer Microstructures for Integrated Circuits. Ph.D. Thesis, Stanford University, 1984; pp 10–120.
12. Harms, T. M.; Kazmierczak, M. J.; Cerner, F. M.; Holke, A.; Henderson, H. T.; Pilchowski, H. T.; Baker, K. In *Experimental Investigation of Heat Transfer and Pressure Drop through Deep Micro channels in a (100) Silicon Substrate,* Proceedings of the ASME, Heat Transfer Division, HTD 351, 1997, pp 347–357.
13. Holland, F. A.; Bragg, R. *Fluid Flow for Chemical Engineers;* Edward Arnold Publishers: London, 1995; pp 1–3.
14. Lee, T. S.; Pejovic, S. Air Influence on Similarity of Hydraulic Transients and Vibrations. *ASME J. Fluid Eng.* **1996,** *118* (4), 706–709.
15. Li, J.; McCorquodale, A. "Modeling Mixed Flow in Storm Sewers", *J. Hydraul. Eng. ASCE. 125* (11), **1999,** 1170–1180.
16. Minnaert, M. On Musical Air Bubbles and the Sounds of Running Water. *Phil. Mag.* **1933,** *16* (7), 235–248.
17. Moeng, C. H.; McWilliams, J. C.; Rotunno, R.; Sullivan, P. P.; Weil, J. "Investigating 2D Modeling of Atmospheric Convection in the PBL." *J. Atm. Sci.* **2004,** *61,* 889–903.
18. Tuckerman, D. B. Pease, R. F. W. High Performance Heat Sinking for VLSI, *IEEE. DEL.* **1981,** *2,* 126–129.
19. Nagiyev, F. B.; Khabeev, N. S. Bubble Dynamics of Binary Solutions. *High Temperature.* **1988,** *27* (3), 528–533.
20. Shvarts, D.; Oron, D.; Kartoon, D.; Rikanati, A.; Sadot, O. "Scaling Laws of Nonlinear Rayleigh-Taylor and Richtmyer-Meshkov Instabilities in Two and Three Dimensions," *C. R. Acad. Sci. Paris IV.* **2000,** *1* (6), 719–726.
21. Cabot W. H.; Cook, A. W.; Miller, P. L.; Laney, D. E.; Miller, M. C.; Childs, H. R. "Large Eddy Simulation of Rayleigh-Taylor Instability." *Phys. Fluids.* **2005,** *17,* 91–106.
22. Cabot, W. University of California, Lawrence Livermore National Laboratory: Livermore, CA. *Phys. Fluids.* **2006,** 94–550.
23. Goncharov, V. N. "Analytical Model of Nonlinear, Single-Mode, Classical Rayleigh-Taylor Instability at Arbitrary Atwood Numbers". *Phys. Rev. Lett.* **2002,** *88,* 134502.
24. Ramaprabhu, P.; Andrews, M. J. "Experimental Investigation of Rayleigh-Taylor Mixing at Small Atwood Numbers." *J. Fluid Mech.* **2004,** *502,* 233.
25. Clark, T. T. "A Numerical Study of the Statistics of a Two-Dimensional Rayleigh-Taylor Mixing Layer". *Phys. Fluids.* **2003,** *15,* 2413.
26. Cook, A. W.; Cabot, W.; Miller, P. L. "The Mixing Transition in Rayleigh-Taylor Instability". *J. Fluid Mech.* **2004,** *511,* 333.

27. Waddell, J. T.; Niederhaus, C. E.; Jacobs, J. W. "Experimental Study of Rayleigh-Taylor Instability: Low Atwood Number Liquid Systems with Single-Mode Initial Perturbations," *Phys. Fluids.* **2001,** *13,* 1263–1273.
28. Weber, S. V.; Dimonte, G.; Marinak, M. M. "Arbitrary Lagrange-Eulerian Code Simulations of Turbulent Rayleigh-Taylor Instability in Two and Three Dimensions." *Laser Part. Beams.* **2003,** *21,* 455.
29. Dimonte, G.; Youngs, D.; Dimits, A.; Weber, S.; Marinak, M. "A Comparative Study of the Rayleigh-Taylor Instability Using High-Resolution Three-Dimensional Numerical Simulations: The Alpha Group Collaboration," *Phys. Fluids.* **2004,** *16,* 1668.
30. Young, Y. N.; Tufo, H.; Dubey, A.; Rosner, R. "On the Miscible Rayleigh-Taylor Instability: Two and Three Dimensions." *J. Fluid Mech.* **2001,** *447,* 377–408.
31. George, E.; Glimm, J. "Self-Similarity of Rayleigh-Taylor Mixing Rates," *Phys. Fluids.* **2005,** *17,* 054101.
32. Oron, D.; Arazi, L.; Kartoon, D.; Rikanati, A.; Alon, U.; Shvarts, D. "Dimensionality Dependence of the Rayleigh-Taylor and Richtmyer-Meshkov Instability Late-Time Scaling Laws". *Phys. Plasmas.* **2001,** *8,* 2883.
33. Nigmatulin, R. I.; Nagiyev, F. B.; Khabeev, N. S. In *Effective Heat Transfer Coefficients of the Bubbles in the Liquid Radial Pulse*, Mater. Second-Union. Conf. Heat Mass Transfer, "Heat Massoob-Men in the Biphasic. with. ". Minsk. 1980, Vol. 5, pp 111–115.
34. Nagiyev, F. B.; Khabeev, N. S. Bubble Dynamics of Binary Solutions. *High Temperature.* **1988,** *27* (3), 528–533.
35. Nagiyev, F. B. In *Damping of the Oscillations of Bubbles Boiling Binary Solutions. Mater*, VIII Resp. Conf. Mathematics and Mechanics, Baku, October 26–29, 1988; pp 177–178.
36. Nagyiev, F. B.; Kadyrov, B. A. Small Oscillations of the Bubbles in a Binary Mixture in the Acoustic Field. *Math. AN Az.SSR Ser. Physicotech. Mater. Sci.* **1986,** *1,* 23–26.
37. Nagiyev, F. B. In *Dynamics, Heat and Mass Transfer of Vapor-Gas Bubbles in a Two-Component Liquid,* Turkey-Azerbaijan Petrol Seminar, Ankara, Turkey, 1993; pp 32–40.
38. Nagiyev, F. B. In *The Method of Creation Effective Coolness Liquids,* Third Baku international Congress, Baku, Azerbaijan Republic, 1995; pp 19–22.
39. Nagiyev, F. B. The Linear Theory of Disturbances in Binary Liquids Bubble Solution. *Dep. VINITI.* **1986,** *405,* 76–79.
40. Nagiyev, F. B. Structure of Stationary Shock Waves in Boiling Binary Solutions. *Math. USSR. Fluid Dyn.* **1989,** *1,* 81–87.
41. Rayleigh, L. On the Pressure Developed in a Liquid During the Collapse of a Spherical Cavity. *Philos. Mag. Ser. 6.* **1917,** *34* (200), 94–98.
42. Perry, R. H.; Green, D. W.; Maloney, J. O. *Perry's Chemical Engineers Handbook;* 7th ed.; McGraw-Hill: New York, 1997; pp 1–61.
43. Nigmatulin, R. I. *Dynamics of Multiphase Media;* Nauka: Moscow, 1987; Vol. 1, pp 12–14.
44. Kodura, A.; Weinerowska, K. *In The Influence of the Local Pipeline Leak on Water Hammer Properties,* Materials of the II Polish Congress of Environmental Engineering: Lublin, 2005; pp 125–133.

45. Kane, J.; Arnett, D.; Remington, B. A.; Glendinning, S. G.; Baz´an, G. "Two-Dimensional Versus Three-Dimensional Supernova Hydrodynamic Instability Growth". *Astrophys. J.* **2000,** *528,* 989.
46. Quick, R. S. "Comparison & Limitations of Various Water hammer Theories". *J. Hyd. Div. ASME.* **1933,** 43–45.
47. Jaeger, C. "Fluid Transients in Hydro-Electric Engineering Practice"; Blackie & Son Ltd., 1977, p. 87–88.
48. Jaime Suárez, A. "Generalized water hammer algorithm for piping systems with unsteady friction" 2005, p. 72–77.

# CHAPTER 5

# NON-REVENUE WATER: SOME PRACTICAL HINTS

KAVEH HARIRI ASLI[1*], SOLTAN ALI OGLI ALIYEV[1], and HOSSEIN HARIRI ASLI[2]

[1]*Department of Mathematics and Mechanics, National Academy of Science of Azerbaijan 'AMEA', Baku, Azerbaijan*

[2]*Civil Engineering Department, Faculty of Engineering, University of Guilan, Rasht, Iran*

[*]*Corresponding author. E-mail: hariri_k@yahoo.com*

## CONTENTS

## ABSTRACT

This work introduces fuzzy logic as a computational method for hydrodynamics instability in water system. Investigation of leakage points in the water distribution network leads to the reduction of the real case amount of non-revenue water (NRW). The relationship classes between the spatial data and non-spatial data are a tool for communication between coordination of water system elements related to NRW data. In this work, the up-to-one-second-signals-detection method based on geographic information system (GIS) and fuzzy logic is incorporated with highly advanced data logger system. The fuzzy logic combination with GIS for rapid data intercommunication through programmable logic control (PLC) system presents a new algorithmic contribution. As a result, this work shows the performance of the fuzzy logic combination with the GIS for rapid data intercommunication which has an important role on water system disaster management.

## 5.1 INTRODUCTION

Theory of fuzzy sets and fuzzy logic was introduced by Professor Lotfali Askar Zadeh in a paper titled "Information and control" in 1965. The theory involved many scientific concepts such as fuzzy sets, fuzzy events, fuzzy numbers, and mathematics and engineering sciences. The most interesting application of fuzzy logic is an interpretation of the structure of intelligent decision-making and human intelligence. This logic clearly shows why classical mathematical logic, with two values including zero and one are not able to explain the biological concepts that form the basis of many smart decisions. More than 20 years later, in 1990, scientists began the industrial use of fuzzy logic. This work presents a new application of fuzzy logic in an area of broad interest to the scientists. In combination with fuzzy logic, the computational performances of a numerical method as a dynamic model are applied for urban water system failure condition in this work. On the other hand fuzzy logic combination with the direct numerical simulation (DNS) and geographic information system (GIS) for rapid data intercommunication through programmable logic control (PLC) system presents a new algorithmic contribution. The approaches proposed to solve the single-phase for the analysis of hydrodynamics instability are the method of characteristics (MOC), finite differences (FD), wave characteristic method (WCM), finite elements (FE), and finite volume (FV). Hence the MOC method as a dynamic model applied for urban water system failure in this work.[1–3]

Fuzzy logic is an approach to computing based on "degrees of truth" rather than the usual "true or false" (1 or 0) logic on which the modern computer is based.

Dr. Lotfi Zadeh introduced the idea of fuzzy logic when he was working on the problem of computer understanding of natural language in the University of California at Berkeley in the 1960s. Natural language (like most other activities in life and indeed the universe) is not easily translated into the absolute terms of 0 and 1. (Whether everything is ultimately describable in binary terms is a philosophical question worth pursuing, but in practice much data we might want to feed to a computer are in some state in between and so, frequently, are the results of computing.)

Fuzzy logic includes 0 and 1 as extreme cases of truth (or "the state of matters" or "fact") but also includes the various states of truth in between so that, for example, the result of a comparison between two things could not be "tall" or "short" but ".38 of tallness."

Fuzzy logic seems closer to the way our brains work. We aggregate data and form a number of partial truths which we aggregate further into higher truths which in turn, when certain thresholds are exceeded, cause certain further results such as motor reaction. It may help to see fuzzy logic as the way reasoning really works and binary logic is simply a special case of it. Classical logic only permits conclusions which are either true or false. For example, the notion that $1 + 1 = 2$ is a fundamental mathematical truth. However, there are also propositions with variable answers, such as one might find when asking a group of people to identify a color. In such instances, the truth appears as the result of reasoning from inexact or partial knowledge in which the sampled answers are mapped on a spectrum. Humans and animals often operate using fuzzy evaluations in many everyday situations. In the case where someone is tossing an object into a container from a distance, the person does not compute exact values for the object weight, density, distance, direction, container height and width, and air resistance to determine the force and angle to toss the object. Instead the person instinctively applies quick "fuzzy" estimates, based upon previous experience, to determine what output values of force, direction, and vertical angle to use to make the toss. Both degrees of truth and probabilities range between 0 and 1 and hence may seem similar at first. For example, let a 100 mL glass contain 30 mL of water. Then we may consider two concepts: empty and full. The meaning of each of them can be represented by a certain fuzzy set. Then one might define the glass as being 0.7 empty and 0.3 full. Note that the concept of emptiness would be subjective and thus would depend on the observer or designer. Another designer might, equally well, design a set membership function where the glass would be considered full for all values down to 50 mL. It is essential to realize that fuzzy logic uses degrees of truth as a mathematical model of vagueness, while probability is a mathematical model of ignorance.[4–22]

The most important propositional fuzzy logics are the following:

Monoidal t-norm-based propositional fuzzy logic (MTL) is an axiomatization of logic where conjunction is defined by a left continuous t-norm and implication is defined as the residuum of the t-norm. Its models correspond to MTL-algebras that are pre-linear commutative bounded integral residuated lattices. Basic propositional fuzzy logic BL is an extension of MTL logic where conjunction is defined by a continuous t-norm, and implication is also defined as the residuum of the t-norm. Its models correspond to BL-algebras. Łukasiewicz fuzzy logic is the extension of basic fuzzy logic BL where standard conjunction is the Łukasiewicz t-norm. It has the axioms of basic fuzzy logic plus an axiom of double negation, and its models correspond to MV-algebras. Gödel fuzzy logic is the extension of basic fuzzy logic BL where conjunction is Gödel t-norm. It has the axioms of BL plus an axiom of idempotence of conjunction, and its models are called G-algebras. Product fuzzy logic is the extension of basic fuzzy logic BL where conjunction is product t-norm. It has the axioms of BL plus another axiom for cancellativity of conjunction, and its models are called product algebras. Fuzzy logic with evaluated syntax (sometimes also called Pavelka's logic), denoted by EVŁ, is a further generalization of mathematical fuzzy logic. This means that each formula has an evaluation. Axiomatization of EVŁ stems from Łukasziewicz fuzzy logic. A generalization of classical Gödel completeness theorem is provable in EVŁ. These extend the above-mentioned fuzzy logics by adding universal and existential quantifiers in a manner similar to the way that predicate logic is created from propositional logic. The semantics of the universal (resp. existential) quantifier in t-norm fuzzy logics is the infimum (resp. supremum) of the truth degrees of the instances of the quantified subformula. The notions of a "decidable subset" and "recursively enumerable subset" are basic ones for classical mathematics and classical logic. Thus the question of a suitable extension of these concepts to fuzzy set theory arises. A first proposal in such a direction was made by E.S. Santos by the notions of fuzzy Turing machine; Markov normal fuzzy algorithm and fuzzy program. Successively, L. Biacino and G. Gerla argued that the proposed definitions are rather questionable and therefore they proposed the following ones. We say that s is decidable if both s and its complement s are recursively enumerable. An extension of such a theory to the general case of the L-subsets is possible. The proposed definitions are well related with fuzzy logic. Indeed, the following theorem holds true (provided that the deduction apparatus of the considered fuzzy logic satisfies some obvious effectiveness property). Theorem. Any axiomatizable fuzzy theory is recursively enumerable. In particular, the fuzzy set of logically true formulas is recursively enumerable in spite of the fact that the crisp set of valid formulas is not recursively enumerable, in general. Moreover, any axiomatizable and complete theory is decidable.[22–31] It is an open question to give supports for a "Church thesis" for fuzzy mathematics, the proposed notion of recursive enumerability for fuzzy subsets is the adequate one. To this aim, an extension of the notions of fuzzy grammar and fuzzy Turing machine should be necessary. Another open question is to start from this notion to find an extension of Gödel's theorems to fuzzy logic. Once fuzzy relations are defined, it is possible to develop fuzzy relational databases. The first fuzzy relational database, FRDB, appeared in Maria Zemankova's dissertation. Later, some other models arose like the Buckles-Petry model, the Prade-Testemale Model, the Umano-Fukami model, and the GEFRED model by J.M. Medina, M.A. Vila, et al. In the context of fuzzy databases, some fuzzy querying languages have been defined, highlighting the SQLf by P. Bosc et al. and the FSQL by J. Galindo et al. These languages define some structures in order to include fuzzy aspects in the SQL statements, like fuzzy conditions, fuzzy comparators, fuzzy constants, fuzzy constraints, fuzzy thresholds, linguistic labels, and so on. Fuzzy logic and probability address different forms of uncertainty. While both fuzzy logic and probability theory can represent degrees of certain kinds of subjective belief, fuzzy set theory uses the concept of fuzzy set membership, i.e., how much a variable is in a set (there is not necessarily any uncertainty about this degree), and probability theory uses the concept of subjective probability, i.e., how probable is it that a variable is in a set (it either entirely is or entirely is not in the set in reality, but there is uncertainty around whether it is or is not). The technical consequence of this distinction is that fuzzy set theory relaxes the axioms of classical probability, which are themselves derived from adding uncertainty, but not degree, to the crisp true/false distinctions of classical Aristotelian logic. Bruno de Finetti argues that only one kind of mathematical uncertainty, probability, is needed, and thus fuzzy logic is unnecessary. However, Bart Kosko shows in Fuzziness vs. Probability that probability theory is a subtheory of fuzzy logic, as questions of degrees of belief in mutually-exclusive set membership in probability theory can be represented as certain cases of non-mutually-exclusive graded membership in fuzzy theory. In that context, he also derives Bayes' theorem from the concept of fuzzy subsethood. Lotfi A. Zadeh argues that fuzzy logic is different in character from probability, and is not a replacement for it. He fuzzified probability to fuzzy probability and also generalized it to possibility theory. More generally, fuzzy logic is one of many different extensions to classical logic intended to deal with issues of uncertainty outside of the scope of classical logic, the inapplicability of probability theory in many domains, and the paradoxes of Dempster–Shafer theory. Leslie Valiant, a winner of the Turing Award, uses the term ecorithms to describe how many less exact systems and techniques like fuzzy logic (and "less robust" logic) can be applied to learning algorithms. Valiant essentially redefines machine learning as evolutionary. Ecorithms and fuzzy logic also have the common property of dealing with possibilities more than probabilities, although feedback and feed forward, basically stochastic "weights," are a feature of both when dealing with, for example, dynamical systems. In general use, ecorithms are algorithms that learn from their more complex environments (Hence Eco) to generalize approximate and simplify solution logic. Like fuzzy logic, they are methods used to overcome continuous variables or systems too complex to completely enumerate or understand discretely or exactly.[32–51] Compensatory fuzzy logic (CFL) is a branch of fuzzy logic with modified rules for conjunction and disjunction. When the truth value of one component of a conjunction or disjunction is increased or decreased, the other component is decreased or increased to compensate. This increase or decrease in truth value may be offset by the increase or decrease in another component. An offset may be blocked when certain thresholds are met. Proponents claim that CFL allows for better computational semantic behaviors.

## 5.2 MATERIALS AND METHODS

In this work DNS was applied for analysis of pressure and flow variation for prediction of water system failure as a branch of fuzzy logic. The DNS and GIS with very fine grid applied for analysis of leakage location and rate due to water system failure. The process for preparation of GIS Ready in order to apply the computational method for this work is as the following process:

- exchange of graphical information from the CAD to GIS space;
- fixed errors in the CAD space;
- convert data from graphical formats DWG to SHP;
- complete description of spatial data layers and fix errors in the GIS (description and location);
- complications from time to time with a good tolerance to Snap;
- toll errors in inappropriate places;
- creating primary and foreign keys for the table side;
- the creation and exchange of good tolerance to the effects of the topology of Spaghetti.

The computational method (eqs 5.1 and 5.2) which was used for prediction of leakage rate and location is as following:

Direct numerical simulation.

- Discretize Navier–Stokes equation on a sufficiently fine grid (Fig. 5.1) for resolving all motions occurring in turbulent flow.
- Not using any models.
- Equivalent to laboratory experiment.
- Relationship between length $\eta$ of the smallest eddies and the length $L$ of the largest eddies,

$$\frac{L}{\eta} \approx (\mathrm{Re}_L)^{\frac{3}{4}}, \tag{5.1}$$

where $R_e$ is Reynolds number, $\eta$ is the length of the smallest eddies, and $L$ is the length of the largest eddies.

The number of elements necessary to discretize the flow field in industrial applications is then:

$$Re\rangle 10^6, \qquad n_{\text{elem}} > 10^3, \tag{5.2}$$

where $n_{\text{elem}}$ is the number of elements necessary to discretize the flow field.

The mechanism of dynamic units can be controlled by using fuzzy logic.[6,7] The fuzzy logic (eqs 5.3–5.5) is a new technology for designing and modeling of a complex system. In fact for considering various factors based on deductive thinking, the values must be defined by the patterning of their words and language based on mathematical formulas which would be very complicated. Such as multi-valued logic, fuzzy logic, and fuzzy sets theory relies. Generalized and extended fuzzy sets of results are conclusive in nature. The definitive collection (Crisp sets) are the same as ordinary sets introduced at the beginning of the classical theory of sets. Adding a definite character to make the distinction that is critical to its innovative concepts is one of the so-called fuzzy logic membership functions (eqs 5.3–5.5) to bring easily in the mind. In the final set, membership function has only two values in the range (in mathematics, the range of a function is equal to the set of all outputs). Yes or no (one or zero), which are the two possible values of classical logic concepts.[52–61] So:

$$\mu_A(x) = \begin{cases} 1 \ldots if \ldots x \in A \\ 0 \ldots if \ldots x \notin A \end{cases}, \tag{5.3}$$

The $\mu_A(x)$ decisive element $x$ in the fuzzy set $\tilde{A}$.

Board membership function of the fuzzy set $\{0,1\}$ is converted to a definitive close range $[0,1]$ for fuzzy logic set.

$$A = \{(x, \mu_A(x)) | \, x \in X\}, \tag{5.4}$$

For example, linguistic variables can be considered in which values such as low, high, low, moderate, or strong can take his place. To have a mathematical language ($P$ = pressure):

$$P\ (\text{pressure}) = \{\text{low, high, low, medium, high}\}, \tag{5.5}$$

The membership $\mu_A(x)$ defines the degree of membership of an element $x$ to a fuzzy set $x$. If the membership degree of an element $x$ is equal to zero and if the member is totally out of membership degree equal to one is a member, the member $x$ is quite complex. If the degree of membership of a member $x$ is between zero and one, this number represents the degree of membership is gradual.[61–71]

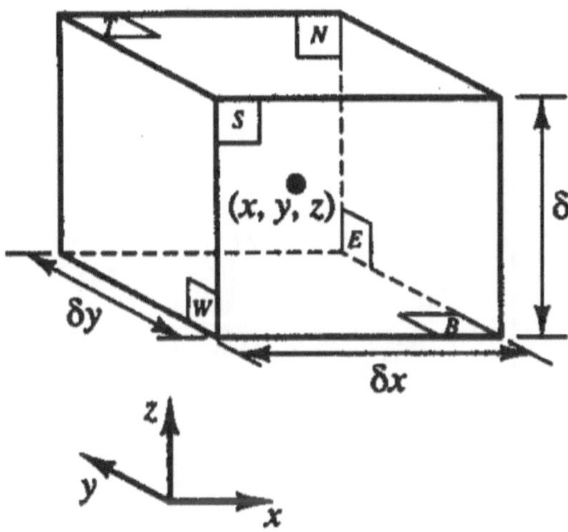


**FIGURE. 5.1** Infinitesimal fluid element.

In this research, fuzzy logic and sightseeing for fast data infrastructure for GIS and GIS Ready to implement and enforce action plans were as follows:

- preparation of a conceptual model for modeling in GIS environment,
- creating a database of suitable land,
- creating ability to track and implement network analysis, network,
- speeding up the investigation of accidents and reduce physical water losses,
- the systematic use of information storage and distribution facilities using GIS,
- saving the declining economy and tracking events,
- creating maps and reports,
- combinations of the above analysis for optimal management.

## 5.3 RESULTS AND DISCUSSION

At this work for hydrodynamics instability in water system (down to 1 s), highly advanced data logger based on GIS Ready and fuzzy logic was used (Figs. 5.2 and 5.3). These data were compared by flow and pressure data which were detected from existent system.

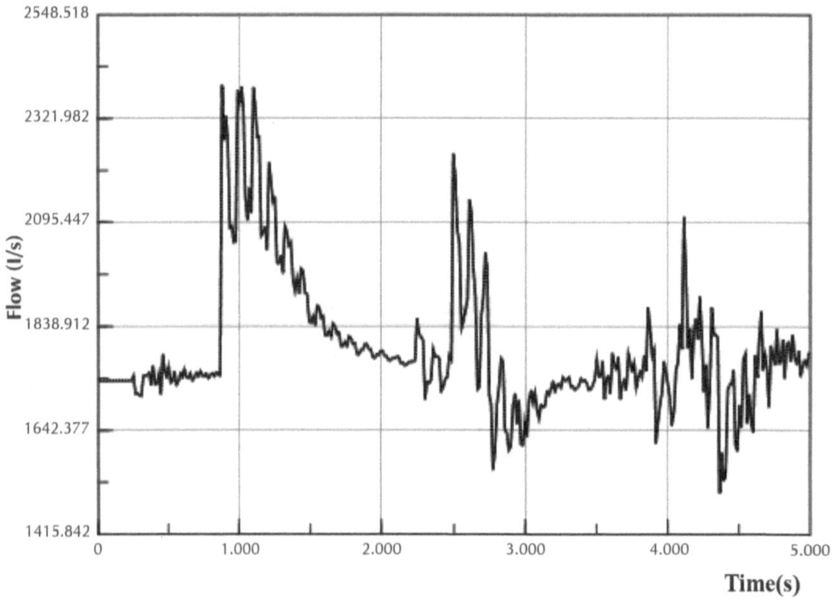


**FIGURE 5.2** Flow variation detection highly advanced data logger by GIS and fuzzy logic.

### *5.3.1 COMPARISON OF PRESENT WORK RESULTS WITH OTHER EXPERT'S RESEARCH*

This work conformed to the results of the works of Kodura and Weinerowska (Fig. 5.3). Comparison of the results of this work with other experts' research results shows significant points which were mentioned in the flowing. The pressure-wave speed as a fundamental parameter for hydraulic transient was modeled at present work. It was determined how the hydrodynamics instability disturbances quickly propagate throughout the system. This affected whether or not different pulses may superpose or cancel each other as they meet at different times and locations. Wave speed was affected by pipe material and bedding, as well as by the presence of fine air bubbles in the fluid. On the other hand, the effects of total transmission flow on the periods of wave oscillations were investigated.[71–84] In this work, the results of experiments and computational numerical analysis are compared with the pressure and flow variation of the work of Kodura and Weinerowska[42] (Fig. 5.3).

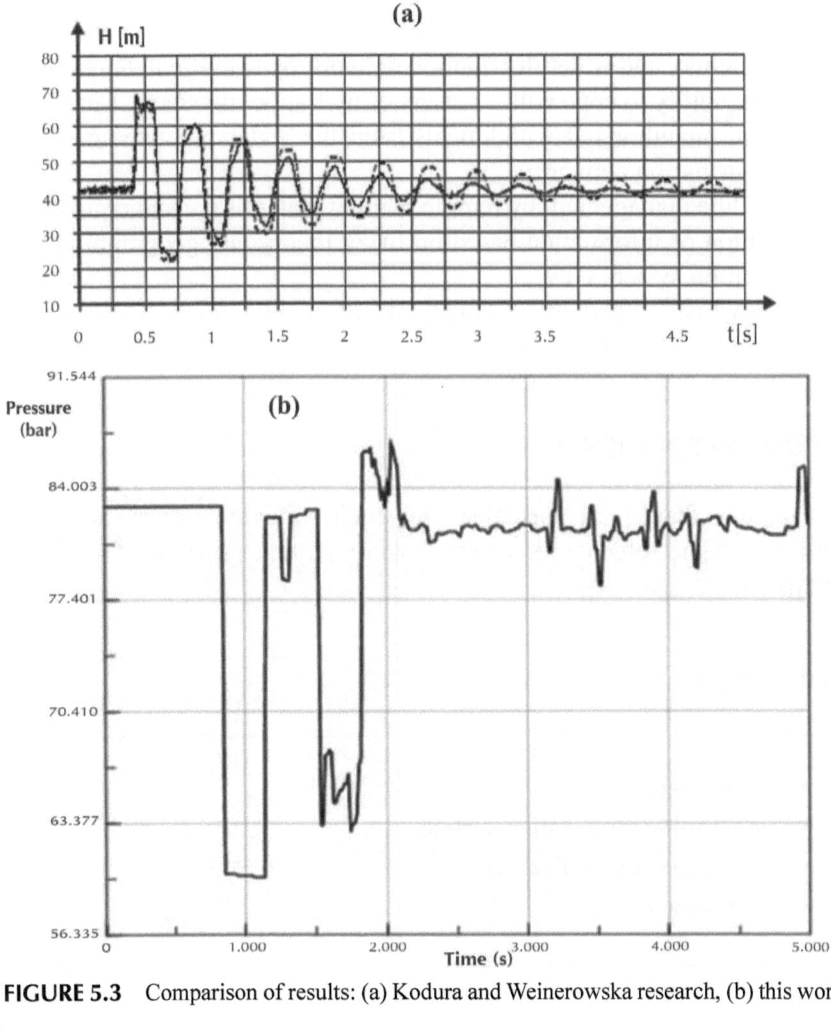


**FIGURE 5.3** Comparison of results: (a) Kodura and Weinerowska research, (b) this work.

## 5.4 CONCLUSION

In this work the hydrodynamics instability in water system was measured by extensometers and was recorded in GIS-based computer's memory. The work showed that the mechanism of dynamic units can be controlled by using fuzzy logic with the best guidance. The results of this work are as follows:

- the scientific management of the data received from various sensors in the hydraulic parameters such as pressure, flow of the water pipeline, and water distribution network by using remote reading;
- the ability to cope with a variety of hydraulic transients, water loss, and unauthorized withdrawals from the water pipeline and water distribution network;
- online management by remote reading water usage and rapid circulation of information based on fuzzy theory and geographic information system GIS;
- governance and management of consumer demand management issues and technical safety.

## ACKNOWLEDGMENTS

The author thanks all specialists for their valuable observations and advice, and the referees for recommendations that improved the quality of this paper.

## KEYWORDS

- **fuzzy logic**
- **geographic information system**
- **data intercommunication**
- **data logger**

## REFERENCES


1. Streeter, V. L.; Wylie, E. B. *Fluid Mechanics;* McGraw-Hill Ltd.: USA, 1979; pp 492–505.
2. Leon Arturo, S. An Efficient Second-Order Accurate Shock-Capturing Scheme for Modeling One and Two-Phase Water Hammer Flows. Ph.D. Thesis, March 29, 2007; pp 4–44.
3. Adams, T. M.; Abdel-Khalik, S. I.; Jeter, S. M.; Qureshi, Z. H. An Experimental Investigation of Single-Phase Forced Convection in Microchannels. *Int. J. Heat Mass Transfer.* **1998,** *41,* 851–857.
4. Peng, X. F.; Peterson, G. P. Convective Heat Transfer and Flow Friction for Water Flow in Microchannel Structure. *Int. J. Heat Mass Transfer.* **1996,** *36,* 2599–2608.
5. Mala, G.; Li, D.; Dale, J. D.; Heat Transfer and Fluid Flow in Microchannels. *J. Heat Transfer.* **1997,** *40,* 3079–3088.
6. Pickford, J. *Analysis of Surge;* Macmillan: London, 1969; pp 153–156.
7. Committee on Pipeline Location of the Pipeline Division, ASCE; *Pipeline Design for Water and Wastewater,* American Society of Civil Engineers: New York, 1975; p 54.
8. Xu, B.; Ooi, K. T.; Mavriplis, C.; Zaghloul, M. E. *Viscous Dissipation Effects for Liquid Flow in Microchannels;* Micorsystems: 2002; pp 53–57.
9. Fedorov, A. G.; Viskanta, R. Three-Dimensional Conjugate Heat Transfer into Microchannel Heat Sink for Electronic Packaging. *Int. J. Heat Mass Transfer.* **2000,** *43,* 399–415.
10. Tuckerman, D. B. Heat Transfer Microstructures for Integrated Circuits. Ph.D. Thesis, Stanford University, 1984; pp 10–120.
11. Harms, T. M.; Kazmierczak, M. J.; Cerner, F. M.; Holke, A.; Henderson, H. T.; Pilchowski, H. T.; Baker, K. In *Experimental Investigation of Heat Transfer and Pressure Drop through Deep Micro channels in a (100) Silicon Substrate,* Proceedings of the ASME, Heat Transfer Division, HTD 351, 1997; pp 347–357.
12. Holland, F. A.; Bragg, R. *Fluid Flow for Chemical Engineers;* Edward Arnold Publishers: London, 1995; pp 1–3.
13. Lee, T. S.; Pejovic, S. Air Influence on Similarity of Hydraulic Transients and Vibrations. *ASME J. Fluid Eng.* **1996,** *118* (4), 706–709.
14. Li, J.; McCorquodale, A. Modeling Mixed Flow in Storm Sewers, *J. Hydraul. Eng. ASCE. 125* (11), **1999,** 1170–1180.
15. Minnaert, M., On Musical Air Bubbles and the Sounds of Running Water. *Phil. Mag.* **1933,** *16* (7), 235–248.
16. Moeng, C. H.; McWilliams, J. C.; Rotunno, R.; Sullivan, P. P.; Weil, J. Investigating 2D Modeling of Atmospheric Convection in the PBL. *J. Atm. Sci.* **2004,** *61,* 889–903.
17. Tuckerman, D. B. Pease, R. F. W. High Performance Heat Sinking for VLSI, *IEEE. DEL.* **1981,** *2,* 126–129.
18. Shvarts, D.; Oron, D.; Kartoon, D.; Rikanati, A.; Sadot, O. Scaling Laws of Nonlinear Rayleigh-Taylor and Richtmyer-Meshkov Instabilities in Two and Three Dimensions, *C. R. Acad. Sci. Paris IV.* **2000,** *1* (6), 719–726.
19. Cabot W. H.; Cook, A. W.; Miller, P. L.; Laney, D. E.; Miller, M. C.; Childs, H. R. Large Eddy Simulation of Rayleigh-Taylor Instability. *Phys. Fluids.* **2005,** *17,* 91–106.

20. Cabot, W. University of California, Lawrence Livermore National Laboratory: Livermore, CA. *Phys. Fluids.* **2006,** 94–550.
21. Goncharov, V. N. Analytical Model of Nonlinear, Single-Mode, Classical Rayleigh-Taylor Instability at Arbitrary Atwood Numbers. *Phys. Rev. Lett.* **2002,** *88,* 134502.
22. Ramaprabhu, P.; Andrews, M. J. Experimental Investigation of Rayleigh-Taylor Mixing at Small Atwood Numbers. *J. Fluid Mech.* **2004,** *502,* 233.
23. Clark, T. T. A Numerical Study of the Statistics of a Two-Dimensional Rayleigh-Taylor Mixing Layer. *Phys. Fluids.* **2003,** *15,* 2413.
24. Cook, A. W.; Cabot, W.; Miller, P. L. The Mixing Transition in Rayleigh-Taylor Instability. *J. Fluid Mech.* **2004,** *511,* 333.
25. Waddell, J. T.; Niederhaus, C. E.; Jacobs, J. W. Experimental Study of Rayleigh-Taylor Instability: Low Atwood Number Liquid Systems with Single-Mode Initial Perturbations, *Phys. Fluids.* **2001,** *13,* 1263–1273.
26. Weber, S. V.; Dimonte, G.; Marinak, M. M. Arbitrary Lagrange–Eulerian Code Simulations of Turbulent Rayleigh-Taylor Instability in Two and Three Dimensions. *Laser Part. Beams.* **2003,** *21,* 455.
27. Dimonte, G.; Youngs, D.; Dimits, A.; Weber, S.; Marinak, M. A Comparative Study of the Rayleigh-Taylor Instability Using High-Resolution Three-Dimensional Numerical Simulations: The Alpha Group Collaboration, *Phys. Fluids.* **2004,** *16,* 1668.
28. Young, Y. N.; Tufo, H.; Dubey, A.; Rosner, R. On the Miscible Rayleigh-Taylor Instability: Two and Three Dimensions. *J. Fluid Mech.* **2001,** *447,* 377–408.
29. George, E.; Glimm, J. Self-Similarity of Rayleigh-Taylor Mixing Rates, *Phys. Fluids.* **2005,** *17,* 054101.
30. Oron, D.; Arazi, L.; Kartoon, D.; Rikanati, A.; Alon, U.; Shvarts, D. Dimensionality Dependence of the Rayleigh-Taylor and Richtmyer-Meshkov Instability Late-Time Scaling Laws. *Phys. Plasmas.* **2001,** *8,* 2883.
31. Nigmatulin, R. I.; Nagiyev, F. B.; Khabeev, N. S. In *Effective Heat Transfer Coefficients of the Bubbles in the Liquid Radial Pulse*, Mater. Second-Union. Conf. Heat Mass Transfer, Heat Massoob-Men in the Biphasic. with Minsk. 1980, vol. 5, pp 111–115.
32. Nagiyev, F. B.; Khabeev, N. S. Bubble Dynamics of Binary Solutions. *High Temperature.* **1988,** *27* (3), 528–533.
33. Nagiyev, F. B. In *Damping of the Oscillations of Bubbles Boiling Binary Solutions. Mater*, VIII Resp. Conf. Mathematics and Mechanics, Baku, October 26–29, 1988; pp 177–178.
34. Nagyiev, F. B., Kadyrov, B. A. Small Oscillations of the Bubbles in a Binary Mixture in the Acoustic Field. *Math. AN Az.SSR Ser. Physicotech. Mater. Sci.* **1986,** *1,* 23–26.
35. Nagiyev, F. B. In *Dynamics, Heat and Mass Transfer of Vapor-Gas Bubbles in a Two-Component Liquid,* Turkey-Azerbaijan Petrol Seminar: Ankara, Turkey, 1993; pp 32–40.
36. Nagiyev, F. B. In *The Method of Creation Effective Coolness Liquids,* Third Baku international Congress, Baku , Azerbaijan Republic, 1995; pp 19–22.
37. Nagiyev, F. B. The Linear Theory of Disturbances in Binary Liquids Bubble Solution. *Dep. VINITI.* **1986,** *405,* 76–79.
38. Nagiyev, F. B. Structure of Stationary Shock Waves in Boiling Binary Solutions. *Math. USSR. Fluid Dyn.* **1989,** *1,* 81–87.

39. Rayleigh, L. On the Pressure Developed in a Liquid During the Collapse of a Spherical Cavity. *Philos. Mag. Ser.* **1917,** *34* (200), 94–98.
40. Perry, R. H.; Green, D. W.; Maloney, J. O. *Perry's Chemical Engineers Handbook;* 7th ed.; McGraw-Hill: New York, 1997; pp 1–61.
41. Nigmatulin, R. I. *Dynamics of Multiphase Media;* Nauka: Moscow, 1987; Vol. 1, pp 12–14.
42. Kodura, A.; Weinerowska, K. In *The Influence of the Local Pipeline Leak on Water Hammer Properties,* Materials of the II Polish Congress of Environmental Engineering: Lublin, 2005; pp 125–133.
43. Kane, J.; Arnett, D.; Remington, B. A.; Glendinning, S. G.; Baz′an, G. Two-Dimensional Versus Three-Dimensional Supernova Hydrodynamic Instability Growth. *Astrophys. J.* **2000,** *528,* 989.
44. Quick, R. S. Comparison & Limitations of Various Water hammer Theories. *J. Hyd. Div. ASME.* **1933,** 43–45.
45. Jaeger, C. Fluid Transients in Hydro-Electric Engineering Practice; Blackie & Son Ltd., 1977, pp 87–88.
46. Jaime Suárez, A. *Generalized Water Hammer Algorithm for Piping Systems with Unsteady Friction.* 2005; pp 72–77.
47. Fok, A.; Ashamalla, A.; Aldworth, G. In *Considerations in Optimizing Air Chamber for Pumping Plants,* Symposium on Fluid Transients and Acoustics in the Power Industry, San Francisco, U.S.A., December, 1978; pp 112–114.
48. Fok, A. In *Design Charts for Surge Tanks on Pump Discharge Lines,* BHRA 3rd International Conference on Pressure Surges, Bedford, England, March, 1980; pp 23–34.
49. Fok, A. In *Water Hammer and its Protection in Pumping Systems,* Hydro Technical Conference, CSCE: Edmonton, May, 1982; pp 45–55.
50. Fok, A. A Contribution to the Analysis of Energy Losses in Transient Pipe Flow. Ph.D. Thesis, University of Ottawa, 1987; pp 176–182.
51. Brunone, B.; Karney, B. W.; Mecarelli, M.; Ferrante, M. Velocity Profiles and Unsteady Pipe Friction in Transient Flow. J. *Water Resour. Planning Manag. ASCE.* **2000,** *126* (4), 236–244.
52. Koelle, E.; Luvizotto Jr. E.; Andrade, J. P. G. In *Personality Investigation of Hydraulic Networks Using MOC – Method of Characteristics,* Proceedings of the 7th International Conference on Pressure Surges and Fluid Transients, Harrogate Durham, UK, 1996; pp 1–8.
53. Filion, Y.; Karney, B. W. In *A Numerical Exploration of Transient Decay Mechanisms in Water Distribution Systems,* Proceedings of the ASCE Environmental Water Resources Institute Conference, American Society of Civil Engineers: Roanoke, Virginia, 2002; p 30.
54. Hamam, M. A.; Mc Corquodale, J. A. Transient Conditions in the Transition from Gravity to Surcharged Sewer Flow. *Canadian J. Civil Eng. Canada.* **1982,** *9* (2), 65–98.
55. Savic, D. A.; Walters, G. A. *Genetic Algorithms Techniques for Calibrating Network Models;* Report No. 95/12, Centre for Systems and Control Engineering: Beijing, 1995; pp 137–146.

56. Walski, T. M.; Lutes, T. L. Hydraulic Transients Cause Low-Pressure Problems. *J. Am. Water Works Assoc.* **1994,** *75* (2), 58.
57. Lee, T. S.; Pejovic, S. Air Influence on Similarity of Hydraulic Transients and Vibrations. *ASME J. Fluid Eng.* **1996,** *118* (4), 706–709.
58. Chaudhry, M. H. *Applied Hydraulic Transients;* Van Nostrand Reinhold Co.: NY. 1979; pp 1322–1324.
59. Parmakian, J. *Water Hammer Analysis;* Dover Publications, Inc.: New York, NY, 1963; pp 51–58.
60. Farooqui, T. A. *Evaluation of Effects of Water Reuse, on Water and Energy Efficiency of an Urban Development Area, Using an Urban Metabolic Framework*; Master of Integrated Water Management Student Project Report, International Water Centre: Australia, 2015.
61. Ferguson, B. C.; Frantzeskaki, N.; Brown, R. R. A Strategic Program for Transitioning to a Water Sensitive City. *Landscape Urban Plan.* **2013,** *117*, 32–45.
62. Kenway, S.; Gregory, A.; McMahon, J. Urban Water Mass Balance Analysis. *J. Indus. Ecol.* **2011,** *15* (5), 693–706.
63. Renouf, M. A.; Kenway, S. J.; Serrao-Neumann, S.; Low Choy, D. *Urban Metabolism for Planning Water Sensitive Cities. Concept for an Urban Water Metabolism Evaluation Framework*; Milestone Report, Cooperative Research Centre for Water Sensitive Cities: Melbourne, 2015.
64. Serrao-Neumann, S.; Schuch, G.; Kenway, S. J.; Low Choy, D. *Comparative Assessment of the Statutory and Non-Statutory Planning Systems*; South East Queensland, Metropolitan Melbourne and Metropolitan Perth, Cooperative Research Centre for Water Sensitive Cities: Melbourne, 2013.
65. Andrews, S.; Traynorp, Guidelines for Assuring Quality of Food and Water Microbiological Culture Media. August, 2004.
66. Andrew, D. Eaton.; Eugene W. Rice.; Lenore S. Clescceri. Standard Methods for the Examination of Water and Waste Water. Part 9000, 2012.
67. Instructions Procedure Rural Water and Wastewater Quality Assurance Test Results of KhorasanRazavi, 2011.
68. Abbassi, B.; Al Baz, I. *Integrated Wastewater Management: A Review. Efficient Management of Wastewater – Its Treatment and Reuse in Water Scarce Countries;* Springer Publishing Co.: Berlin, Germany, 2008.
69. Andreasen, P. *Chemical Stabilization. Sludge into Biosolids – Processing, Disposal and Utilization;* IWA Publishing: London, 2001.
70. CIWE; *The Chartered Institution of Water and Environmental Management Sewage Sludge: Stabilization and Disinfection – Handbooks of UK Wastewater Practice;* CIWEM Publishing: Lavenham, Suffolk, 1996.
71. Halalsheh, M.; Wendland, C. *Integrated Anaerobic-Aerobic Treatment of Concentrated Sewage. Efficient Management of Wastewater – Its Treatment and Reuse in Water Scarce Countries;* Springer Publishing Co.: Berlin, Germany, 2008.
72. ISWA; *Handling, Treatment and Disposal of Sludge in Europe;* Situation Report 1, ISWA Working Group on Sewage Sludge and Water Works: Copenhagen, 1995.
73. Matthews, P. *Agricultural and Other Land Uses. Sludge into Biosolids – Processing, Disposal and Utilization*; IWA Publishing: London, 2001.
74. IWK Sustainability Report, 2012–2013.

75. Novák, V.; Perfilieva, I.; Močkoř, J. *Mathematical Principles of Fuzzy Logic*; Kluwer Academic: Dodrecht, 1999; ISBN 0-7923-8595-0.
76. Fuzzy Logic; *Stanford Encyclopedia of Philosophy;* Bryant University: USA, 2006-07-23. Retrieved 2008-09-30.
77. Zadeh, L. A. Fuzzy Sets. *Inf. Control.* **1965,** *8* (3), 338–353. doi:10.1016/s0019-9958(65)90241-x.
78. Pelletier Francis, J. Review of Metamathematics of Fuzzy Logics. The *Bull. Symb. Logic.* **2000,** *6* (3), 342–346. JSTOR 421060.
79. Zadeh, L. A., et al. *Fuzzy Sets, Fuzzy Logic, Fuzzy Systems;* World Scientific Press: Hackensack, NJ, 1996; ISBN 981-02-2421-4.
80. Kosko, B. *Fuzzy Thinking: The New Science of Fuzzy Logic;* Hyperion: NY, 1994.
81. Bansod Nitin, A.; Marshall K.; Patil, S. H. *Soft Computing- A Fuzzy Logic Approach. Soft Computing;* Allied Publishers: New Delhi, 2005; p 73.
82. Novák, V. Are Fuzzy Sets a Reasonable Tool for Modeling Vague Phenomena? *Fuzzy Sets Syst.* **2005,** *156,* 341–348. doi:10.1016/j.fss.2005.05.029.
83. Valiant, L. *Probably Approximately Correct: Nature's Algorithms for Learning and Prospering in a Complex World New York;* Basic Books: New York, 2013; ISBN 978-0465032716.
84. Cejas, J. *Compensatory Fuzzy Logic;* Revista de Ingeniería Industrial: La Habana, 2011; ISSN 1815-5936.

CHAPTER 6

# THREE-DIMENSIONAL HEAT TRANSFER AND WATER FLOW MODELING

KAVEH HARIRI ASLI[1*], SOLTAN ALI OGLI ALIYEV[1], and HOSSEIN HARIRI ASLI[2]

[1]*Department of Mathematics and Mechanics, Azerbaijan National Academy of Sciences 'AMEA', Baku, Azerbaijan*

[2]*Civil Engineering Department, Faculty of Engineering, University of Guilan, Rasht, Iran*

**Corresponding author. E-mail: hariri_k@yahoo.com*

## CONTENTS

## ABSTRACT

A finite volume (FV) method is used to solve the conjugate heat transfer through the heat sinks. In this work the flow and heat transfer development regions inside the tubes are considered. The numerical results are then compared with the available experimental data. The effects of liquid velocity through channels and their effects on heat transfer and pressure drop along microchannels are investigated. Finally, the effects of aspect ratio on heat dissipation and pressure drop in microtubes are predicted.

## 6.1 INTRODUCTION

A search on the Internet provided the needed information in the form of a downloadable subroutine. The program uses FORTRAN's capability to deal with complex numbers, avoiding tedious computations to arrive at the real and imaginary portions of complicated complex expressions. The user inputs the three parameters involved in the Joukowsky transformations, as well as the angle of attack of the airfoil. The program then generates the shape of the airfoil, as well as the velocity and pressure distributions on its surface. The program is an example of the generation of a potential flow about a halfbody consisting of a source in a uniform stream. After specifying source strength and uniform stream velocity, the body shape and surrounding streamlines are found. The computation is complicated by the fact that on a streamline it is not possible to solve directly for $y$ as a function of $x$. An iteration procedure must be used. The appropriate source strength on the panel is computed so as to make the velocity tangent to the body. The finite volume method (FVM) for three-dimensional heat transfer and water flow modeling is used to solve the continuity, momentum, and energy equations.

A very brief description of the method used is given here:[1–27]

- FVM uses integral form of conservation (transport) equation.
- Domain subdivided in control volumes (CV).
- Surface and volume integrals approximated by numerical quadrature.
- Interpolation used to express variable values at CV faces in terms of nodal values.

- It results in an algebraic equation per CV.
- Suitable for any type of grid.
- Conservative by construction.
- Commercial codes: CFX, Fluent, Phoenics, Flow3D.
- Most flows in practice are turbulent.
- With increasing Re, smaller eddies.
- Very fine grid necessary to describe all length scales.
- Even the largest supercomputer does not have (yet) enough speed and memory to simulate turbulent flows of high Re.

Computational methods for turbulent flows:

- Direct numerical simulation (DNS)
- Large Eddy simulation (LES)
- Reynolds-averaged Navier–Stokes (RANS)

Direct numerical simulation_

- Discretize Navier–Stokes equation on a sufficiently fine grid for resolving all motions occurring in turbulent flow.
- Not using any models.
- Equivalent to laboratory experiment.

Large Eddy simulation

- Only large eddies are computed.
- Small eddies are modeled, subgrid-scale (SGS) models.

Reynolds-averaged Navier–Stokes

- Variables decomposed in a mean part and a fluctuating part.
- Navier–Stokes equations averaged over time.
- Turbulence models are necessary.

## 6.2 MATERIALS AND METHODS

The characteristics of the computational methods are depicted in Figure 6.1. The width and thickness of microchannels are $W_1$ and $H_1$, respectively.

The thickness of the silicon substrate is $H_2 - H_1$; and the total length of the microchannels is $L$. The heat supplies by a $1 \times 1$ cm$^2$ heat source located at the entrance of the channels and were centered on the whole channel heat sink. A uniform heat flux of $q$ is provided to heat the microchannels. The heat is removed by flowing water through channels. The inlet temperature of the cooling water is 20°C. The analysis is performed for five different cases.

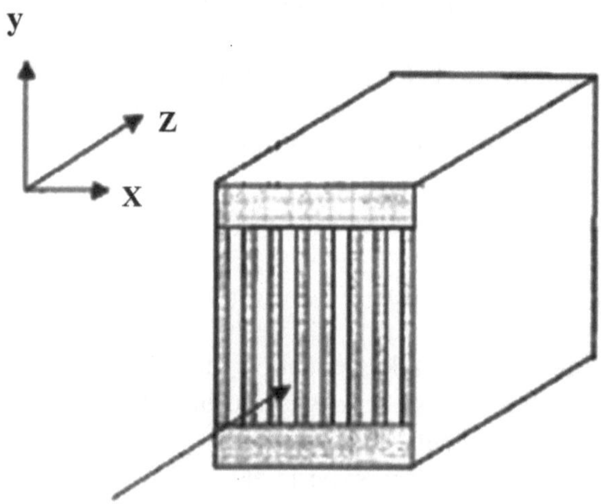


**FIGURE 6.1** A sample microtube.

The dimensions related to each case are given in Table 6.1. By these dimensions, there will be 150 microchannels for cases 0 and 1 and 200 microchannels for cases 2–4.

For the second section, case 2 was considered as the base geometry. The water velocity changed from 50 to 400 cm/s.

In the third section, the Tuckerman's geometries were solved by the unique Reynolds of 150. As performing a numerical method for the whole microtubes heat sink is hard; a certain computational domain is considered (Fig. 6.2).

To prevent the various boundary condition effects, the computational domain is taken at the center of the heat sink to have the quoted uniform heat flux in Table 6.1 This is because there is very little spreading of heat toward the heat sinks. There is also some geometrical symmetry which simplifies the[28–35] computation. So only a semi-channel and semi-silicon

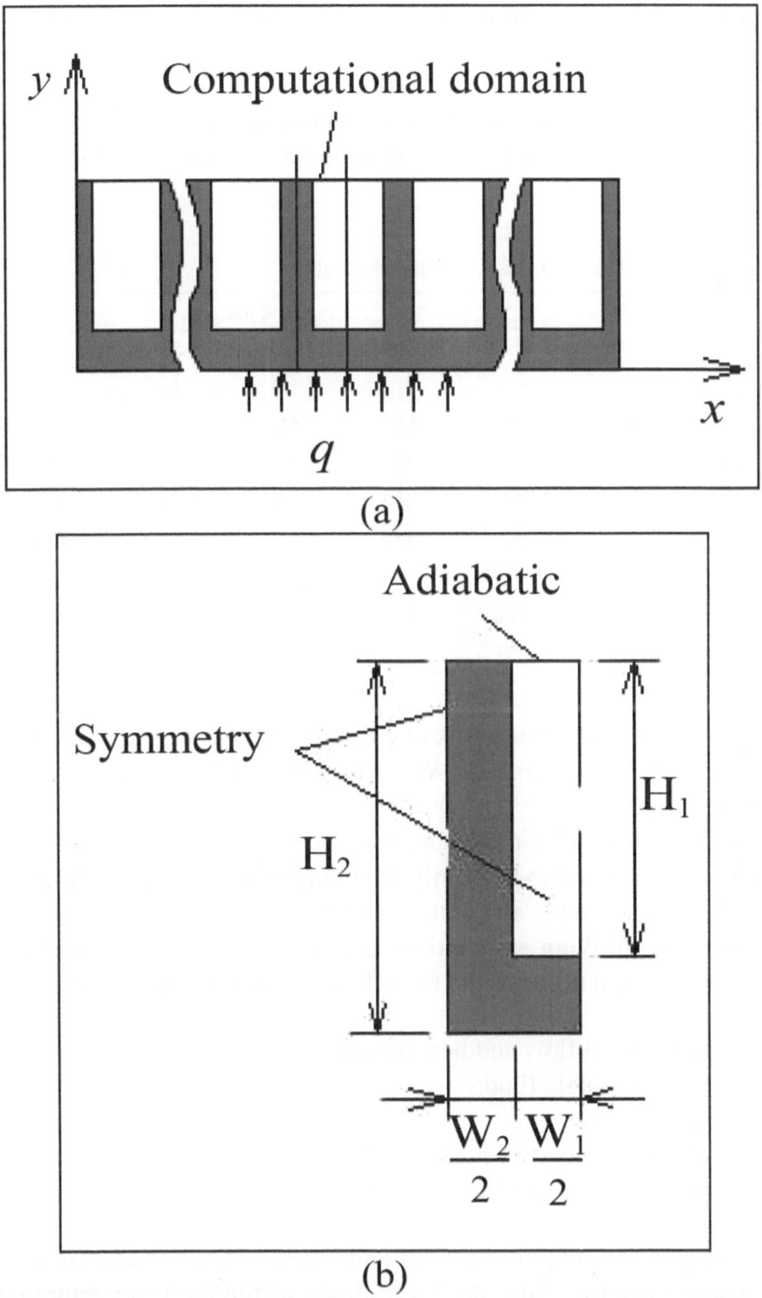


**FIGURE 6.2** Dimensions and computational domain.

substrate will be considered and the results will be the same for the other half. The whole substrate is made of silicon with thermal conductivity (k) of 148 W/m K). At the top of the channel $y = H_2$, there is a Pyrex plate to make an adiabatic condition (its thermal conductivity is two orders lower than that of silicon). There are two different boundary conditions at the bottom.

**TABLE 6.1** Four Different Cases of Microchannels.

| | Case | | | | |
|---|---|---|---|---|---|
| | **0** | **1** | **2** | **3** | **4** |
| $L$ (cm) | 2 | 2 | 1.4 | 1.4 | 1.4 |
| $W_1$ (μm) | 64 | 64 | 56 | 55 | 50 |
| $W_2$ (μm) | 36 | 36 | 44 | 45 | 50 |
| $H_1$ (μm) | 280 | 280 | 320 | 287 | 302 |
| $H_2$ (μm) | 489 | 489 | 533 | 430 | 458 |
| $\dot{Q}$ (cm$^3$/s) | 1.277 | 1.86 | 4.7 | 6.5 | 8.6 |
| $q$ (W/cm$^2$) | 34.6 | 34.6 | 181 | 277 | 790 |
| Number of channels | 150 | 150 | 200 | 200 | 200 |

For $z < L_h$ a uniform heat flux of $q$ is imposed over the heat sink and the rest is assumed to be adiabatic. Water flows through the channel from the entrance in $z$ direction.[36–55]

The transverse velocities of the inlet are assumed to be zero. The axial velocity is considered to be evenly distributed through the whole channel. The velocities at the top and bottom of tubes are zero.

Several simplifying assumptions are incorporated before establishing the governing equations for the fluid flow and heat transfer in a unit cell:

1. steady fluid flow and heat transfer;
2. incompressible fluid;
3. laminar flow;
4. negligible radiative heat transfer;
5. constant solid and fluid properties.

In the last assumption the solid and liquid properties are assumed to be constant because of the small variations within the temperature range tested.

Based on the above assumptions, the governing differential equations and nomenclatures used to describe the fluid flow and heat transfer in a unit cell are expressed as follows:

$A$ = flow area, $T$ = temperature, $c_p$ = specific heat, $T_{\mathrm{in}}$ = inlet temperature, $D_{\mathrm{h}}$ = hydraulic diameter, $T_{\mathrm{max}}$ = maximum temperature, EDL = electric double layer, $w$ = velocity ($z$ direction), $H_1$ = channel height, $W_1$ = channel width, $H_2$ = total height, $W_2$ = substrate width, $k$ = thermal conductivity, $L$ = heat sink length, $\Phi$ = viscous dissipation terms in energy equation, $L_{\mathrm{h}}$ = heated length, $p$ = pressure, $\mu$ = viscosity, $P$ = wetted perimeter, $\rho$ = density, $q$ = heat flux subscripts, $Q$ = total average volumetric flow rate, $R$ = thermal resistance , in = inlet, Re = Reynolds number, max = maximum.

Continuity equation

$$\frac{\partial w}{\partial z} = 0. \tag{6.1}$$

Momentum equation

$$\rho \frac{\partial (w_i w_j)}{\partial z_j} = -\frac{\partial p}{\partial z_i} + \mu \frac{\partial}{\partial z_j}\left(\frac{\partial w_i}{\partial z_j} + \frac{\partial w_j}{\partial z_i}\right) \tag{6.2}$$

$$-\frac{2}{3}\mu \frac{\partial}{\partial z_i}\left(\frac{\partial w_k}{\partial z_k}\right).$$

Energy equation

$$\rho c_p \frac{\partial w_j T}{\partial z_j} = k \frac{\partial^2 T}{\partial x_j^{\,2}} + \mu \Phi \tag{6.3}$$

where

$$\Phi = 2\left[\left(\frac{\partial u}{\partial x}\right)^2 + \left(\frac{\partial v}{\partial y}\right)^2 + \left(\frac{\partial w}{\partial z}\right)^2\right] + \left(\frac{\partial u}{\partial y} + \frac{\partial v}{\partial x}\right)^2$$

$$+\left(\frac{\partial u}{\partial z} + \frac{\partial w}{\partial x}\right)^2 + \left(\frac{\partial v}{\partial z} + \frac{\partial w}{\partial y}\right)^2. \tag{6.4}$$

The FVM is used to solve the continuity, momentum, and energy equations. A very brief description of the method used is given here.

In this method the domain is divided into a number of control volumes such that there is one control volume surrounding each grid point. The grid point is located in the center of a control volume. The governing equations are integrated over the individual control volumes to construct algebraic equations for the discrete dependent variables such as velocities, pressure, and temperature. The discretization equation then expresses the conservation principle for a finite control volume, just as the partial differential equation expresses it for an infinitesimal control volume. In this study, a solution is deemed converged when the mass imbalance in the continuity equation is less than $10^{-6}$.

As the thermal specifications and flow characteristics are of the great importance in design of microchannel heat sinks, the results are concentrated in these fields.

For solving the equations several grid structures were used. The grid density of 120 × 40 × 20 in *z*, *y*, and *x* directions is considered to be appropriate.

The thermal resistance is calculated as follows:

$$R(z) = \frac{T_{\text{max}}(z) - T_{\text{in}}}{q}. \tag{6.5}$$

In eq (6.5), $R(z)$ $T_{\text{max}}(z)$, $T_{\text{in}}$ and $q$ the thermal resistance at $z$ (cm) from the entrance, the inlet water temperature and the heat flux at the heating area.

In addition to thermal resistance, the Reynolds number is calculated as:

$$\text{Re} \equiv \frac{\rho w_{\text{ave}} D_{\text{h}}}{\mu}, \tag{6.6}$$

where

$$D_{\text{h}} \equiv \frac{4A}{P} = \frac{2H_1W_1}{H_1 + W_1}. \tag{6.7}$$

First, the temperature distribution, thermal resistances, and pressure drops based on the numerical results are demonstrated. Then the numerical results are compared with Tuckerman's experiments data.

### 6.2.1 TEMPERATURE DISTRIBUTION AND THERMAL RESISTANCE

The temperature distributions at four *x–y* cross-sections along the channel are shown in Figure 6.3 for case 0. The four sections, $z = 1, 3, 6,$ and 9 mm are all in the heated area. As all heat exchangers, isotherms are the closest at the entrance of the channels. This means that heat transfer rate is the highest at the entrance and it decreases along the channel.

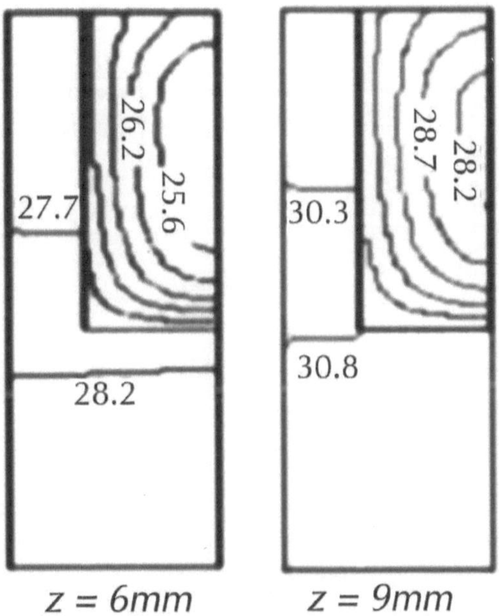


**FIGURE 6.3** Temperature distribution along the tube for case 0.

The thermal resistance along the tube is shown in Figure 6.3 for cases 0 and 1. It can be seen that the thermal resistance has increased by increasing the *z* value and attained the maximum value at $z = 9$ mm. It is consistent with Tuckerman's experiments. It may be seen that the thermal resistance increased linearly through the channel except the entry region and near its peak. The sharper slopes in the experimental data of Figure 6.4(a) and the numerical prediction in Figure 6.4(b) are evidently due to the entrance region effects. So the flow may be considered thermally fully developed with a proper precision. The maximum thermal resistance is occurred in

$z = 9$ mm which is consistent with Tuckerman's experiments. For the other cases the results are given in Table 6.2. The numerical values of resistances are predicted well.[56–65]

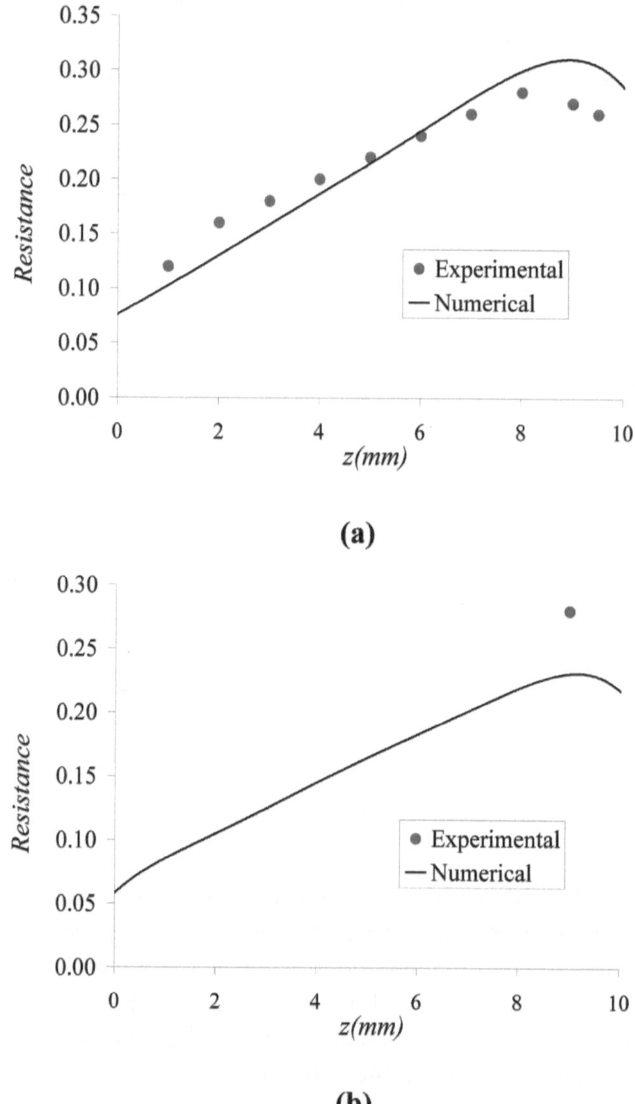


**FIGURE 6.4** Numerical and experimental thermal resistances for: (a) case 0, and (b) case 1.

Pressure drop is linear along the channel. Figure 6.5(a) shows the pressure drop for case 4. The pressure drop increases by increasing the inlet velocity. The slope of the pressure line in the entrance of the channel is maximum. This is due to the entry region effects. The velocity field will be

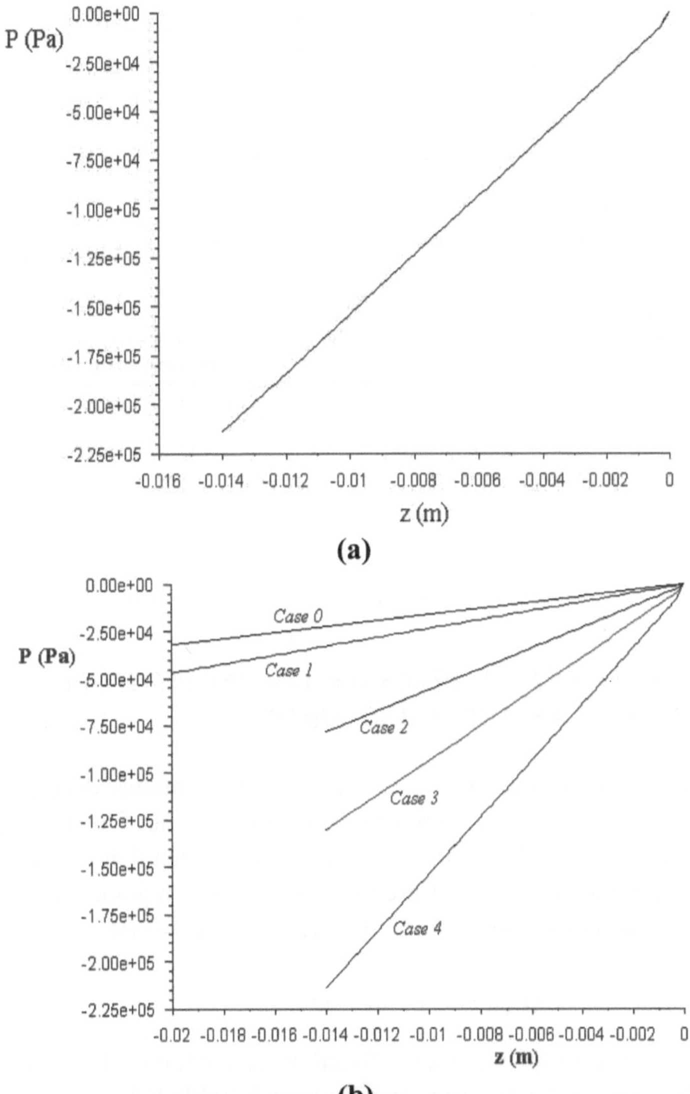


**FIGURE 6.5** Pressure drop: (a) case 4; (b) all cases.

fully developed after a small distance from the entrance. So the assumption of fully developed flow is acceptable. Figure 6.5(b) shows the pressure drop for all five cases. These amounts are tabulated in Table 6.2.

**TABLE 6.2** Thermal Resistance Comparison.

| Case | $q$ (W/cm$^2$) | $R$ (cm$^2$/K/W) | | Error (%) |
|---|---|---|---|---|
| | | Experimental | Numerical | |
| 0 | 34.6 | 0.277 | 0.253 | 5.8 |
| 1 | 34.6 | 0.280 | 0.246 | 12.1 |
| 2 | 181 | 0.110 | 0.116 | 5.0 |
| 3 | 277 | 0.113 | 0.101 | 8.1 |
| 4 | 790 | 0.090 | 0.086 | 3.94 |

**TABLE 6.3** Pressure Drop in Five Cases.

| Case | Pressure Drop (bar) |
|---|---|
| 0 | 0.322 |
| 1 | 0.469 |
| 2 | 0.784 |
| 3 | 1.302 |
| 4 | 2.137 |

### 6.2.2 THE EFFECT OF VELOCITY ON TEMPERATURE DISTRIBUTION AND PRESSURE DROP

The effects of velocity on the temperature rise and pressure drop in microtubes are displayed in Figures 6.6 and 6.7. The amount of heat dissipation increases by increasing the velocity. But it can be seen that the amount of decrease in temperature falls drastically by increase in velocity.

The following function can predict the temperature rise:

$$(T_{max} - T_{in}) = 265.67w^{-0.4997}. \tag{6.8}$$

The pressure drop is a linear function of velocity. The amounts of temperature rise and pressure drop are given in Table 6.4.

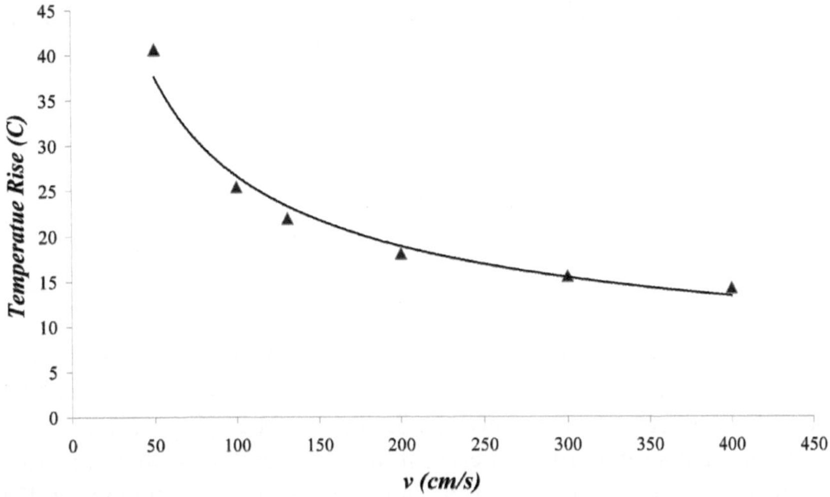


**FIGURE 6.6** Temperature rise for different velocities.

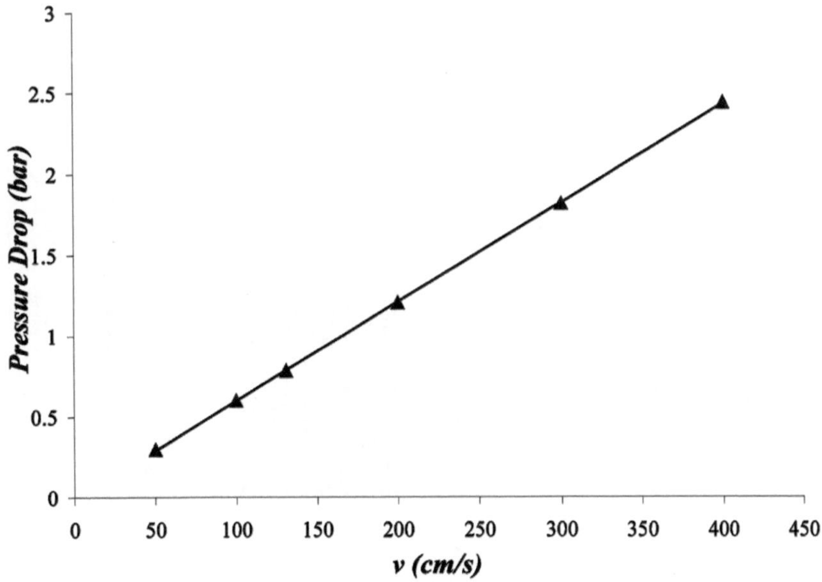


**FIGURE 6.7** Pressure drop for different velocities.

**TABLE 6.4** Temperature Rise and Pressure Drop for Second Section.

| Velocity (cm/s) | Re | Temp. Rise (°C) | Pressure Drop (bar) |
|---|---|---|---|
| 50 | 47 | 40.6 | 0.298 |
| 100 | 95 | 25.4 | 0.598 |
| 131 | 124 | 21.9 | 0.784 |
| 200 | 190 | 18.0 | 1.206 |
| 300 | 285 | 15.5 | 1.817 |
| 400 | 380 | 14.2 | 2.439 |

Aspect ratio is an important factor in microtube design. In this section the inlet heat flux of 181 W/cm$^2$ imposed over the heat sinks and the hydraulic diameter changes between 85.8 and 104.2. (Fig. 6.8) show the maximum temperature of each case. It may be seen that with an identical heat flux, the heat dissipation of the largest aspect ratio is the lowest. But this case has the minimum pressure drop too (Fig. 6.9). By increasing the amount of aspect ratio the ability of heat dissipation increases.[66–76] The amounts of temperature rise and pressure drop of all geometries are given in Table 6.5.

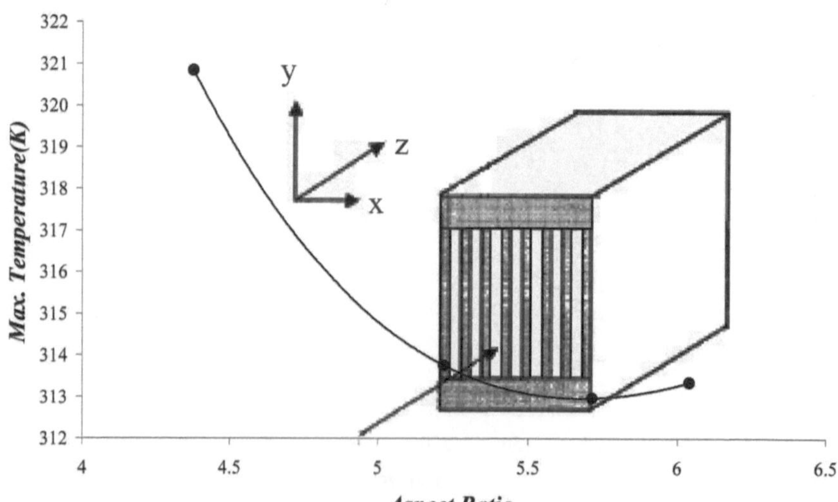


**FIGURE 6.8** Temperature with respect to aspect ratio.

**TABLE 6.5** Temperature Rise and Pressure Drop for Third Section.

| Case | Aspect Ratio | Hydraulic Diameter (μm) | Velocity (m/s) | Temp. Rise (°C) | Pressure Drop (bar) |
|---|---|---|---|---|---|
| 1 | 4.375 | 104.19 | 1.447 | 27.84 | 0.98 |
| 2 | 5.714 | 95.32 | 1.58 | 19.97 | 0.95 |
| 3 | 5.218 | 92.31 | 1.633 | 20.76 | 1.03 |
| 4 | 6.040 | 85.80 | 1.76 | 20.35 | 1.31 |

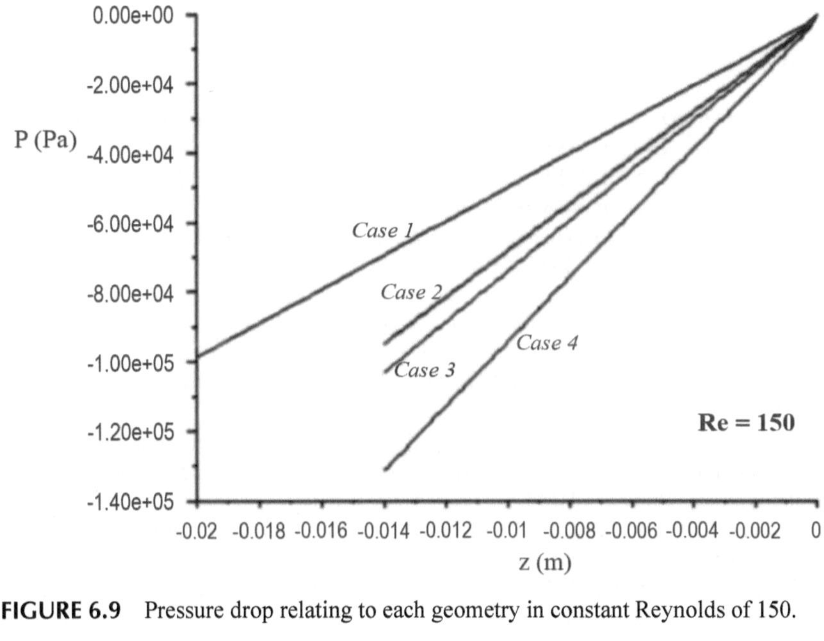


**FIGURE 6.9** Pressure drop relating to each geometry in constant Reynolds of 150.

## 6.3 CONCLUSION

The third geometry has the minimum temperature rise. The temperature rises of second and forth geometries are approximately alike. But the pressure drop of these two cases has a noticeable difference. The pressure drop increases with increase in aspect ratio.

## KEYWORDS

- **temperature distribution**
- **thermal resistances**
- **pressure drops**
- **three-dimensional heat transfer**
- **microchannel**

## REFERENCES

1. Tuckerman, D. B.; Pease, R. F. W. High-Performance Heat Sinking for VLSI. *IEEE Electron Device Lett.* **1981,** *5,* 126–129.
2. Lee, P. S.; Suresh, G. V.; Liu, D. Investigation of Heat Transfer in Rectangular Microchannels. *Int. J. Heat Mass Transf.* **2005,** *9,* 1688–1704.
3. Hsieh, S. S.; Lin, C. Y.; Huang, C. F.; Tsai, H. H. Liquid Flow in a Micro-Channel. *J. Micromech. Microeng.* **2004,** *14,* 436–445.
4. Liu, J. T.; Peng, X. F.; Yan, W. M. Numerical Study of Fluid Flow and Heat Transfer in Microchannel Cooling Passages. *Int. J. Heat Mass Transf. 50* (9–10), 1855–1864.
5. Harms, T. M.; Kazmierczak, M. J.; Gerner, F. M. Developing Convective Heat Transfer in Deep Rectangular Microchannels. *Int. Heat Fluid Flow.* **1999,** *20* (2), 149–157.
6. Agostini, B.; Bontemps, A.; Thonon , B. Effect of Geometrical and Thermophysical Parameters on Heat Transfer Measurements in Small Diameter Channels. *Heat Transf. Eng.* **2006,** *27* (1), 14–24.
7. Hwang, Y. W.; Kim, M. S. The Pressure Drop in Microtubes and the Correlation Development. *Int. J. Heat Mass Transf.* **2006,** *49* (11–12), 1804–1812.
8. Hrnjak, P.; Tu, X. Single Phase Pressure Drop in Microchannels. *Int. J. Heat Fluid Flow.* **2007,** *28* (1), 2–14.
9. Rosa, P.; Karayiannis, T.; Collins, M. Single-Phase Heat Transfer in Microchannels: The Importance of Scaling Effects. *Appl. Therm. Eng.* **2009,** *29* (17–18), 3447–3468.
10. Kandlikar, S.; Lu, Z.; Domigan, W.; White, A.; Benedict, M. Measurement of Flow Maldistribution in Parallel Channels and its Application to Ex-Situ and In-Situ Experiment's in PEMFC Water Management Studies. *Int. J. Heat Mass Transf.* **2009,** *52* (7–8), 1741–1752.
11. Tonomura, O.; Tanaka, S.; Noda, M.; Kano, M.; Hasebe, S.; Hashimoto, I. CFD-Based Optimal Design of Manifold in Plate-Fin Microdevices. *Chem. Eng. J.* **2004,** 101 (1–3), 397–402.
12. Balaji, S.; Lakshminarayanan, S. Improved Design of Microchannel Plate Geometry for Uniform Flow Distribution. *Can. J. Chem. Eng.* **2006,** 84 (6), 715–721.

13. Pan, M.; Shao, X.; Liang, L. Analysis of Velocity Uniformity in a Single Microchannel Plate with Rectangular Manifolds at Different Entrance Velocities. *Chem. Eng. Technol.* **2013,** *36* (6), 1067–1074.
14. Huang, C. Y.; Wu, C. M.; Chen, Y. N.; Liou, T. M. The Experimental Investigation of Axial Heat Conduction Effect on the Heat Transfer Analysis in Microchannel Flow. *Int. J. Heat Mass Transf.* **2014,** *70,* 169–173.
15. Fedorov, A. G.; Viskanta, R. Three-Dimensional Conjugate Heat Transfer in the Microchannel Heat Sink for Electronic Packaging. *Int. J. Heat Mass Transf.* **2000,** *43* (3), 399–415.
16. Mansoor, M. M.; Wong, K. C.; Siddique, M. Numerical Investigation of Fluid Flow and Heat Transfer under High Heat Flux Using Rectangular Micro-Channels. *Int. Commun. Heat Mass Transf.* **2012,** 39 (2), 291–297.
17. Mirmanto, S. T.; Kenning, D. B. R.; Karayiannis, T. G.; Lewis, J. S. Pressure Drop and Heat Transfer Characteristics for Single Phase Developing Flow of Water in Rectangular Microchannels. *J. Phys. Conf. Ser.* **2012,** *395* (conference 1).
18. Tuckerman, D. B.; Pease, R. F. W. High-Performance Heat Sinking for VLSI. *IEEE Electron Device Lett.* **1981,** *2* (5), 126–129.
19. Wu, P. Y.; Little, W. A. Measurement of Friction Factor for Flow of Gases in Very Fine Channels Used for Microminiature Joule Thompson Refrigerators. *Cryogenics.* **1983,** *24* (8), 273–277.
20. Harms, T. M.; Kazmierczak, M. J.; Cerner, F. M.; Holke, A.; Henderson, H. T.; Pilchowski, H. T.; Baker, K. In *Experimental Investigation of Heat Transfer and Pressure Drop through Deep Micro Channels in a (100) Silicon Substrate*, Proceedings of the ASME. Heat Transfer Division, HTD 351, 1997, pp 347–357.
21. Fedorov, A.G.; Viskanta, R. Three-Dimensional Conjugate Heat Transfer into Microchannel Heat Sink for Electronic Packaging. *Int. J. Heat Mass Transf.* **2000,** *43* (3), 399–415.
22. Qu, W.; Mala, G. M.; Li, D. Heat Transfer for Water Flow in Trapezoidal Silicon Microchannels. *Int. J. Heat Mass Transf.* **1993,** *21,* 399–404.
23. Choi, S. B.; Barren, R. F.; Warrington, R. O. Fluid Flow and Heat Transfer in Microtubes, ASME DSC 40, **1991,** 89–93.
24. Adams, T. M.; Abdel-Khalik, S. I. Jeter, S. M.; Qureshi, Z. H. An Experimental Investigation of Single-Phase Forced Convection in Microchannels. *Int. J. Heat Mass Transf.* **1998,** *41 (6–7),* 851–857.
25. Peng, X. F.; Peterson, G. P. Convective Heat Transfer and Flow Friction for Water Flow in Microchannel Structure. *Int. J. Heat Mass Transf.* **1996,** *39* (12), 2599–2608.
26. Mala, G. M.; Li, D.; Dale, J. D. Heat Transfer and Fluid Flow in Microchannels. *Int. J. Heat Mass Transf.* **1997,** *40* (13), 3079–3088.
27. Xu, B.; Ooi, K. T.; Mavriplis, C.; Zaghloul, M. E. Viscous Dissipation Effects for Liquid Flow in Microchannels, Micorsystems, 2002, pp 53–57.
28. Iaccarinno, G.; Ooi, A.; Durbin, P.; Behnia, M. Conjugate Heat Transfer Predictions in Two-dimensional Ribbed Passages. *Int. J. Heat Fluid Flow* 23 (2002), pp 340–345.
29. Tiselj, I.; Hetsroni, G.; Mavko, B.; Mosyak, A.; Pogrebnyak, E.; Segal, Z. Effect of Axial Conduction on the Heat Transfer in Microchannels. *Int. J. Heat Mass Transf.* **2004,** 47 (12–13), 2551–2565.

30. Moharana, M. K.; Singh, P. K.; Khandekar, S. Optimum Nusselt Number for Simultaneously Developing Internal Flow under Conjugate Conditions in a Square Microchannel. *J. Heat Transfer.* **2012,** *134* (7), 1–10.
31. Bansode, A. S.; Patel, S.; Kumar, T. R.; Muralidhar, B.; Sundararajan, T.; Das, S. K. Numerical Simulation of Effects of Flow Maldistribution on Heat and Mass Transfer in a PEM Fuel Cell Stack. *Heat Mass Transfer*. **2007,** *43* (10), 1037–1047.
32. Phillips, R. J. Forced-Convection, Liquid-Cooled, Microchannel Heat Sinks. MS. Thesis, Department of Mechanical Engineering, Cambridge, 1987, pp 70.
33. Bejan, A. *Convection Heat Transfer;* John Wiley & Sons, Inc.: New Jersey, 22013.
34. Dittus, F. W.; Boelter, L. M. K. *Heat Transfer in Automobile Radiators of Tubular Type;* Publications in Engineering. University of California: Berkeley, California,1930; pp 443–461.
35. Sahara, A. M.; Özdemira, M. R.; Fayyadhb, E. M.; Wissinka, J.; Mahmouda, M. M.; Karayiannisa, T. G. Single Phase Flow Pressure Drop and Heat Transfer in Rectangular Metallic Microchannels. *Appl. Therm. Eng.* **2016,** *93,* 1324–1336.
36. Nagiyev, F. B. Dynamics, Heat and Mass Transfer of Vapor-Gas Bubbles in a Two-Component Liquid. Turkey-Azerbaijan petrol semin., Ankara, Turkey, 1993, pp 32–40.
37. Nagiyev, F. B. The Method of Creation Effective Coolness Liquids, Third Baku international Congress. Baku, Azerbaijan Republic, 1995, 19–22.
38. Nagiyev, F. B. The Linear Theory of Disturbances in Binary Liquids Bubble Solution. Dep. In VINITI, 1986, pp 76–79.
39. Nagiyev, F. B. Structure of Stationary Shock Waves in Boiling Binary Solutions. Math. USSR, Fluid Dynamics, 1989, pp 81–87.
40. Rayleigh, L. On the Pressure Developed in a Liquid during the Collapse of a Spherical Cavity. *Philos. Mag. Ser. 6.* **1917,** *34* (200), 94–98.
41. Perry, R. H.; Green, D. W.; Maloney, J. O. In *Perry's Chemical Engineers Handbook,* 7th Ed.; Perry, R. H, Green, D. W, Eds.; McGraw-Hill Professional Publishing: New York, 1997; 1–61.
42. Nigmatulin, R. I. *Dynamics of Multiphase Media;* Izdatel'stvo Nauka: Moscow, 1987; pp 12–14.
43. Kodura, A.; Weinerowska, K. The Influence of the Local Pipeline leak on water hammer properties, Materials of the II Polish Congress of Environmental Engineering, Lublin, 2005, 125–133.
44. Kane, J.; Arnett, D.; Remington, B. A.; Glendinning, S. G.; Baz'an, G. Two-Dimensional versus Three-Dimensional Supernova Hydrodynamic Instability Growth. *Astrophys. J.* **2000,** *528* (2), 528–989.
45. Quick, R. S. Comparison and Limitations of Various Water hammer Theories. *J. Hyd. Div.* **1933,** 43–45.
46. Jaeger, C. In *Fluid Transients in Hydro-Electric Engineering Practice;* Blackie and Son Ltd.: Glasgow, Scotland, 1977; pp 87–88.
47. Jaime, S. A. Generalized Water Hammer Algorithm for Piping Systems with Unsteady Friction. E-theses, HKU School of Professional and Continuing Education, 2005.
48. Fok, A.; Ashamalla, A.; Aldworth, G. Considerations in Optimizing Air Chamber for Pumping Plants, Symposium on Fluid Transients and Acoustics in the Power Industry, San Francisco, U.S.A. Dec., 1978, 112–114.

49. Fok, A. *Design Charts for Surge Tanks on Pump Discharge Lines, BHRA 3rd Int. Conference on Pressure Surges*; Bedford, England, 1980; pp 23–34.
50. Fok, A. *Water Hammer and its Protection in Pumping Systems, Hydro technical Conference;* CSCE: Edmonton, 1982; pp 45–55.
51. Fok, A. A contribution to the Analysis of Energy Losses in Transient Pipe Flow, Ph.D. Thesis, University of Ottawa, 1987.
52. Brunone, B.; Karney, B. W.; Mecarelli, M.; Ferrante, M. Velocity Profiles and Unsteady Pipe Friction in Transient Flow. *J. Water Res. Plan. Manag. ASCE.* **2000,** *126* (4), 236–244.
53. Koelle, E.; Luvizotto Jr. E.; Andrade, J. P. G. In *Personality Investigation of Hydraulic Networks using MOC – Method of Characteristics.* Proceedings of the 7th International Conference on Pressure Surges and Fluid Transients, Harrogate Durham, UK, 1996, pp 1–8.
54. Filion, Y.; Karney, B. W. In *A Numerical Exploration of Transient Decay Mechanisms in Water Distribution Systems.* Proceedings of the ASCE Environmental Water Resources Institute Conference, American Society of Civil Engineers, Roanoke, Virginia, 2002, p 30.
55. Hamam, M. A.; McCorquodale, J. A. Transient Conditions in the Transition from Gravity to Surcharged Sewer Flow. *Canadian J. Civil Eng.* **1982,** *9* (2), 65–98.
56. Savic, D. A.; Walters, G. A. Genetic Algorithms Techniques for Calibrating Network Models, Report No. 95/12, Centre for Systems and Control Engineering, 1995, 137–146.
57. Walski, T. M.; Lutes, T. L. Hydraulic Transients Cause Low-Pressure Problems. *J. Am. Water Works Assoc.* **1994,** *75* (2), 58.
58. Lee, T. S.; Pejovic, S. Air Influence on Similarity of Hydraulic Transients and Vibrations. *J. Fluids Eng.* **1996,** *118* (4), 706–709.
59. Chaudhry, M. H. *Applied Hydraulic Transients;* Van Nostrand Reinhold Co.: New York, 1979; 1322–1324.
60. Parmakian, J. *Water Hammer Analysis;* Dover Publications, Inc.: New York, 1963; pp 51–58.
61. Tuckerman, D. B.; Pease, R. F. W. High Performance Heat Sinking for VLSI. *IEEE Electron Device Lett.* **1981,** *2* (5), 126–129.
62. Farooqui, T. A. Evaluation of Effects of Water Reuse, on Water and Energy Efficiency of an Urban Development Area, using an Urban Metabolic Framework. Master of Integrated Water Management Student Project Report, International Water Centre.
63. Ferguson, B. C.; Frantzeskaki, N.; Brown, R. R. A Strategic Program for Transitioning to a Water Sensitive City. *Landsc. Urban Plan.* **2013,** *117*, 32–45.
64. Kenway, S.; Gregory, A.; McMahon, J. Urban Water Mass Balance Analysis. *J. Ind. Ecol.* **2011,** *15* (5), 693–706.
65. Renouf, M. A.; Kenway, S. J.; Serrao-Neumann, J.; Low Choy, D. (2015). *Urban Metabolism for Planning Water Sensitive Cities. Concept for an Urban Water Metabolism Evaluation Framework. Milestone Report.* Melbourne: Cooperative Research Centre for Water Sensitive Cities.
66. Serrao-Neumann, S.; Schuch, G.; Kenway, S. J.; Low Choy, D. Comparative Assessment of the Statutory and Non-Statutory Planning Systems: South East Queensland,

Metropolitan Melbourne and Metropolitan Perth. Melbourne: Cooperative Research Centre for Water Sensitive Cities, 2013.

67. Andrews, S.; Traynor, P. *Guidelines for Assuring Quality of Food and Water Microbiological Culture Media;* August 2004.
68. Andrew, D. Eaton; Eugene, W. Rice; Lenore S. Clescceri. Standard Methods for the Examination of Water and Waste Water. Part 9000, 2012.
69. Instructions Procedure Rural Water and Wastewater Quality Assurance Test Results of KhorasanRazavi, 2011.
70. Abbassi, B.; Al Baz, I. Integrated Wastewater Management: A Review. In *Efficient Management of Wastewater–Its Treatment and Reuse in Water Scarce Countries.* Springer Publishing Co.: Heidelberg, 2008.
71. Andreasen, P. Chemical Stabilization. In *Sludge into Biosolids – Processing, Disposal and Utilization;* Spinosa, L., Vesilind, P. A., Eds.; IWA Publishing: UK, 2001; pp 392.
72. CIWEM – The Chartered Institution of Water and Environmental Management. Sewage Sludge: Stabilization and Disinfection – Handbooks of UK Wastewater Practice. CIWEM Publishing, 1996.
73. Halalsheh, M.; Wendland, C. Integrated Anaerobic-Aerobic Treatment of Concentrated Sewage. In *Efficient Management of Wastewater – Its Treatment and Reuse in Water Scarce Countries;* Baz, I. A., Otterpohl, R., Wendland, C., Eds.; Springer Publishing Co.: Berlin, Heidelberg, 2008; pp 177–186.
74. ISWA. Handling, Treatment and Disposal of Sludge in Europe. Situation Report 1. Copenhagen. ISWA Working Group on Sewage Sludge and Water Works, 1995.
75. Matthews, P. Agricultural and Other Land Uses. Sludge into Biosolids – Processing, Disposal and Utilization. IWA Publishing, 2001.
76. IWK Sustainability Report 2012–2013.

## CHAPTER 7

# NON-LINEAR MODELING FOR NON-REVENUE WATER

KAVEH HARIRI ASLI[1*], SOLTAN ALI OGLI ALIYEV[1], and HOSSEIN HARIRI ASLI[2]

[1]*Department of Mathematics and Mechanics, Azerbaijan National Academy of Sciences AMEA, Baku, Azerbaijan*

[2]*Civil Engineering Department, Faculty of Engineering, University of Guilan, Rasht, Iran*

[*]*Corresponding author. E-mail: hariri_k@yahoo.com*

## CONTENTS

## ABSTRACT

Present research was defined for investigation of leakage points in the water distribution network in order to reduce the real case amount of non-revenue water (NRW). In this research, the Relationship Class between the spatial data and non-spatial data is a tool for communication between coordination of water system elements related to NRW data. Therefore the relationship class for water system elements is affected according to the type of connection (simple or complex) and this method provides the access to the leakage points. Hence in this research the hydraulic model of non-revenue water for analysis of water loss was defined in compliance with geographical information system (GIS) and by WATERGEMS8.2 software under management of ArcGIS9-ArcMap9.3 software. The results of this research lead to reduction of the amount of NRW.

## 7.1 INTRODUCTION

Water loss may be defined as the water that has been obtained from a source and put into a supply and distribution system is lost via leaks or is allowed to escape or is taken there from for no useful purpose. "Water loss" is usually considered as "leakage" and "water loss reduction" is referred to as "leakage control."

For example, old iron mains still form the majority of mains and they are the worst culprits for leakage. They suffer from both external and internal corrosion attacks which progressively weaken them. Iron mains can then crack and leak, or holes are formed due to the corrosion process. Once the leakage occurs, which may be finally precipitated by an increase in pressure, flow, or temporary change, then it will worsen. This may occur steadily, or rapidly degenerate into a large burst. Cases of subterranean caverns beneath metallic roadways are known where the escaping water hollows out a void by its pressure jet.

A properly designed distribution system should prevent some vulnerability to leakage at the outset. Such design would assess the need for cathodic protection of steel and ductile iron mains. It would ensure that all mains with unrestrained flexible joints had appropriately sized and positioned concrete thrust blocks at all changes of direction and blank ends. All mains and services should be laid with the correct amount of cover to the surface, and appropriately distanced from other underground services. The use of marker tape sited 300 mm above the main will alert excavation to the presence of the main, thus preventing incidental damage and ensuing leakage. Where plastic pipe is used, such tape should have a metallic strip incorporated to assist with location equipment. Correct sizing of mains at the outset, considering such factors as peak flow, fire fighting requirements, and future development, will prevent the temptation to “force” more water through by increasing pressures at a later date. Oversized mains also need to be avoided, particularly from a water quality point of view.

There is no substitute for good workmanship of the initial installation in preventing future leakage. Pipe handling, bedding, laying, jointing, and backfilling must be of a high standard. Extra care should be given to repair work, as a repair does represent a potential weakness to the integrity of the system.

It is obvious that all materials used in the distribution system must comply with relevant standards for long-term usefulness, be of a high quality, be appropriate to the surrounding conditions, and be of the correct operational capabilities. It should also be ensured that the same standards apply to repair materials, and that poor substitutes are not used for permanent repairs. High pressure equals high leakage. This factor is very important in leakage control, and will be discussed in more detail in a subsequent section.

This may seem obvious but it is very important to remember that a small increase in the size of the leak has a big effect in terms of volume leaked. The longer a leak is left to run, the bigger the hole will get. A speedy location and repair of leaks is essential to reduce waste levels.

A leak running for a long time can waste just as much water as catastrophic trunk main burst which is repaired quickly. Time before discovery, time to detect, and time to repair are the major components. Leakage will only be reduced by sustained, determined detection, and rapid repair. “Find and Fix Fast” is an appropriate axiom.

Severe pressures can be generated by the rapid operation of isolating valves, thus precipitating bursts and leakage. Ironically, rapid re-charging of a system following leakage repair work can cause further damage and leakage. Valve closures and mains re-charging work should therefore be carried out in a steady, controlled manner. This is particularly relevant when mains scraping and relining is taking place. Similar care must be taken during mains flushing, swabbing, and air scouring.

The aging process cannot be stopped, and increasing leakage is indicative of deteriorating structural condition. It should therefore be recognized that a realistic and consistent level of renewal of the infrastructure is an essential part of leakage strategy development. This may be achieved by targeted mains relining (where iron pipes are in use, and corrosion is mostly internal), or by targeted mains replacement. The former has little, if any, impact on leakage rates from those mains, whereas the latter should eradicate it for a substantial period of time if done well. The modern techniques of mains replacement have substantially cut excavation and backfilling costs. It is essential that the renewal of service pipes is included in such work for the greatest benefit.

Leakage grows with time, and without action to curb it, would grow to a point where supplies were unsustainable. Passive control, that is, the repair of bursts and leakage showing on the surface, and the elimination of poor pressure and flow complaints, is the minimum possible response. This is required to prevent damage to persons and property, and to maintain supplies to customers. The actual leakage level reached will depend on how quickly low pressure and flow will be experienced and other factors, such as how quickly leakage appears on the surface and is reported.

For any given area in the distribution system there will be a characteristic growth rate. This characteristic growth rate will be affected by changes in the physical elements of the system, such as rehabilitation of mains, renewal of service pipes, and changes in pressure. Sooner or later, leakage control must be associated with a program of mains renewal in order to maintain the supply/demand balance. However, improvement of mains and services is expensive and clearly, for the system as whole, is very much a long-term strategy. Reduction in pressure is also effective in reducing both the volume of leakage and its rate of growth, although there is some doubt whether the latter effect persists in the long term. The scope for pressure reduction is, of course, limited, given that adequate supplies to customers must be maintained.[1–13]

## 7.2 MATERIALS AND METHODS

To reduce the natural level of leakage at any pressure, a program of leakage detection must be planned, co-ordinated, and implemented. In this work conclusions were drawn on the basis of experiments and calculations for the pipeline with a local leak. Hence, the most important effects that were observed are as follows: The pressure wave speed generated by phenomenon was influenced by some additional factors. Therefore the ratio of local leakage and discharge from the leak location was mentioned. The effect of total discharge from the pipeline and its effect on the values of wave oscillations period were studied. The outflow from the leak affected the value of wave celerity. The pipeline was equipped with the valve at the end of the main pipe, which was joined with the closure time register. The pressure characteristics were measured by extensometers, as similar as in the work of Kodura and Weinerowska. Simultaneously it was recorded in the computer's memory. The supply of the water at the system was realized with the use of the reservoir which enabled inlet pressure stabilization. In this work, positive pressure in the pipeline was introduced with local leaks in two scenarios: first, with the outflow from the leak to the overpressure reservoir; and second, with free outflow from the leak to atmospheric pressure, with the possibility of sucking air in the negative pressure phase. The bubbles in non-linear dynamics fluid state act as a separator gate for two distinct parts of the flow at upstream and downstream of separator gate. It causes high surge wave velocity at one of these distinct parts. Therefore the compressed air builds high pressure flow which can destroy the water pipeline.

Obtained pressure heads by the steady and unsteady friction model are shown in Figure 7.1. Comparison showed similarity in this work with the results of the work of Chaudhry.

*Parmakian (1955) and Streeter and Wylie (1983)***.** MOC-based scheme was most popular because this scheme provided the desirable attributes of accuracy, numerical efficiency, and programming simplicity in the work of Parmakian (1955) and Wylie and Streeter (1983). The CFD software package which was used in this work applied Numerical solutions of the non-linear Navier–Stokes equations by method of characteristics (MOC).

*Arturo S. Leon (2006–2007)*. Comparison showed similarity between present work results and the results of the work of Arturo S. Leon (2007). The results are as following:

1. The proposed formulation maintains the conservation property of FV schemes and introduced no unphysical perturbations into the computational domain.
2. Numerical tests were performed for smooth (i.e., flows that do not present discontinuities) and sharp transients.

3. The high efficiency of the proposed scheme was important for real-time control (RTC) of water hammer flows in large networks.

In this work, changes at system boundaries (sudden changes) created a transient pressure pulse. In this regard, model design needed to find the relation between many variables in accordance to fluid transient. Therefore, a computational technique was presented and the results were compared by field tests. In this work after closing the valves on the horizontal pipe of constant diameter, which moves the liquid with an average speed $V_o$, a liquid layer, located directly at the gate, immediately stops, and then successively terminate movement of the liquid layers (turbulence, counter flows) to increase with time away from the gate. In this work the air was sucked into the pipeline. Pressure wave velocity was recorded in fast transient up to 5 ms (in this work 1 s). The assessment procedure was used to analyze the collected data which were obtained at real system.[14–42]

It is defined as the combination of momentum equation and continuity equation for determining the velocity and pressure in a one-dimensional flow system. The solving of these equations produces a theoretical result that usually corresponds quite closely to actual system measurements.

$$P\Delta A-(P+\frac{\partial P}{\partial S}.\Delta S)\Delta A-$$

$$W.\sin\theta-\tau.\Delta S.\pi.d=\frac{W}{g}.\frac{dV}{dt}.$$

Both sides are divided by $m$ and with assumption:

$$-\frac{1}{\partial}.\frac{\partial \mathrm{P}}{\partial S}-\frac{\partial Z}{\partial S}-\frac{4\tau}{\gamma D}=\frac{1}{g}.\frac{dV}{dt}, \qquad (7.1)$$

$$\Delta A=\frac{\Pi.D^2}{4}.$$

If fluid diameter assumed equal to pipe diameter, then:

$$\frac{-1}{\gamma}.\frac{\partial \mathrm{P}}{\partial S}-\frac{\partial Z}{\partial S}-\frac{4\tau_o}{\gamma.D}, \qquad (7.2)$$

$$\tau_o=\frac{1}{8}\rho.f.V^2,$$

$$\frac{1}{-\gamma}.\frac{\partial \mathrm{P}}{\partial S}-\frac{\partial Z}{\partial S}-\frac{f}{D}.\frac{V^2}{2g}=\frac{1}{g}.\frac{dV}{dt}, \tag{7.3}$$

$$V^2=V\,|V\,|,\frac{dV}{dt}+\frac{1}{\rho}.\frac{\partial \mathrm{P}}{\partial S}+g\frac{dZ}{dS}+\frac{f}{2D}V|V|=0. \tag{7.4}$$

(Euler equation)

For finding $V$ and $P$ we need to “conservation of mass law”:

$$\rho \mathrm{A} V-\left[\rho A V-\frac{\partial}{\partial S}(\rho A V)dS\right]=\frac{\partial}{\partial t}(\rho A dS)-\frac{\partial}{\partial S}(\rho A V)dS=\frac{\partial}{\partial t}(\rho A dS) \tag{7.5}$$

$$-\left(\rho \mathrm{A}\frac{\partial V}{\partial S}dS+\rho V\frac{\partial \mathrm{A}}{\partial S}dS+\mathrm{A}V\frac{\partial \rho}{\partial S}dS\right)=\rho \mathrm{A}\frac{\partial}{\partial t}(dS)+\rho dS\frac{\partial \mathrm{A}}{\partial t}+\mathrm{A}dS\frac{\partial p}{\partial t}, \tag{7.6}$$

$$\frac{1}{\rho}\left(\frac{\partial \rho}{\partial t}+V\frac{\partial \rho}{\partial S}\right)+\frac{1}{\mathrm{A}}\left(\frac{\partial \mathrm{A}}{\partial t}+V\frac{\partial \mathrm{A}}{\partial S}\right)+\frac{1}{dS}.\frac{\partial}{\partial t}(dS)+\frac{\partial V}{\partial S}=\circ$$

With $\frac{\partial \rho}{\partial t}+V\frac{\partial \rho}{\partial S}=\frac{d\rho}{dt}$ & $\frac{\partial \mathrm{A}}{\partial t}+V\frac{\partial \mathrm{A}}{\partial S}=\frac{d\mathrm{A}}{dt}$

$$\frac{1}{\rho}\cdot\frac{d\rho}{dt}+\frac{1}{\mathrm{A}}\cdot\frac{d\mathrm{A}}{dt}+\frac{\partial V}{\partial S}+\frac{1}{dS}.\frac{1}{dt}(dS)=\circ, \tag{7.7}$$

$K=\left(\frac{d\rho}{\left(\frac{d\rho}{\rho}\right)}\right)$ (Fluid module of elasticity) then:

$$\frac{1}{\rho}.\frac{d\rho}{dt}=\frac{1}{k}.\frac{d\rho}{dt}, \tag{7.8}$$

Put eq 7.7 into eq 7.8, then:

$$\frac{\partial V}{\partial S}+\frac{1}{k}.\frac{d\rho}{dt}+\frac{1}{\mathrm{A}}\cdot\frac{d\mathrm{A}}{dt}+\frac{1}{dS}.\frac{d}{dt}(dS)=\circ,$$

$$\rho\frac{\partial V}{\partial S}+\frac{d\rho}{dt}\rho\left[\frac{1}{k}+\frac{1}{\mathrm{A}}.\frac{d\mathrm{A}}{d\rho}+\frac{1}{dS}\cdot\frac{d}{d\rho}(dS)\right]=\circ, \tag{7.9}$$

$$\rho\left[\frac{1}{k}+\frac{1}{\mathrm{A}}.\frac{d\mathrm{A}}{dt}+\frac{1}{dS}\cdot\frac{d}{d\rho}(dS)\right]=\frac{1}{C^{2}}.$$

Then $C^{2}\frac{\partial V}{\partial S}+\frac{1}{\rho}.\frac{d\rho}{dt}=\circ,$

(Continuity equation)

Partial differential equations 7.4 and 7.10 are solved by method of characteristics "MOC":

$$\frac{dp}{dt}=\frac{\partial p}{\partial t}+\frac{\partial p}{\partial S}\cdot\frac{dS}{dt}, \tag{7.11}$$

$$\frac{dV}{dt}=\frac{\partial V}{dt}+\frac{\partial V}{\partial S}\cdot\frac{dS}{dt}, \tag{7.12}$$

Then:

$$\left|\begin{array}{l}\frac{\partial V}{\partial t}+\frac{1}{\rho}\frac{\partial p}{\partial S}+g\frac{dz}{dS}+\\ \frac{f}{2D}V|V|=\circ,\\ C^{2}\frac{\partial V}{\partial S}+\frac{1}{\rho}\frac{\partial P}{\partial t}=\circ.\end{array}\right. \qquad (7.13), (7.14)$$

By linear combination of eqs 7.13 and 7.14

$$\begin{array}{l}\lambda\left(\frac{\partial V}{\partial t}+\frac{1}{\rho}\frac{\partial p}{\partial S}+g.\frac{dz}{dS}+\frac{f}{2D}V|V|\right)+\\ C^{2}\frac{\partial V}{\partial S}+\frac{1}{\rho}\frac{\partial p}{\partial t}=\circ,\end{array} \tag{7.15}$$

$$\begin{array}{l}(\lambda\frac{\partial V}{\partial t}+C^{2}\frac{\partial V}{\partial S})+(\frac{1}{\rho}.\frac{\partial \rho}{\partial t}+\frac{\lambda}{\rho}.\frac{\partial P}{\partial S})+\\ \lambda g.\frac{dz}{dS}+\frac{\lambda f}{2D}V|V|=\circ,\end{array} \tag{7.16}$$

$$\lambda\frac{\partial V}{\partial t}+C^2\frac{\partial V}{\partial S}=\lambda\frac{dV}{dt}\Rightarrow\lambda\frac{dS}{dt}=C^2, \tag{7.17}$$

$$\begin{aligned}&\frac{1}{\rho}.\frac{\partial p}{\partial t}+\frac{\lambda}{\rho}.\frac{\partial\rho}{\partial S}=\frac{1}{\rho}.\frac{d\rho}{dt}\Rightarrow\\&\frac{\lambda}{\rho}=\frac{1}{\rho}.\frac{dS}{dt}\end{aligned}, \tag{7.18}$$

$$\left|\frac{C^2}{\lambda}=\lambda\text{ (By removing }\frac{dS}{dt}\text{)},\ \lambda=\pm C.\right.$$

For λ = ±*C* from eq 7.18 we have:

$$\mathrm{C}\ \frac{dV}{dt}+\frac{1}{\rho}.\frac{dp}{dt}+C.g.\frac{dz}{dS}+C.\frac{f}{2D}V|V|=\circ.$$

By dividing both sides by "*C*":

$$\begin{aligned}&\frac{dV}{dt}+\frac{1}{c.\rho}\frac{dP}{dt}+g.\frac{dz}{dS}+\\&\frac{f}{2D}V|V|=\circ.\end{aligned}, \tag{7.19}$$

For λ = –*C* from by eq 7.16:

$$\begin{aligned}&\frac{dV}{dt}-\frac{1}{c.\rho}\frac{dp}{dt}+g\frac{dZ}{dS}+\\&\frac{f}{2D}V|V|=\circ\end{aligned} \tag{7.20}$$

If $\rho=\rho.g\,(H-Z)$
From eqs 7.9 and 7.10:

$$\left|\begin{aligned}&\frac{dV}{dt}+\frac{g}{c}\cdot\frac{dH}{dt}+\frac{f}{2D}V|V|=\circ\\&if:\frac{dS}{dt}=C,\end{aligned}\right., \tag{7.21), (7.22}$$

$$\left| \begin{array}{l} \frac{dV}{dt}+\frac{g}{c}.\frac{dH}{dt}+\frac{f}{2D}V|V|=\circ, \\ if:\frac{dS}{dt}=-C, \end{array} \right. \quad (7.23), (7.24)$$

The method of characteristics is a finite difference technique by which pressures (Fig. 7.2) were computed along the pipe for each time step eqs 7.1–7.33.

Calculation automatically sub-divided the pipe into sections (intervals) and selected a time interval for computations.

If $f = 0$, then eq 7.23 will be (Fig. 7.1):

$$\frac{dV}{dt}-\frac{g}{c}.\frac{dH}{dt}=\circ,$$

or

$$dH=\left(\frac{C}{g}\right)dV \text{ (Zhukousky)}, \quad (7.25)$$

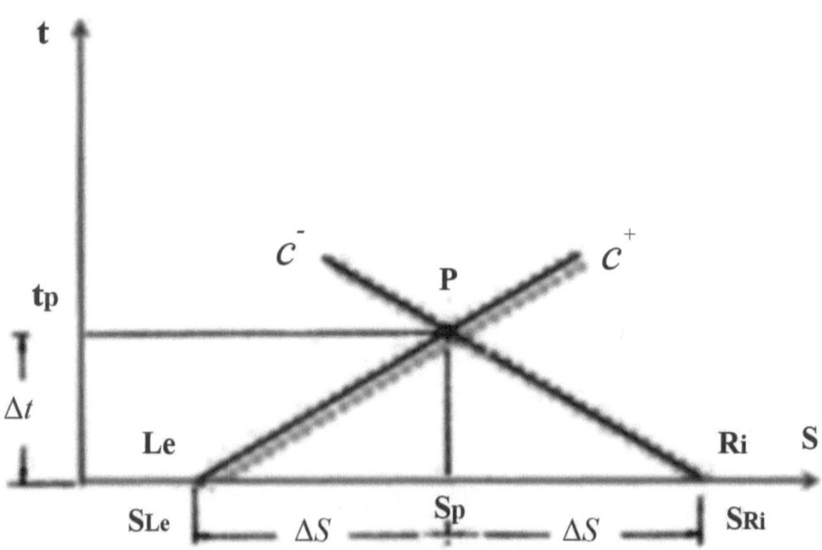


**FIGURE 7.1** Intersection of characteristic lines with positive and negative slope.

If the pressure at the inlet of the pipe and along its length is equal to $p_0$, then slugging pressure undergoes a sharp increase:

$\Delta p$: $p = p_0 + \Delta p$.

The Zhukousky formula is as flowing:

$$\Delta p = \left(\frac{C.\Delta V}{g}\right). \tag{7.26}$$

The speed of the shock wave is calculated by the formula:

$$C = \sqrt{\frac{g.\dfrac{E_W}{\rho}}{1+\dfrac{d}{t_W}\cdot\dfrac{E_W}{E}}}. \tag{7.27}$$

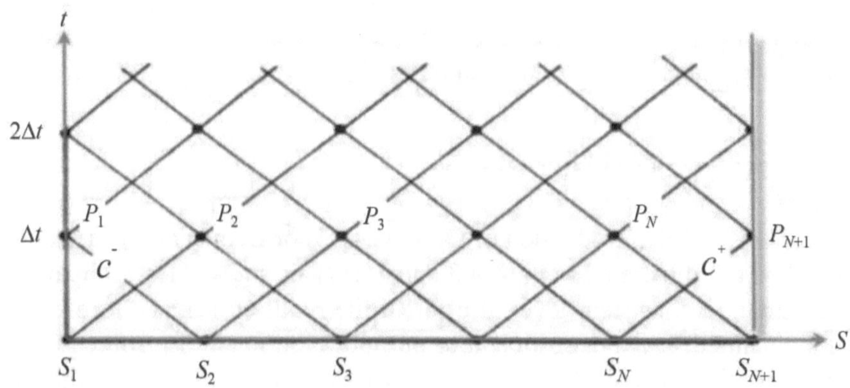


**FIGURE 7.2** Set of characteristic lines intersection for assumed pipe.

By finite difference method of water hammer: $T_p - 0 = \Delta t$

$$c^{+}:(V_P - V_{Le})/(T_P - \circ) + (\tfrac{g}{c})(H_P - H_{Le})/(T_P - \circ) + fV_{Le}|V_{Le}|/2D) = \circ, \quad (7.28)$$

$$c^{-}:(V_P - V_{Ri})/(T_P - \circ) + (\tfrac{g}{c})(H_P - H_{Ri})/(T_P - \circ) + fV_{Ri}|V_{Ri}|/2D) = \circ, \quad (7.28)$$

$$c^{-}:(V_P - V_{Ri})/(T_P - \circ) + (\tfrac{g}{c})(H_P - H_{Ri})/(T_P - \circ) + fV_{Ri}|V_{Ri}|/2D) = \circ, \quad (7.29)$$

$$c^{+}:(V_{p}-V_{Le})+(\frac{g}{c})(H_{p}-H_{Le})+(f.\Delta t)(f.V_{Le}|V_{Le}|/2D)=\circ, \quad (7.30)$$

$$c^{-}:(V_{p}-V_{Ri})+(\frac{g}{c})(H_{p}-H_{Ri})+(f.\Delta t)(fV_{Ri}|V_{Ri}|/2D)=\circ, \quad (7.31)$$

$$V_{p}=\frac{1}{2}\left[(V_{Le}+V_{Ri})+\frac{g}{c}\left(H_{Le}-H_{Ri}\right)-(f.\Delta t/2D)(V_{Le}|V_{Le}|-VR_{i}|V_{Ri}|)\right] \quad (7.32)$$

$$H_{p}=\frac{1}{2}\left[\frac{c}{g}(V_{Le}+V_{Ri})+(H_{Le}-H_{Ri})-\frac{c}{g}(f.\Delta t/2D)(V_{Le}|V_{Le}|-V_{Ri}|V_{Ri}|)\right]. \quad (7.33)$$

*VLe' V Ri HLe' HRi' L*, *D* are initial conditions parameters.

They are applied for solution at steady state condition. The calculation starts with pipe length "*N*" divided by "*N*" parts:

$$\Delta S=\frac{L}{N}\ \&\ \Delta t=\frac{\Delta s}{C}.$$

Equations 7.28 and 7.29 are solved for the range $P_2$ through *H*, therefore *H* and *V* are found for internal points.[43–64]

## 7.3 RESULTS AND DISCUSSION

The effects of the introduction of various levels or frequencies of leakage detection are illustrated in Figs. 7.3 and 7.4). This shows clearly the need to maintain a consistent level of effort if the required leakage level is to be maintained. It is not sufficient to put in a high level of resource for a short period, as any slackening of effort will lead to an increase in leakage over a period of time. Given that no two water distribution systems are identical in terms of physical or economic characteristics, it is not possible to determine the most appropriate leakage control policy in a general manner. The best policy for any given system will depend on its particular characteristics.

The economic balance of searching for and repairing leakage, and of controlling it to an acceptable level, is a complex issue. Typically a leakage percentage of below 10 or even 15% may not be economic to pursue. In other words, the effect of hunting down, identifying, and repairing the leakage costs more than the value of the water saved. These remarks need to be heavily qualified, however. For instance, a modern housing estate

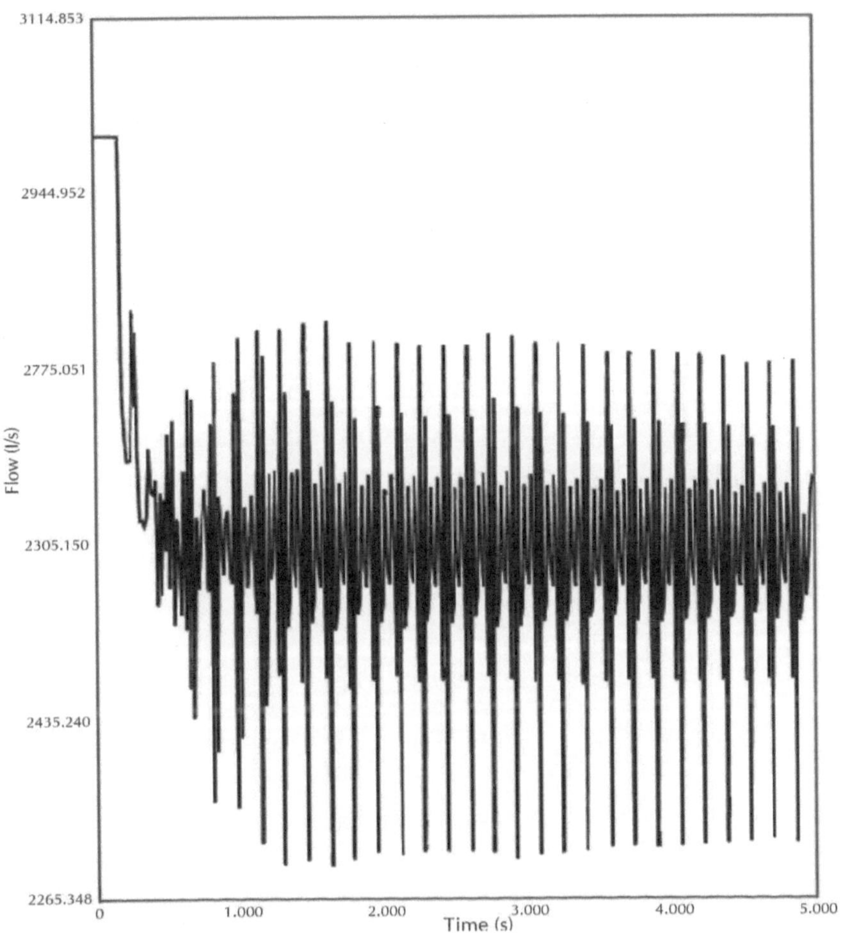


**FIGURE 7.3** Graph to show growth of leakage with time.

could have a serious problem with 10% leakage whilst an old area with a stubborn leakage of 30%, say, may require a mains renewal scheme. Each area will have its own intrinsic economic leakage level. It was further suggested that of the "acceptable leakage," the quantity of leakage which was undetectable was approximately 30 1/prop/day. Figure 7.5 suggests the relative percentages of leaks caused by different types of bursts, and the possible water quantity lost through them. In addition to the volume of water lost, its scarcity and marginal cost per mega liter are vital factors.

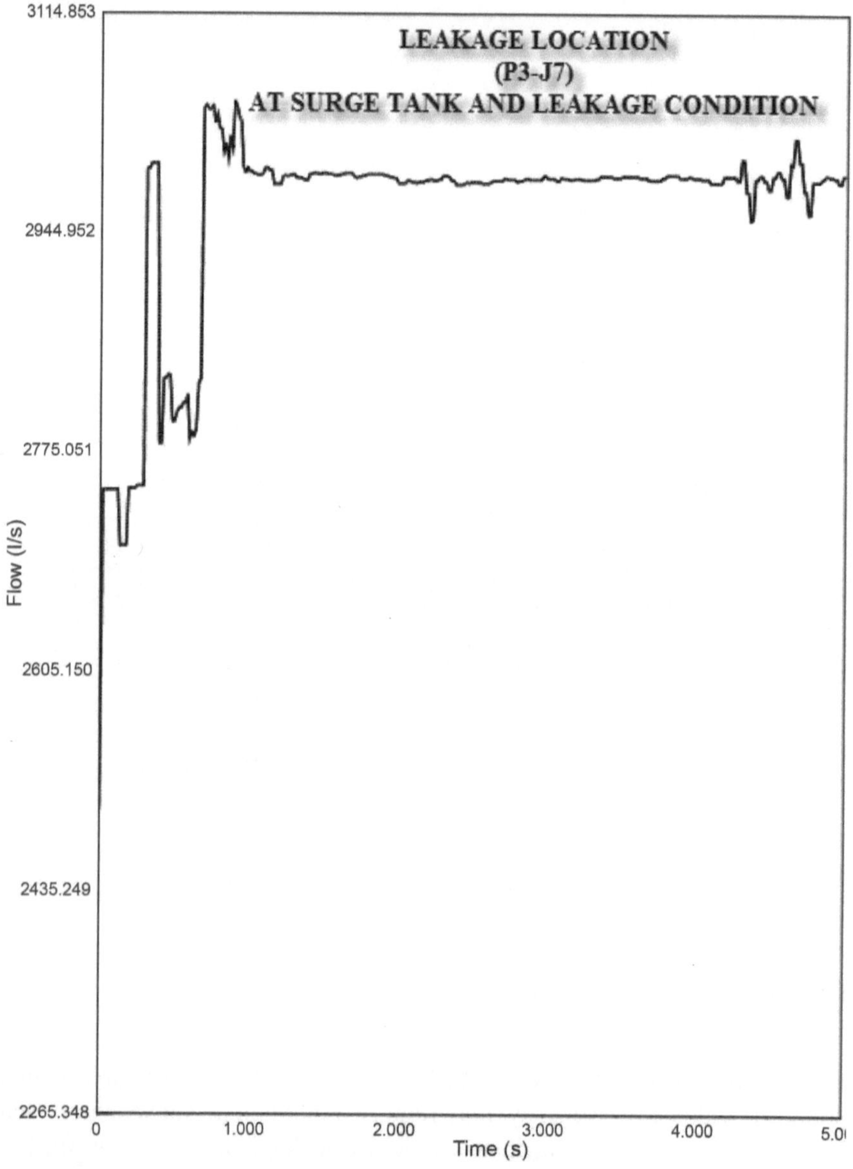


**FIGURE 7.4** Graph to show growth of leakage with time.

In an area of rising demand, needing to promote, build, and commission a new source, intensive leakage control activity would be essential. In an area which relied upon pumped supplies with high electricity costs, a high degree of leakage control would make sense, and have priority over an area with plentiful supplies fed by gravity. In selecting the required leakage reduction approach, there are two policies which may be adopted.

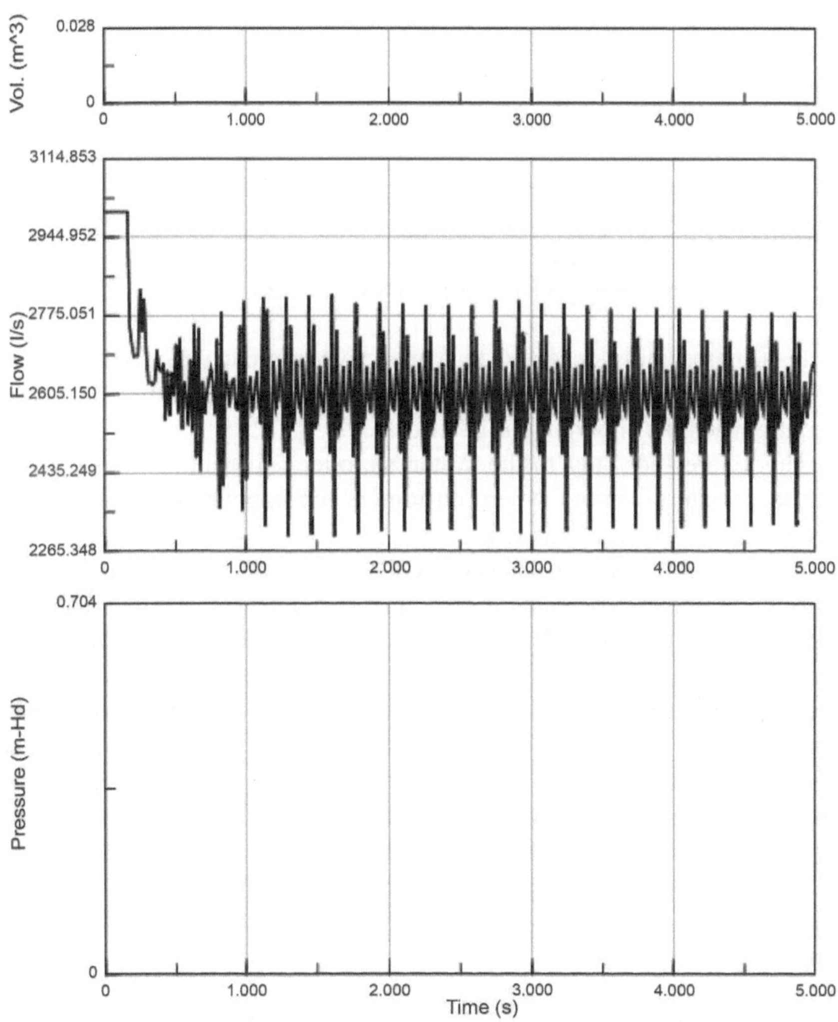


**FIGURE 7.5** Graph to show growth of leakage with time.

This is a procedure whereby water loss is only tackled when leakage is visible or when problems are reported from the public. The adoption of this policy minimizes the day-to-day operating costs of leakage detection, but increases the risk of water being wasted. It results in an ever-increasing upward trend in the annual supply of water, since leaks can remain undetected for many years until they reach such a magnitude that urgent action has to be taken due to customer complaints. It is, however, a perfectly feasible policy to adopt, providing it is "politically" acceptable, and may be carried out with full instrumentation to allow rapid location of leaks, although this will give rise to modest maintenance costs.

Such a policy is only applicable if:

- the revenue costs of leakage detection are high;
- the costs of production are low, and there is ample capacity to supply all foreseen demands;
- bursts are readily visible and easily repaired.

It is increasingly accepted that an active approach of searching for leakage is preferential in cost/benefit terms to a passive approach of only reacting when the situation has deteriorated.

This relates not only to the Water Supplier's distribution system but also to private pipework where customers are encouraged to carry out repairs on any leakage detected.

Active control would usually involve the monitoring of flows in a distribution network by using a system of permanently installed distribution meters. If unexpectedly high flows of water are observed, these are immediately investigated; leakage detection teams being carefully directed to ensure that leakage is maintained within defined criteria (such criteria being prepared using an acceptable cost/benefit basis). It is obvious that monitoring which does not initiate further action is unproductive. It will also be unproductive if, when further action is worthwhile, resources are not available to proceed with location of the leakage.

An active policy requires expenditure on meter installations, etc., and the day-to-day operating costs of leakage detection teams. The following benefits should be achieved:

- It minimizes leakage and hence reduces the loss of water in monetary terms.
- It results in an overall reduction of water demand.

- Limited water resources are conserved for legitimate use, and rationing, etc., is avoided.
- It reduces operating costs (savings on electrical power and chemical treatment costs).
- Work is planned (rather than acting in response to emergency).
- Dangerous leakage is minimized (e.g., freezing water on highway).
- Customer perception is improved.
- Capital expenditure requirements on treatment works, reservoirs, and mains are reduced.

Because of their potential, it is worth noting that active leakage detection in the future is likely to be increasingly employed acoustic loggers, some permanently installed. This could result in larger meter areas, and hence fewer district meters. A well-managed active leakage detection policy ensures that the cost of the leakage detection teams and the repayments of the capital necessary to establish the system is exceeded by the value of the water saved.

It is applicable if:

- The cost of water production is high.
- The sources of water have limited capacity and cannot meet normal and/or foreseen demand.
- Bursts are "invisible" due to the strata.
- The quantity of water being put into supply is increasing at an unacceptable rate.

Leakage reduction and control is a long-term activity, and should be regarded as a part of good distribution management. Occasional short bursts of effort are unlikely to produce lasting results because distribution systems continue to deteriorate for one reason or another. If only the obvious leaks are repaired, leakage levels will still increase, as will consumer problems. The development of a long-term leakage control strategy is therefore essential if water supply and distribution systems are to be effectively managed. Such development needs to be flexible, with occasional reviews to ensure that the strategy adopted is the most appropriate one for the situation. Cost/benefit analysis is important in this regard.

"Active" leakage control (i.e., finding and repairing leaks before their presence becomes obvious or generates problems) has been found to be a cost-effective method of reducing water supply deficiencies. A planned approach should result in lower complaint costs, and lower repair and maintenance costs. The establishment of controllable, manageable areas (District Meter Areas or DMAs) within a distribution system, whose demands are easily monitored, has been found to be extremely helpful for effective leakage control and supply management. It forces plans to be updated, locating mains and buried fittings. It introduces new valves to give better operational control. It locates illegal connections and identifies malfunctioning meters and public supplies. In the very process of this setting-up work, leakages and wastages are found and repaired. It enforces good housekeeping. Regard has to be given, however, to the minimization of dead-ends and their associated quality problems.

Leakage reduction requires a dedicated core of highly trained, specialist personnel using appropriate "state-of-the-art" equipment and techniques. Local knowledge is essential together with an understanding of the day-to-day operation of the distribution system and demand patterns. Support can be obtained from specialist agencies/contractors, given precise briefs and targets. Personnel motivation, good communication, and synchronization of activities and continuous feedback of decisions/results cannot be over-emphasized. This is vital for understanding, efficiency, and success. Everyone should be included, from planners to repair teams. The organization of leakage control personnel can vary widely. Distribution personnel can either be organized as a specialist team, or be integrated into general distribution system operational duties, and spend only part of their time on leakage control. It is generally accepted that to properly pursue active leakage control and to meet agreed monitoring/detection frequencies, it is necessary to set up specialist teams. However, general operational duties cannot and should not be entirely divorced from leakage control. Technical support is required for design and modification of district metering, computer systems support, compilation of base data for DMAs, production of reports, overall performance monitoring, production of drawings, system records updating, and for problem solving. Clerical support is required for computer input and administrative duties such as serving notices relating to private pipe repairs. Skilled and knowledgeable technical support is crucial if the mass of data now regularly available are to be handled and analyzed to the best advantage for the leakage reduction effort. Good leakage control depends upon good and progressively improving data. To achieve this it is necessary to establish and keep an audit trail of data, building from individual DMAs and their component data up to the regional total figures for the Water Supplier. These can be collected in the two data streams of:[65–75]

- Aggregated night-lines/"bottom up" calculation
- Total integrated flow/ "top down" calculation.

## 7.4 CONCLUSION

Generally, water leakage of pipelines is manifested as a hydro-machines phenomenon which come from the non-revenue water (NRW). In this work, cycles of increased and decreased pressure iterate at intervals equal to time for dual-path shock wave length of the pipeline from the valve prior to the pipeline. Thus, the hydraulic impact of the liquid in the pipeline performed oscillatory motion. The cause of oscillatory motion was the hydraulic resistance and viscosity. It absorbed the initial energy of the liquid for overcoming the friction.

This work focused on the effects of the NRW due to the surge wave velocity in water pipeline. It showed that Eulerian-based computational model is an accurate model. Hence in order to present the importance of NRW on water saving, it was compared with the models for laboratory, computational and field tests experiments. This procedure showed that the Eulerian-based model for water transmission line. On the other hand, this idea was included in the proper analysis to provide a dynamic response to the shortcomings of the system. It also performed the design protection equipments to manage the NRW phenomena and determine the operational procedures to avoid water loss. Consequently, the results of this work will help to reduce the risk of system damage or failure at the water pipeline.

## KEYWORDS

- **non-revenue water**
- **geography information system**
- **\relationship classes**
- **hydraulic model**

## REFERENCES


1. Streeter, V. L.; Wylie, E. B. *Fluid Mechanics;* McGraw-Hill Ltd.: USA, 1979; pp 492–505.
2. Leon, A. S. An Efficient Second-Order Accurate Shock-Capturing Scheme for Modeling One and Two-Phase Water Hammer Flows. Ph.D. Thesis, March 2007.
3. Adams, T. M.; Abdel-Khalik, S. I.; Jeter, S. M.; Qureshi, Z. H. An Experimental Investigation of Single-Phase Forced Convection in Microchannels. *Int. J. Heat Mass Transf.* **1998,** *41* (6–7), 851–857.
4. Peng, X. F.; Peterson, G. P. Convective Heat Transfer and Flow Friction for Water Flow in Microchannel Structure. *Int. J. Heat Mass Transf.* **1996,** *39* (12), 2599–2608.
5. Mala, G.; Li, D.; Dale, J. D. Heat Transfer and Fluid Flow in Microchannels. *Int. J. Heat Transf.* **1997,** *13*, 3079–3088.
6. Xu, B.; Ooi, K. T.; Mavriplis, C.; Zaghloul, M. E. Viscous Dissipation Effects for Liquid Flow in Microchannels. *Micorsystems.* **2002,** 53–57.
7. Pickford, J. *Analysis of Surge;* Macmillan: London, 1969; pp 153–156.
8. American Society of Civil Engineers, Task Committee on Engineering Practice in the Design of Pipelines. *Pipeline Design for Water and Wastewater;* American Society of Civil Engineers: New York, 1975; pp 128.
9. Fedorov, A. G.; Viskanta, R. Three-Dimensional Conjugate Heat Transfer into Microchannel Heat Sink for Electronic Packaging. *Int. J. Heat Mass Transf.* **2000,** *43,* 399–415.
10. Tuckerman, D. B. Heat Transfer Microstructures for Integrated Circuits. Ph.D. thesis, Stanford University, 1984.
11. Harms, T. M.; Kazmierczak, M. J.; Cerner, F. M.; Holke, A.; Henderson, H. T.; Pilchowski, H. T.; Baker, K. In *Experimental Investigation of Heat Transfer and Pressure Drop through Deep Micro channels in a (100) Silicon Substrate,* Proceedings of the ASME, Heat Transfer Division, HTD 351, 1997, 347–357.
12. Holland, F. A.; Bragg, R. *Fluid Flow for Chemical Engineers;* Edward Arnold Publishers: London, 1995; pp 1–3.
13. Lee, T. S, Pejovic, S. Air Influence on Similarity of Hydraulic Transients and Vibrations. *J. Fluid Eng.* **1996,** *118* (4), 706–709.
14. Li, J.; McCorquodale, A. Modeling Mixed Flow in Storm Sewers. *J. Hydraul. Eng.* **1999,** *125* (11), 1170–1180.
15. Minnaert, M. On Musical Air Bubbles and the Sounds of Running Water. *Phil. Mag.* **1933,** *16* (7), 235–248.
16. Moeng, C. H.; McWilliams, J. C.; Rotunno, R.; Sullivan, P. P.; Weil, J. Investigating 2D Modeling of Atmospheric Convection in the PBL. *J. Atm. Sci.* **2004,** *61,* 889–903.
17. Tuckerman, D. B.; Pease, R. F. W. High Performance Heat Sinking for VLSI. *IEEE Electron Device Lett.* **1981,** *2* (5), 126–129.
18. Nagiyev, F. B.; Khabeev, N. S. Dynamics of Bubble in Binary Solutions. *High Temp.* **1988,** *27* (3), 528–533.
19. Shvarts, D.; Oron, D.; Kartoon, D.; Rikanati, A.; Sadot, O. Scaling Laws of Nonlinear Rayleigh-Taylor and Richtmyer-Meshkov Instabilities in Two and Three Dimensions. *C. R. Acad. Sci. Series IV Phys.* **2000,** *6,* 719–726.

20. Cabot, W. H.; Cook, A. W.; Miller, P. L.; Laney, D. E.; Miller, M. C.; Childs, H. R. Large Eddy Simulation of Rayleigh-Taylor Instability. *Phys. Fluids.* **2005,** *17,* 91–106.
21. Cabot, W. *Phys. Fluids*, **2006,** 94–550.
22. Goncharov, V. N. Analytical Model of Nonlinear, Single-Mode, Classical Rayleigh-Taylor Instability at Arbitrary Atwood Numbers. *Phys. Rev. Lett.* **2002,** *88* (13), 10–15.
23. Ramaprabhu, P.; Andrews, M. J. Experimental Investigation of Rayleigh-Taylor Mixing at Small Atwood Numbers. *J. Fluid Mech.* **2004,** *502,* 233–271.
24. Clark, T. T. A Numerical Study of the Statistics of a Two-Dimensional Rayleigh-Taylor Mixing Layer. *Phys. Fluids.* **2003,** *15,* 2413.
25. Cook, A. W.; Cabot, W.; Miller, P. L. The Mixing Transition in Rayleigh-Taylor Instability. *J. Fluid Mech.* **2004,** *511,* 333–362.
26. Waddell, J. T.; Niederhaus, C. E.; Jacobs, J. W. Experimental Study of Rayleigh-Taylor Instability: Low Atwood Number Liquid Systems with Single-Mode Initial Perturbations. *Phys. Fluids.* **2001,** *13,* 1263–1273.
27. Weber, S. V.; Dimonte, G.; Marinak, M. M. *Arbitrary Lagrange-Eulerian Code Simulations of Turbulent Rayleigh-Taylor Instability in Two and Three Dimensions;* Laser and Particle Beams: 21, 2003, pp 455–461.
28. Dimonte, G.; Youngs, D.; Dimits, A.; Weber, S.; Marinak, M. A Comparative Study of the Rayleigh-Taylor Instability using High-Resolution Three-Dimensional Numerical Simulations: The Alpha Group Collaboration. *Phys. Fluids.* **2004,** *16,* 1668.
29. Young, Y. N.; Tufo, H.; Dubey, A.; Rosner, R. On the Miscible Rayleigh-Taylor Instability: Two and Three Dimensions. *J. Fluid Mech.* **2001,** *447,* 377–408.
30. George, E.; Glimm, J. Self-Similarity of Rayleigh-Taylor Mixing Rates. *Phys. Fluids.* **2005,** *17,* 054101.
31. Oron. D.; Arazi, L.; Kartoon, D.; Rikanati, A.; Alon, U.; Shvarts, D. Dimensionality Dependence of the Rayleigh-Taylor and Richtmyer-Meshkov Instability Late-Time Scaling Laws. *Phys. Plasmas.* **2001,** *8,* 2883.
32. Nigmatulin, R. I.; Nagiyev, F. B.; Khabeev, N. S. In *Effective Heat Transfer Coefficients of the Bubbles in the Liquid Radial Pulse,* Mater. Second-Union. Conf. Heat Mass Transf, "Heat massoob-men in the biphasic. with." Minsk, 1980, Vol. 5, 111–115.
33. Nagiyev, F. B. In *Damping of the Oscillations of Bubbles Boiling Binary Solutions,* Mater. VIII Resp. Conf. Mathematics and Mechanics, Baku, October 26–29, 1988, 177–178.
34. Nagyiev, F. B.; Kadyrov, B. A. Small Oscillations of the Bubbles in a Binary Mixture in the Acoustic Field. *Math. AN Az.SSR Ser. Physicotech. Mate. Sci.* **1986,** 1, 23–26.
35. Nagiyev, F. B. Dynamics, Heat and Mass Transfer of Vapor-Gas Bubbles in a Two-Component Liquid. Turkey-Azerbaijan petrol semin., Ankara, Turkey, 1993, 32–40.
36. Nagiyev, F. B. The Method of Creation Effective Coolness Liquids, Third Baku international Congress. Baku, Azerbaijan Republic, 1995, 19–22.
37. Nagiyev, F. B. The Linear Theory of Disturbances in Binary Liquids Bubble Solution. *Dep. In VINITI.* **1986,** 405, 76–79.
38. Nagiyev, F. B. Structure of Stationary Shock Waves in Boiling Binary Solutions. Math. USSR, Fluid Dynamics, 1989, № 1, 81–87.
39. Lord Rayleigh, O. M. F. R. S. On the Pressure Developed in a Liquid during the Collapse of a Spherical Cavity. *Philos. Mag. Ser. 6.* **1917,** *34* (200), 94–98.

40. Perry. R. H.; Green, D. W.; Maloney, J. O. In *Perry's Chemical Engineers Handbook,* 7th Ed.; Perry, R. H., Green, D. W., Eds.; McGraw-Hill Professional Publishing: New York, 1997; pp 1–61.
41. Nigmatulin, R. I. *Dynamics of Multiphase Media;* Izdatel'stvo Nauka: Moscow, 1987; pp 12–14.
42. Kodura, A.; Weinerowska, K. The Influence of the Local Pipeline Leak on Water Hammer Properties, Materials of the II Polish Congress of Environmental Engineering, Lublin, 2005, 125–133.
43. Kane, J.; Arnett, D.; Remington, B. A.; Glendinning, S. G.; Baz'an, G. Two-Dimensional versus Three-Dimensional Supernova Hydrodynamic Instability Growth. *Astrophys. J.* **2000,** *528* (2), 528–989.
44. Quick, R. S. Comparison and Limitations of Various Water hammer Theories. *J. Hyd. Div.* **1933,** 43–45.
45. Jaeger, C. *Fluid Transients in Hydro-Electric Engineering Practice;* Blackie and Son Ltd.: Glasgow, 1977; pp 87–88.
46. Jaime, S. A. Generalized Water Hammer Algorithm for Piping Systems with Unsteady Friction. E-theses, HKU School of Professional and Continuing Education, 2005.
47. Fok, A.; Ashamalla, A.; Aldworth, G. "Considerations in Optimizing Air Chamber for Pumping Plants", Symposium on Fluid Transients and Acoustics in the Power Industry, San Francisco, CA, 1978, 112–114.
48. Fok, A. *Design Charts for Surge Tanks on Pump Discharge Lines, BHRA 3rd Int. Conference on Pressure Surges;* Bedford: England, 1980; pp 23–34.
49. Fok, A. *Water Hammer and its Protection in Pumping Systems, Hydro technical Conference;* CSCE: Edmonton, 1982; pp 45–55.
50. Fok, A. A Contribution to the Analysis of Energy Losses in Transient Pipe Flow. Ph.D. Thesis, University of Ottawa, 1987.
51. Brunone, B.; Karney, B. W.; Mecarelli, M.; Ferrante, M. Velocity Profiles and Unsteady Pipe Friction in Transient Flow. *J. Water Res. Plan. Manag. ASCE.* **2000,** *126* (4), 236–244.
52. Koelle, E.; Luvizotto Jr. E.; Andrade, J. P. G. In *Personality Investigation of Hydraulic Networks using MOC – Method of Characteristics,* Proceedings of the 7th International Conference on Pressure Surges and Fluid Transients. Harrogate Durham, UK, 1996.
53. Filion, Y.; Karney, B. W. In *A Numerical Exploration of Transient Decay Mechanisms in Water Distribution Systems,* Proceedings of the ASCE Environmental Water Resources Institute Conference, American Society of Civil Engineers, Roanoke, Virginia, 2002.
54. Hamam, M. A.; McCorquodale, J. A. Transient Conditions in the Transition from Gravity to Surcharged Sewer Flow. *Canadian J. Civil Eng.* **1982,** *9* (2), 65–98.
55. Savic, D. A.; Walters, G. A. *Genetic Algorithms Techniques for Calibrating Network Models*; Report No. 95/12, Centre for Systems and Control Engineering, 1995, 137–146.
56. Walski, T. M.; Lutes, T. L. Hydraulic Transients Cause Low-Pressure Problems. J. Am. Water Works Assoc. **1994,** *75* (2), 58.
57. Lee, T. S.; Pejovic, S. Air Influence on Similarity of Hydraulic Transients and Vibrations. *J. Fluids Eng.* **1996,** *118* (4), 706–709.
58. Chaudhry, M. H. *Applied Hydraulic Transients;* Van Nostrand Reinhold Co.: New York, 1979; pp 1322–1324.

59. Parmakian, J. *Water Hammer Analysis;* Dover Publications, Inc.: New York, 1963; pp 51–58.
60. Tuckerman, D. B.; Pease, R. F. W. High Performance Heat Sinking for VLSI. *IEEE Electron device lett.* **1981,** *DEL-2,* 126–129.
61. Farooqui, T. A. *Evaluation of Effects of Water Reuse, on Water and Energy Efficiency of an Urban Development Area, Using an Urban Metabolic Framework;* Master of Integrated Water Management Student Project Report. International Water Centre, 2015.
62. Ferguson, B. C.; Frantzeskaki, N.; Brown, R. R. A Strategic Program for Transitioning to a Water Sensitive City. *Landsc. Urban Plan.* **2013,** *117*, 32–45.
63. Kenway, S.; Gregory, A.; McMahon, J. Urban Water Mass Balance Analysis. *J. Ind. Ecol.* **2011,** *15* (5), 693–706.
64. Renouf, M. A.; Kenway, S. J.; Serrao-Neumann, S.; Low Choy, D. *Urban metabolism for Planning Water Sensitive Cities. Concept for an Urban Water Metabolism Evaluation Framework. Milestone Report.* Melbourne: Cooperative Research Centre for Water Sensitive Cities, 2015.
65. Serrao-Neumann, S.; Schuch, G.; Kenway, S. J.; Low Choy, D. Comparative Assessment of the Statutory and Non-Statutory Planning Systems: South East Queensland, Metropolitan Melbourne and Metropolitan Perth. Melbourne: Cooperative Research Centre for Water Sensitive Cities, 2013.
66. Andrews, S.; Traynorp. P. *Guidelines for Assuring Quality of Food and Water Microbiological Culture Media;* August 2004.
67. Andrew, D. E.; Rice, E. W.; Clescceri, L. S. Standard Methods for the Examination of Water and Waste Water. Part 9000, 2012.
68. Instructions Procedure Rural Water and Wastewater Quality Assurance Test Results of KhorasanRazavi, 2011.
69. Abbassi, B.; Al Baz, I. *Integrated Wastewater Management: A Review. Efficient Management of Wastewater – Its Treatment and Reuse in Water Scarce Countries*; Springer Publishing Co.: Berlin, Heidelberg, 2008.
70. Andreasen, P. In *Chemical Stabilization. Sludge into Biosolids – Processing, Disposal and Utilization;* Spinosa, L., Vesilind, P. A., Eds.; IWA Publishing: UK, 2001; pp 392.
71. CIWEM – The Chartered Institution of Water and Environmental Management. Sewage Sludge: Stabilization and Disinfection – Handbooks of UK Wastewater Practice. CIWEM Publishing, 1996.
72. Halalsheh, M.; Wendland, C. Integrated Anaerobic-Aerobic Treatment of Concentrated Sewage. In *Efficient Management of Wastewater – its Treatment and Reuse in Water Scarce Countries;* Baz, I. A., Otterpohl, R., Wendland, C., Eds.; Springer Publishing Co.: Berlin, Heidelberg, 2008; pp 177–186.
73. ISWA. Handling, Treatment and Disposal of Sludge in Europe. Situation Report 1. Copenhagen. ISWA Working Group on Sewage Sludge and Water Works, 1995.
74. Matthews, P. Agricultural and Other Land Uses. Sludge into Biosolids – Processing, Disposal and Utilization. IWA Publishing, 2001.
75. IWK Sustainability Report, 2012–2013.

## CHAPTER 8

# UNACCOUNTED-FOR WATER OF A WATER SYSTEM: SOME COMPUTATIONAL ASPECTS

KAVEH HARIRI ASLI[1*], SOLTAN ALI OGLI ALIYEV[1], and HOSSEIN HARIRI ASLI[2]

[1]*Department of Mathematics and Mechanics, Azerbaijan National Academy of Sciences AMEA, Baku, Azerbaijan*

[2]*Civil Engineering Department, Faculty of Engineering, University of Guilan, Rasht, Iran*

[*]*Corresponding author. E-mail: hariri_k@yahoo.com*

## CONTENTS

## ABSTRACT

Non-revenue water (NRW) is water that has been produced and is "lost" before it reaches the customer. In this work the combination of direct numerical simulation (DNS) and geographic information system (GIS) presents a new algorithmic contribution for recognition of NRW. As the main idea, this work presents the role of rapid data intercommunication related to the intelligent water system pressure management for reduction of NRW. Hence the programmable logic control (PLC) systems incorporated with highly advanced flow and pressure sensors and equipped with data logger system were applied in this work. Pressure management is a major element in the leakage management strategy. Pressure reduction is probably the simplest and most immediate way of reducing leakage within the distribution system. As a result, this work shows the performance of the web GIS for water system data intercommunication and calibration which has an important role on management of urban water system failure. This work reveals that the future of the urban water cycle depends on the joint efforts of cities, utilities, and industries.


## 8.1 INTRODUCTION

It is being argued that the level of non-revenue water (NRW) should be as low as possible, independent of economic or financial considerations, in order to minimize the risk of drinking water loss in the water system failure. The first national study on the NRW was published in 1980 setting down a methodology for the assessment of economic leakage levels in the U.K. As a result of droughts, a number of countries initiated major leakage management programs based on water- and energy-saving assessments. In this chapter, a dynamic model based on geo-reference data for water system failure condition shows the computational performance of a numerical method. The combination of direct numerical simulation (DNS) with geographic information system (GIS) through programmable logic control (PLC) system presents a new algorithmic contribution. During a failure condition and NRW analysis, the fluid and system boundaries can be either elastic or inelastic. Both branches of water system failure theory stem from the same governing equations. Among the approaches proposed to solve the single-phase (pure liquid) equations are the method of characteristics (MOC), finite differences (FD), wave characteristic method

(WCM), finite elements (FE), and finite volume (FV). One difficulty that commonly arises relates to the selection of an appropriate level of time step to use for the analysis.[13] In this work, geo-reference data of water system have an important role between computational speed and accuracy.

This work will illustrate how CFD model can predict the behavior of a water distribution system through time using an extended period simulation (EPS). An EPS can be conducted for any duration you specify. System conditions are computed over the given duration at a specified time increment. Some of the types of system behaviors that can be analyzed using an EPS include how tank levels fluctuate, when pumps are running, whether valves are open or closed, and how demands change throughout the day. Water demand in a distribution system fluctuates over time. For example, residential water use on a typical weekday is higher than average in the morning before people choose work, and is usually highest in the evening when residents are preparing dinner, washing clothes, etc. This variation in demand over time can be modeled using demand patterns. Demand patterns are multipliers that vary with time and are applied to a given base demand, most typically the average daily demand. In this lesson, you will be dividing the single fixed demands for each junction node in Lesson 1 into two individual demands with different demand patterns. One demand pattern will be created for residential use, and another for commercial use. You will enter demand patterns at the junction nodes through the junction editors.

One of the many project tools is water management by GIS. Scenarios allow you to calculate multiple "What If?" situations in a single project file. You may wish to try several designs and compare the results, or analyze an existing system using several different demand alternatives and compare the resulting system pressures. A scenario is a set of alternatives, while alternatives are groups of actual model data. Scenarios and alternatives are based on a parent/child relationship where a child scenario or alternative inherits data from the parent scenario or alternative. In construction of the water distribution network, one must define the characteristics of the various elements, enter demands and demand patterns, and perform steady-state simulation and EPS. In this work, these scenarios were needed to test four "What If?" situations for water distribution system. These "What If?" situations would involve changing demands and pipe sizes. At the end, all of the results were compared using the Scenario Comparison tool. The first need is to set up the required data sets, or alternatives. An alternative is a group of data that describes a specific part of the model.

The setup is a different physical or demand alternative for each design trial which should be evaluated. Each alternative will contain different pipe size or demand data. Base alternatives are alternatives that do not inherit data from any other alternative. Child alternatives can be created from the base alternative. A child alternative inherits the characteristics of its parent, but specific data can be overridden to be local to the child. A child alternative can, in turn, be the parent of another alternative. Alternatives are the building blocks of a scenario. A scenario is a set of one of the types of alternatives, plus all of the calculation information needed to solve a model. Just as there are base, parent, and child alternatives, there are also base, parent, and child scenarios. The difference is that instead of inheriting model data, scenarios inherit sets of alternatives. To change the new scenario, change one or more of the new scenario's alternatives. For this aim, an engineer will create a new scenario for each different set of conditions that need to be evaluated. There is always a default Base Scenario that is composed of the base alternatives. Initially, only the Base is available, because any new scenarios are not created. It needs to further examine what is going on in the system as a result of the fire flow, and find solutions to any problems that might have arisen in the network as a result. It can be review output tables to quickly see what the pressures and velocities are within the system, and create new alternatives and scenarios to capture the necessary modifications. [1–17] An important feature in all water distribution modeling software is the ability to present results clearly. The CFD model reporting features include

- *Reports*, which display and print information on any or all elements in the system.
- *Element Tables (FlexTables)*, for viewing, editing, and presentation of selected data and elements in a tabular format.
- *Profiles*, to graphically show, in a profile view, how a selected attribute, such as hydraulic grade, varies along an interconnected series of pipes.
- *Contouring*, to show how a selected attribute, such as pressure, varies throughout the distribution system.
- *Element Annotation*, for dynamic presentation of the values of user-selected variables in the plan view.
- *Color Coding*, which assigns colors based on ranges of values to elements in the plan view. Color coding is useful in performing quick diagnostics on the network.

The three required spatial and non-spatial attributes of a CFD model map feature files are as follows:

- *Main File*, the main file is a binary file with an extension of .SHP. It contains the spatial attributes associated with the map features. For example, a polyline record contains a series of points, and a point record contains x and y coordinates.
- *Index File*, the index file is a binary file with an extension of .SHX. It contains the byte position of each record in the main file.
- *Database File*, The database file is a dBase III file with an extension of .DBF. It contains the non-spatial data associated with the map features. All three files must have the same file name with the exception of the extension, and be located in the same directory. Listed below are the files you will be importing. Only the main files are listed; however, corresponding .SHX and .DBF are presented as well.

    - PresJunc.shp
    - PresPipe.shp
    - PRV.shp
    - Pump.shp
    - Reservoi.shp
    - Tank.shp.

After performing the conversion, the Drawing Review window lets you navigate to and fix any problems that may be encountered. [18–27]

There are three scenarios for Darwin Designer to optimize the setup of a pipe network:

- existing System representing current system conditions,
- future Condition representing the system expansion layout,
- optimization base representing the scenario for Designer base.

There are two design tasks:

- New pipes to be sized.
- Old pipes need to be rehabilitated by applying possible actions including cleaning pipe, relining pipe, and leaving the pipe as it is (no action or do thing to a pipe).

The design criteria are:

- minimum pressure at all demand junction,
- maximum pressure at all demand junction,
- filling each tank to or above the initial tank level.

The Darwin Designer need to consider two ways of accomplishing a cost-effective design: create new or parallel pipes and rehabilitate existing pipes. Clearly, the new subdivision will get new pipes. And, as you can design an appropriate size for these new pipes, there is no need for parallel pipes and there are no existing pipes on which to perform rehabilitation. With that in mind, you would create a parallel pipe option for all existing pipes. This parallel pipe option should include a variety of sizes so Darwin Designer has flexibility to choose the most efficient size. Additionally, the pipe sizes must include a 0 diameter, which lets Darwin Designer calculate the efficiency of the system with the pipe absent (without installing the parallel pipe). There are four options in this tutorial for existing pipe:

- Install parallel pipe.
- Clean existing pipe.
- Reline existing pipe.
- Take no action.

Pressure management is a major element in a leakage management strategy. Pressure reduction is probably the simplest and most immediate way of reducing leakage within the distribution system. Its benefits are immediate. Even where already practiced, it is likely to be worthwhile to re-examine and reset equipment and schemes to take advantage of progressive technical developments, and local system alterations. Pressure management can be accomplished in a number of ways and not just via the installation of a new pressure reduction valve (PRV). In fact, the generation of pressure almost always costs money, so reducing pressure by means of a PRV is intrinsically inefficient. The following options should be considered first:

- Re-zoning the area supplied to match input head to topography and minimize system losses. This may include boosting to a smaller, critical area, reinforcing or reconditioning mains to allow low pressure zones to be extended, or transferring demand zones to an alternative source with a lower overall head. Network analysis could greatly facilitate this investigation.

- Matching pump output curves to closely match distribution demands. This could include resizing pumps to match known demands, or staged or variable speed pumping, or closed loop control using flow or pressure signals.
- Installation of break pressure tanks. These generally have a higher capital cost and are a potential contamination risk. On account of this they are no longer used in the U.K. Having considered these three options, mechanical pressure control devices, typically PRVs, provide the next stage in a pressure control strategy.

Pressure control can

- reduce leakage;
- educe pressure-related consumption such as hand-washing, car washing, etc.;
- reduce the frequency of bursts, at least in the immediate future—subsequent savings in repair costs can exceed those due to reduced leakage;
- stabilize pressure, decreasing the possibility of pipe work movement and fatigue type failures, and possibly eliminating certain household plumbing problems;•provide a more constant service to customers—large diurnal pressure variations may give customers an impression of a poorly managed service, and unnecessary high pressures raise customers' expectations and perceptions of what is adequate;
- enable a company to standardize on pipes and fittings which have a lower pressure rating, and are therefore cheaper; and
- assist demand management when flow restriction is necessary, i.e., during drought.

Some examples of the problems that can potentially arise, and their consequences, are listed below. Some of these can be designed out of the system. In correctly configured systems this is typically a result of restrictions and blockages of individual supplies. Flow and pressure tests at the property affected will reveal the location of the problem which can then be dealt with in the normal way. Partly closed stop taps and valves are a typical problem. Poor pressures may also be the result of pipe work simply being undersized, perhaps through corrosion. They may also occur by the setting up of the PRV area, severing the normal interlinking of the system. This should be assessed beforehand at the area design stage.

Noise can be a problem close to PRV installations. Noise is usually associated with small valve openings and may be associated with cavitation problems. Attention to pipe work detail and valve settings can reduce noise levels but it is best avoided by correct selection and sitting. Noise through a PRV does create difficulties for leak detection work in the vicinity because of its interference. [28–32] Blockages can occur as a result of mains material becoming trapped in the PRV. This may result in failure of the control and actuating mechanism and loss of pressure control, leading to excessively high or low pressures. Attention to the maintenance of filters and correct flushing are necessary to avoid blockages in distribution systems which are prone to solid contamination. It is generally recommended that planned preventative maintenance be carried out on a half-yearly basis. Valves without close mechanical tolerances are less susceptible to this type of failure. Strainers upstream of the PRV will also help Closing of valves between the PRV and a remote pressure monitoring point will result in the PRV attempting to rectify the apparent loss of pressure at the remote point. Typically this occurs when valves are shut in the course of a routine repair. The results of exposing the system to maximum pressures at moderate flows will usually be a series of burst mains. This situation should be avoided by ensuring that Inspectors, in particular, are aware of pressure control systems and follow appropriate procedures before closing critical valves. Network models can also be used to simulate valve closures prior to operation on site to help understand how the system will react.

Under certain circumstances surges of pressure and flow can cause unpredictable PRV behavior. This can result, with certain valves, in the piston exceeding its travel and jamming in the fully open or closed position. Usually, the surges which cause this type of failure result from valve or pump operations which should be examined to minimize the risk. The provision of "stops" to limit the travel in mechanical systems can be helpful.

Ordnance survey data alone are insufficient in planning an area from a topographical point of view—a tall building survey should be undertaken. In areas where existing flats rely upon a high pressure mains supply, pressure reduction may only be possible if the Supplier is willing to bear costs of pumping and plumbing modifications. Where small boosters are already feeding multistory buildings, the lowering of pressures may cause the boosters to operate more regularly, which in turn lead to rapid payback of investment, reduction in leakage, reduced incidence of burst mains, and reduced customer complaints.

A PRV can be defined as a mechanical device which will give a reduced outlet (downstream) pressure for a range of flow rates and upstream pressures.

All PRVs have certain features in common. These are a means of controlling the flow (the valve), a means of sensing the pressure differential between the inlet and the outlet, and a means of actuating the valve. A variety of more or less sophisticated means of providing these features have been developed by manufacturers.

The two principal categories of PRV are fixed outlet and flow-modulated, each with several variations. Generally, fixed outlet characteristics maintain approximately the same value of downstream pressure over a range of flow rates. The pressure has to be set so that level of service (LOS) pressure is maintained at the target point for the maximum design flow rate. The resultant average zone night pressure (AZNP) will be at a higher value than a flow-modulated pressure in a similar system since in the latter case pressures can be optimized for minimum demand. In reality, some fixed outlet valves are not always capable of maintaining a constant outlet pressure, particularly at low flow when some rise in outlet pressure can be experienced. A "pilot" can assist in providing the necessary variable throttling effect to keep a constant outlet pressure as inlet pressures and flows vary. Two pilots with a timed changeover can give a "day" and "night" setting of outlet pressure. Flow-modulated PRVs vary the outlet pressure in such a manner that a constant head can be maintained at a target point in the distribution system for a range of flow rates and inlet pressures. The mechanism for regulating the outlet pressure may be mechanical or electronic or a combination of both. "Look-up" tables or telemetry may be involved in the outlet pressure control. Generally speaking, where head losses across the target area exceed 10 m (night time/no flow pressure minus day-time peak flow pressure) flow-modulated devices will provide greater net benefit (in spite of the extra cost), and are to be preferred.[33–41] Because of advances in control practice and communications, control systems for PRVs are becoming more complex and more effective. The valves are now fitted primarily to reduce leakage and to some extent pressure-dependent consumption, rather than the traditional reason of protecting the downstream infrastructure.

Water Suppliers must be seen to be operating efficiently and effectively, and must be able to justify the cost with their level of leakage and works designed to manage leakage, particularly to their customers who want to see their costs minimized. Leakage is often seen as synonymous with waste, and reducing leakage is seen as a means of saving money. Water lost through leakage has a value and so reducing the level of leakage offers benefits. However, eliminating leakage completely is impracticable and the cost of reducing it to low levels may exceed the cost of producing the water saved. Conversely, when little effort is expended on active leakage control, leakage levels will raise to levels where the cost of the water lost predominates. Water Suppliers must therefore strike a balance between the cost of reducing leakage and the value of the water saved. The level of leakage at which it would cost more to make further reductions than to produce the water from another source is what is known as the economic level of leakage (ELL). Operating at ELLs means that the total cost to the customer of supplying water is minimized, and Suppliers are operating efficiently. This means that leakage reduction should be pursued to the point where the long-run marginal cost of leakage control is equal to the long-run marginal benefit of the water saved. The latter depends on the long-run marginal costs of augmenting supplies by alternative means, including an assessment of the environmental benefits. The ELL is not fixed for all time. It depends on a wide range of factors, which will vary over time. For example, the cost of detecting and repairing leaks will fall as new technology is introduced. This will cause the ELL to fall. Conversely, if total demand falls to a point where there is a large surplus of water, it may not be economic to reduce leakage, unless the water can be sold to other Suppliers. Once the economic optimum level is known, this can be compared to the present level of leakage, and the Supplier can then set targets for leakage control in conjunction with other corporate policies on customer metering, mains rehabilitation, resource development, and pressure control. To do so they will need to appraise the investment required for these various different supply and demand management solutions, and the benefits which are expected to accrue. Due to the complexity of the issues, it is not possible to generalize to provide standardized formulas for setting leakage targets. Even if the same leakage policy is pursued, it is likely to be uneconomical to set the same target leakage levels for areas of differing characteristics. Thus there is a need to examine each system to determine the most appropriate method of leakage control and to plan the required capital investment, manpower, and revenue resource. However, any Supplier who is prepared to commit resources to collecting the required data, and to carry out the analysis and appraisals, will develop a greater understanding of the factors which are important to target setting. They will also be less likely to have unrealistic or uneconomic targets imposed on them from outside, or fall into the trap of setting leakage targets themselves without full consideration of the practicalities of achieving them, or the economic consequences. There are many possible ways of setting a leakage target. These can include targets based on minimum night flows, areas with excess pressure, areas with expensive water, or the most urbanized areas. The setting of economic targets, i.e., a level of leakage which provides the most economic mix of leakage-related costs, is independent of variations in physical factors such as property density, pressure, etc., and can provide clear information upon which sound management decisions may be based. However, it is recognized that there may be social, environmental, and political factors which dictate the target leakage level, as well as economic ones related to the Suppliers' own operating environment.[42–53] This has given rise to a broader fluid mechanics concept of "the most appropriate leakage target," being described as "that level of leakage which, over a long term planning horizon, provides the least cost combination of demand management and resource development, whilst adequately providing a low risk of security of supply to customers, and not unduly over-abstracting water from the environment."

## 8.2 MATERIAL AND METHODS

This chapter will explore each aspect of creating a fully functional water distribution model from disparate GIS data sources, from the initial creation of the network and the incorporation of loading and elevation data to the customized display of calculated results. At the beginning of each part, pressure-dependent demands (PDD) are used to simulate situations where a change in pressure affects the pressure management and quantity of water used. In order to simulate a situation where pressure significantly drops, the near source is taken out of service and the behavior with and without consideration of PDD is made. The starter file consists of a model with two non-PDD scenarios, Steady NoPD and EPSNoPDD. The demands have been loaded and the diurnal demand function has been created.[54–62]

In this work DNS and GIS with very fine grid applied for analysis of NRW at leakage location due to water system failure. An experimental and computational method (1–4) was used for prediction of leakage rate and location related to NRW by the following process (Figs. 8.1 and 8.2):

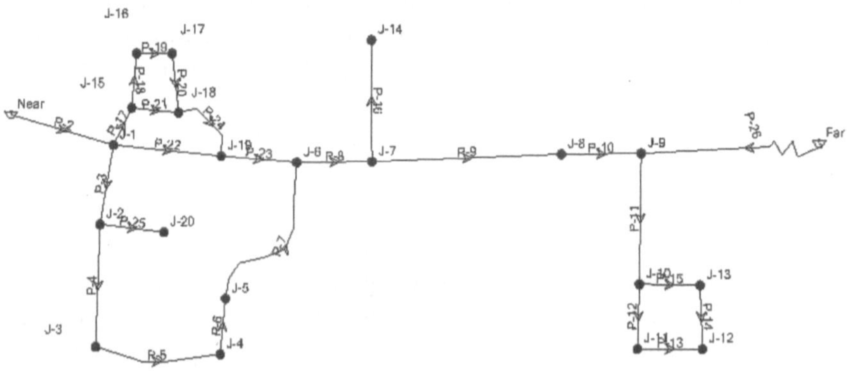


**FIGURE 8.1** Water distribution demands model.

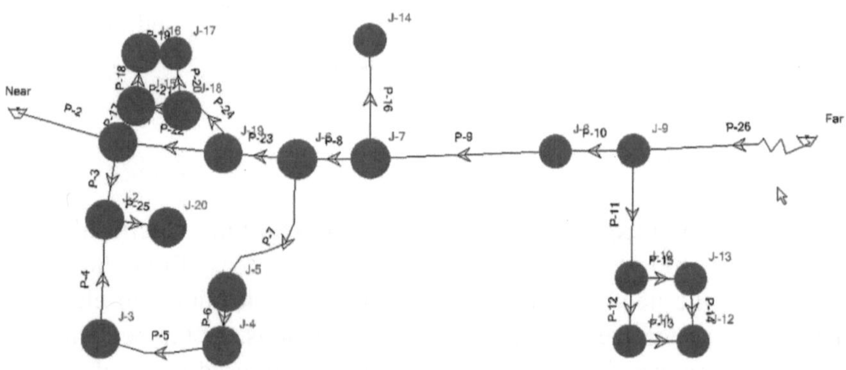


**FIGURE 8.2** Water distribution demands model from disparate GIS data sources.

— Convert data from graphical formats drawing (DWG) to Shapefile Format (SHP).

— Complete description of spatial data layers and fix errors in the GIS.

Discretize Navier–Stokes equation was formulated on a sufficiently fine grid for resolving all motions occurring in water system failure:

$$(\partial \rho / \partial t) + div(\rho \vec{u}) = 0, \text{(Continuity equation)} \tag{8.1}$$

$$(\partial \rho u / \partial t) + div(\rho u \vec{u}) = -(\partial p / \partial x) + div(\mu grad u) + q_x, \text{ X (momentum equation)} \tag{8.2}$$

$$(\partial \rho v / \partial t) + div(\rho v \vec{u}) = -(\partial p / \partial y) + div(\mu grad v) + q_y, \text{ Y (momentum equation)} \tag{8.3}$$

$$(\partial \rho w / \partial t) + div(\rho w \vec{u}) = -(\partial p / \partial z) + div(\mu grad w) + q_z, \text{ Z (momentum equation).} \tag{8.4}$$

In computational method by geo-reference data due to water system failure, the smaller the time step, the longer the run time but the greater the numerical accuracy. First, to calculate many boundary conditions, such as obtaining the head and discharge at the junction with geo-reference data of pipes, it is necessary that the time step be common to all pipes. The second constraint arises from the nature of the MOC. The MOC requires that ratio of the distance $\Delta x$ to the time step $\Delta t$ be equal to the wave speed in each pipe. In other words, the Courant number should ideally be equal to 1 and must not exceed 1 for stability reasons. For most urban systems, having as they do a variety of different pipes materials and coordinates with a range of wave speeds and lengths, it is impossible to satisfy exactly the Courant requirement in all pipes with a reasonable value of $\Delta t$. Faced with this challenge, this work used geo-reference data for prediction of physical NRW rate and location related in ways of relaxing the numerical constraints. The method of wave-speed adjustment changes of water network geo-reference data by the web GIS to satisfy exactly the Courant condition. In the other hand, this work prepared a conceptual model for physical NRW in GIS base. The main results of this work are:[63–67]

— Create ability to track and implement network analysis.

— Save the declining economy and track events for optimal management.
— Speed up the investigation of accidents which reduced physical water losses.

There is similarity between these results (Fig. 8.3) with the work of Leon.[2]

*Laboratory Models*—a scale model have been built to reproduce transients observed in a prototype (real) system, typically for forensic or steam system investigations. This research lab model has recorded flow and pressure data (Table 8.1). The model is calibrated using one set of data and, without changing parameter values, it is used to match a different set of results. If successful, it is considered valid for these cases. (Fig. 8.3).

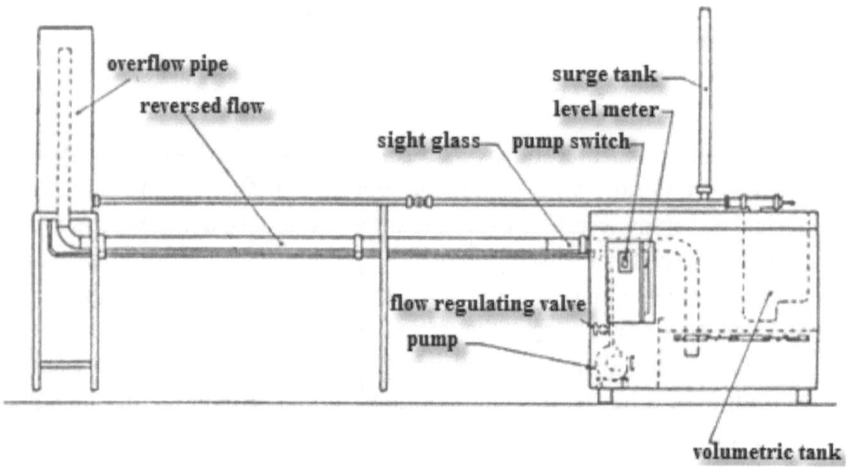


**FIGURE 8.3** Research laboratory model for laboratory experiments.

The model has been calibrated and final-checked by Laboratory Models with the flowing specifications (Table 8.1).

*Field Tests*—Field tests have provided key modeling parameters such as the pressure-wave speed or pump inertia. Water pipeline has equipped with advanced flow and pressure sensors, high-speed data loggers and "PLC." Hence fast transients, down to 5 ms, have been recorded. Methods such as inverse transient calibration and leak detection in calculation of unaccounted-for water "UFW" has used such data. Field tests have been formed on actual systems with flow and pressure data records. These comparisons require threshold and span calibration of all sensor groups, multiple simultaneous datum and time base checks, and careful test planning and interpretation.[6–8]

**TABLE 8.1** Research Laboratory Model Technical Specifications.

| Laboratory Model Technical Specifications | Notation | Value | Dimension |
|---|---|---|---|
| Pipe diameter | d | 22 | mm |
| Surge tank cross section area | A | $1.521 \times 10^{-3}$ | $m^2$ |
| Pipe cross section area | a | $0.3204 \times 10^{-3}$ | $m^2$ |
| Pipe thickness | t | 0.9 | mm |
| Fluid density | ρ | 1000 | $kg/m^3$ |
| Volumetric coefficient | K | 2.05 | $GN/m^2$ |
| Fluid power | P | * | * |
| Fluid force | F | * | * |
| Friction loss | hf | * | * |
| Frequency | W | * | * |
| Fluid velocity | v | * | m/s |
| Max fluctuation | Ymax | * | * |
| Flow rate | q | * | $m^3/s$ |
| Pipe length | L | * | m |
| Period of motion | T | * | * |
| Surge tank and reservoir elevation difference | y | * | m |
| Surge wave velocity | C | * | m/s |

*Laboratory experiments and Field Tests results.[11]

*Regression Equations*—There is a relation between two or many of Physical Units variables. For example, there is a relation between volume of gases and their internal temperatures. The main approach in this research is investigation of relation between *P*-surge pressure (m) as a function "Y," and several factors, as variables "X," such as the following: $\rho$, density(kg $m^{-3}$); $C$, velocity of surge wave (m $s^{-1}$); $g$, acceleration of gravity (m $s^{-2}$); $\Delta V$, changes in velocity of water (m $s^{-1}$); $d$, pipe diameter (m); $T$, pipe thickness (m); $E_p$, pipe module of elasticity (kg $m^{-2}$); $E_w$, module of elasticity of water (kg $m^{-2}$); $C_1$, pipe support coefficient; $T$, time (s), $T_p$, pipe thickness (m).[12] There are two cases for Modeling (5–12):

1. The combined elasticity of both the water and the pipe walls is characterized by the pressure wave speed (Arithmetic method—a combination of Joukowski formula and Allievi formula):

$$H_2 - H_1 = \frac{C}{g}(V_2 - V_1) = \rho C(V_2 - V_1), \text{ (Joukowski formula)} \quad (8.5)$$

$$C = 1/\left[\rho\left((1/k) + (dC_1/Ee)\right)\right]1/2. \text{ (Allievi formula)} \quad (8.6)$$

With combination of Joukowski formula and Allievi formula:

$$H_2 - H_1 = 1/[\rho( \quad ((1/k) + (d.C1/E.e))]1/2(V_2 - V_1)/g. \quad (8.7)$$

Hence, pressure or surge pressure ($\Delta H$) is a function of independent variables (***X***) such as:

$$\Delta H \approx \rho, K, d, C_1, Ee, V, g. \quad (8.8)$$

2. The MOC based on a finite difference technique where pressures are computed along the pipe for each time step,

$$V_p = 1/2((V_{Le} + V_{ri}) + (g/c)(H_{Le} - H_{ri}) \\ -(f\Delta t/2D)(V_{Le}|V_{Le}| + V_{ri}|V_{ri}|)), \quad (8.9)$$

$$H_p = 1/2(C/g(V_{Le} - V_{ri}) + (H_{Le} + H_{ri}) - C/g \\ (f\Delta t/2D)(V_{Le}|V_{Le}||V_{ri}||V_{ri}|)), \quad (8.10)$$

Hence, pressure or surge pressure ($\Delta H$) is a function of independent variables (***X***) such as:

$$\Delta H \approx f, T, C, V, g, D. \quad (8.11)$$

For a model definition in this research relation between surge pressure (m), as a function "Y," and several factors, as variables "X," has been investigated; then CFD software evaluates transient flow data as a function of parameters: $\rho, K, d, C_1, Ee, V, f, T, C, g, T_p$. Regression software has fitted the function curve and provides regression analysis. For simplicity and in the curve-fitting process of function by regression software, assumed that

$H$ is dependent variable and $V$ is the only independent variable. Hence other variables assumed constant:[68–74]

## 8.3 RESULTS AND DISCUSSION

Like all fluid research, however, data are obtained at a finite number of locations and generalizing the findings requires assumptions, with uncertainties spreading across the system (Table 8.2).

**TABLE 8.2** Model Summary and Parameter Estimates.

| Model | | Un-Standardized Coefficients | Standardized Coefficients | | t | Sig. |
|---|---|---|---|---|---|---|
| | | | Std. Error | Beta | | |
| 1 | (Constant) | 28.762 | 29.73 | – | 0.967 | 0.346 |
| | flow | 0.031 | 0.01 | 0.399 | 2.944 | 0.009 |
| | Distance | −0.005 | 0.001 | −0.588 | −4.356 | 0 |
| | Time | 0.731 | 0.464 | 0.117 | 1.574 | 0.133 |
| 2 | (Constant) | 14.265 | 29.344 | – | 0.486 | 0.632 |
| | Flow | 0.036 | 0.01 | 0.469 | 3.533 | 0.002 |
| | Distance | −0.004 | 0.001 | −0.52 | −3.918 | 0.001 |
| 3 | (Constant) | 97.523 | 1.519 | – | 64.189 | 0 |
| 4 | (Constant) | 117.759 | 2.114 | – | 55.697 | 0 |
| | Distance | −0.008 | 0.001 | −0.913 | −10.033 | 0 |
| 5 | (Constant) | 14.265 | 29.344 | – | 0.486 | 0.632 |
| | flow | 0.036 | 0.01 | 0.469 | 3.533 | 0.002 |
| | distance | −0.004 | 0.001 | −0.52 | −3.918 | 0.001 |

Regression equation defined in stage 1 has been accepted, because its coefficients are meaningful:

$$\text{pressure} = 28.762 + .031\ \text{Flow} - .005\ \text{Distance} + .731\ \text{Time} \quad (8.12)$$

In worst cases, tests can lead to physically doubtful conclusions limited by the scope of the test program. Neither laboratory models nor field testing can substitute for the careful and correct application of a proven hydraulic transient computer model. If a system is faced with large changes in velocity and pressure in short time periods, then transient analysis is required.

Regression modeling results has been compared with Research Field Tests Model.

Assumption: $p = f(V, T, L)$, where $V$, the velocity (flow), $T$, the time, and $L$, the distance, are the most important requested variables. Regression software fitted the function curve (Fig. 8.4) and has provided regression analysis. Results are shown in Tables 8.1 and 8.3.

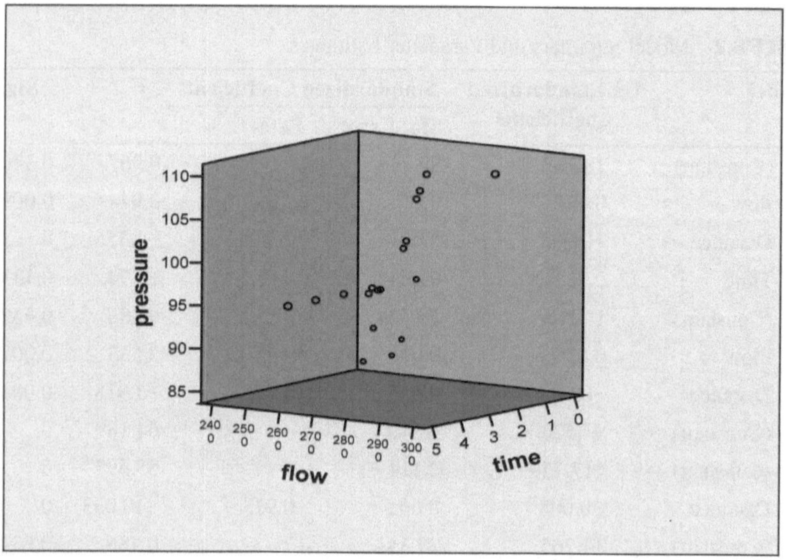


**FIGURE 8.4** Scatter diagram for lab tests (Research Field Tests Model).

**TABLE 8.3** Output Data Table Min. and Max. Head Compared to Equation of Regression Software "SPSS."

| End Point | Max. Press. (mH) | Min. Press. (mH) | Max. Head (m) | Min. Head (m) |
|---|---|---|---|---|
| P2:J6 | 124.1 | 97.0 | 160.4 | 133.4 |
| P2:J7 | 123.7 | 96.7 | 160.0 | 133.0 |
| P3:J7 | 123.7 | 96.7 | 160.0 | 133.0 |
| P3:J8 | 121.9 | 95.1 | 158.0 | 131.2 |
| P4:J3 | 126.7 | 99.5 | 162.2 | 135.0 |
| P4:J4 | 125.1 | 97.8 | 162.3 | 135.0 |

**TABLE 8.3** *(Continued)*

| End Point | Max. Press. (mH) | Min. Press. (mH) | Max. Head (m) | Min. Head (m) |
|---|---|---|---|---|
| P5:J4 | 125.1 | 97.8 | 162.3 | 135.0 |
| P5:J26 | 125.1 | 97.8 | 162.3 | 135.0 |
| P6:J26 | 125.1 | 97.8 | 162.3 | 135.0 |
| P6:J27 | 124.1 | 97.0 | 160.5 | 133.5 |
| P7:J27 | 124.1 | 97.0 | 160.5 | 133.5 |
| P7:J6 | 124.1 | 97.0 | 160.4 | 133.4 |
| P8:J8 | 121.9 | 95.1 | 158.0 | 131.2 |
| P8:J9 | 119.9 | 93.1 | 157.9 | 131.1 |
| P9:J9 | 119.9 | 93.1 | 157.9 | 131.1 |
| P9:J10 | 117.4 | 90.8 | 155.7 | 129.2 |
| P10:J10 | 117.4 | 90.8 | 155.7 | 129.2 |
| P10:J11 | 115.8 | 89.4 | 154.4 | 128.0 |
| P11:J11 | 115.8 | 89.4 | 154.4 | 128.0 |
| P11:J12 | 112.6 | 86.4 | 152.2 | 126.0 |
| P12:J12 | 112.6 | 86.4 | 152.2 | 126.0 |
| P12:J13 | 109.1 | 83.1 | 150.1 | 124.1 |
| P13:J13 | 109.1 | 83.1 | 150.1 | 124.1 |
| P13:J14 | 107.5 | 81.5 | 149.8 | 123.8 |
| P14:J14 | 107.5 | 81.5 | 149.8 | 123.9 |
| P14:J15 | 100.9 | 60.9 | 146.1 | 106.1 |
| P15:J15 | 100.9 | 60.9 | 146.1 | 106.1 |
| P15:J16 | 102.3 | 62.3 | 145.8 | 105.8 |
| P16:J16 | 102.3 | 62.3 | 145.8 | 105.8 |
| P16:J17 | 99.3 | 59.2 | 144.3 | 104.3 |
| P17:J17 | 99.3 | 59.2 | 144.3 | 104.3 |
| P17:J18 | 101.2 | 61.1 | 144.2 | 104.1 |
| P18:J18 | 101.2 | 61.1 | 144.2 | 104.1 |
| P18:J19 | 101.8 | 61.7 | 144.1 | 104.0 |
| P19:J19 | 101.8 | 61.7 | 144.1 | 104.0 |
| P19:J20 | 99.8 | 59.8 | 144.0 | 104.0 |
| P20:J20 | 99.8 | 59.8 | 144.0 | 104.0 |
| P20:J21 | 98.3 | 58.1 | 140.9 | 100.6 |
| P21:J21 | 98.3 | 58.1 | 140.9 | 100.6 |
| P21:J22 | 94.7 | 54.4 | 139.3 | 99.0 |
| P22:J22 | 94.7 | 54.4 | 139.3 | 99.0 |

**TABLE 8.3** *(Continued)*

| End Point | Max. Press. (mH) | Min. Press. (mH) | Max. Head (m) | Min. Head (m) |
|---|---|---|---|---|
| P22:J23 | 68.4 | 28.1 | 138.3 | 98.0 |
| P23:J23 | 68.4 | 28.1 | 138.3 | 98.0 |
| P23:J24 | 56.3 | 15.9 | 138.1 | 97.7 |
| P24:J24 | 56.3 | 15.9 | 138.1 | 97.7 |
| P24:J28 | 42.4 | 0.0 | 137.6 | 95.2 |
| P25:J28 | 42.4 | 0.0 | 137.6 | 95.2 |
| P25:N1 | 16.7 | 16.7 | 112.6 | 112.6 |
| P0:J1 | 0.0 | 0.0 | 40.6 | 40.6 |
| P0:J2 | 27.5 | 2.1 | 63.0 | 37.6 |
| P1:J2 | 27.5 | 2.1 | 63.0 | 37.6 |
| P1:J3 | 27.5 | 2.1 | 63.0 | 37.6 |

Elapsed time: 12 s.

## 8.4 CONCLUSION

In this chapter, Laboratory Model results are similar to a scale model which can be built to reproduce transients observed in a prototype (real) system. Therefore it is always a good idea to check extreme transient pressures for any system with large changes in elevation, long pipelines with large diameters (i.e., mass of water) and initial (e.g., steady-state) velocities in excess of 1 m/s. In some cases, hydraulic transient forces can result in cracks or breaks, even with low steady-state velocities.

## KEYWORDS

- **direct numerical simulation**
- **geographic information system**
- **non-revenue water**
- **pressure management**

## REFERENCES

1. Streeter, V. L.; Wylie, E. B. *Fluid Mechanics;* McGraw-Hill Ltd.: USA, 1979; pp 492–505.
2. Leon, A. S. An Efficient Second-Order Accurate Shock-Capturing Scheme for Modeling One and Two-Phase Water Hammer Flows. Ph.D. Thesis, March 2007.
3. Adams, T. M.; Abdel-Khalik, S. I.; Jeter, S. M.; Qureshi, Z. H. An Experimental Investigation of Single-Phase Forced Convection in Microchannels. *Int. J. Heat Mass Transfer.* **1998,** *41,* 851–857.
4. Peng, X. F.; Peterson, G. P. Convective Heat Transfer and Flow Friction for Water Flow in Microchannel Structure. *Int. J. Heat Mass Transfer.* **1996,** *36,* 2599–2608.
5. Mala, G.; Li, D.; Dale, J. D. Heat Transfer and Fluid Flow in Microchannels. *J. Heat Transfer.* **1997,** *40,* 3079–3088.
6. Xu, B.; Ooi, K. T.; Mavriplis, C.; Zaghloul, M. E. Viscous Dissipation Effects for Liquid Flow in Microchannels. *Micorsystems.* **2002,** 53–57.
7. Pickford, J. *Analysis of Surge;* Macmillan: London, 1969; pp 153–156.
8. American Society of Civil Engineers, Task Committee on Engineering Practice in the Design of Pipelines. *Pipeline Design for Water and Wastewater;* American Society of Civil Engineers: New York, 1975; p 128 .
9. Fedorov, A. G.; Viskanta, R.; Three-Dimensional Conjugate Heat Transfer into Microchannel Heat Sink for Electronic Packaging. *Int. J. Heat Mass Transfer.* **2000,** *43,* 399–415.
10. Tuckerman, D. B. Heat Transfer Microstructures for Integrated Circuits. Ph.D. Thesis, Stanford University, 1984.
11. Harms, T. M.; Kazmierczak, M. J.; Cerner, F. M.; Holke, A.; Henderson, H. T.; Pilchowski, H. T.; Baker, K. In *Experimental Investigation of Heat Transfer and Pressure Drop through Deep Micro channels in a (100) Silicon Substrate,* Proceedings of the ASME, Heat Transfer Division, HTD 351, 1997; pp 347–357.
12. Holland, F. A.; Bragg, R. *Fluid Flow for Chemical Engineers;* Edward Arnold Publishers: London, 1995; pp 1–3.
13. Lee, T. S; Pejovic, S. Air Influence on Similarity of Hydraulic Transients and Vibrations. *J. Fluids Eng.* **1996,** *118* (4), 706–709.
14. Li, J.; McCorquodale, A. Modeling Mixed Flow in Storm Sewers. *J. Hydraul. Eng.* **1999,** *125* (11), 1170–1180.
15. Minnaert, M. On Musical Air Bubbles and the Sounds of Running Water. *Phil. Mag.* **1933,** *16* (7), 235–248.
16. Moeng, C. H.; McWilliams, J. C.; Rotunno, R.; Sullivan, P. P.; Weil, J. Investigating 2D Modeling of Atmospheric Convection in the PBL. *J. Atm. Sci.* **2004,** *61,* 889–903.
17. Tuckerman, D. B.; Pease, R. F. W. High Performance Heat Sinking for VLSI. *IEEE Electron Device Lett.* **1981,** *DEL-2,* 126–129.
18. Nagiyev, F. B.; Khabeev, N. S. Bubble Dynamics of Binary Solutions. *High Temp.* **1988,** *27* (3), 528–533.
19. Shvarts, D.; Oron, D.; Kartoon, D.; Rikanati, A.; Sadot, O. Scaling Laws of Nonlinear Rayleigh-Taylor and Richtmyer-Meshkov Instabilities in Two and Three Dimensions. *C. R. Acad. Sci. Series IV Phys.* **2000,** *6,* 719–726.

20. Cabot, W. H.; Cook, A. W.; Miller, P. L.; Laney, D. E.; Miller, M. C.; Childs, H. R. Large Eddy Simulation of Rayleigh-Taylor Instability. *Phys. Fluids.* **2005,** *17,* 91–106.
21. Cabot, W. *Phys. Fluids.* **2006,** 94–550.
22. Goncharov, V. N. Analytical Model of Nonlinear, Single-Mode, Classical Rayleigh-Taylor Instability at Arbitrary Atwood Numbers. *Phys. Rev. Lett.* **2002,** *88* (13), 10–15.
23. Ramaprabhu, P.; Andrews, M. J. Experimental Investigation of Rayleigh-Taylor Mixing at Small Atwood Numbers. *J. Fluids Mech.* **2004,** *502,* 233–271.
24. Clark, T. T. A Numerical Study of the Statistics of a Two-Dimensional Rayleigh-Taylor Mixing Layer. *Phys. Fluids.* **2003,** *15,* 2413.
25. Cook, A. W.; Cabot, W.; Miller, P. L. The Mixing Transition in Rayleigh-Taylor Instability. *J. Fluid Mech.* **2004,** *511,* 333–362.
26. Waddell, J. T.; Niederhaus, C. E.; Jacobs, J. W. Experimental Study of Rayleigh-Taylor Instability: Low Atwood Number Liquid Systems with Single-Mode Initial Perturbations. *Phys. Fluids.* **2001,** *13,* 1263–1273.
27. Weber, S. V.; Dimonte, G.; Marinak, M. M. Arbitrary Lagrange-Eulerian Code Simulations of Turbulent Rayleigh-Taylor Instability in Two and Three Dimensions. *Laser Part. Beams.* **2003,** *21,* 455.
28. Dimonte, G.; Youngs, D.; Dimits, A.; Weber, S.; Marinak, M. A Comparative Study of the Rayleigh-Taylor Instability using High-Resolution Three-Dimensional Numerical Simulations: The Alpha Group Collaboration. *Phys. Fluids.* **2004,** *16,* 1668.
29. Young, Y. N.; Tufo, H.; Dubey, A.; Rosner, R. On the Miscible Rayleigh-Taylor Instability: Two and Three Dimensions. *J. Fluids Mech.* **2001,** *447,* 377–408.
30. George, E.; Glimm, J. Self-Similarity of Rayleigh-Taylor Mixing Rates. *Phys. Fluids.* **2005,** *17,* 054101.
31. Oron. D.; Arazi, L.; Kartoon, D.; Rikanati, A.; Alon, U.; Shvarts, D. Dimensionality Depéndence of the Rayleigh-Taylor and Richtmyer-Meshkov Instability Late-Time Scaling Laws. *Phys. Plasmas.* **2001,** *8,* 2883.
32. Nigmatulin, R. I.; Nagiyev, F. B.; Khabeev, N. S. In *Effective Heat Transfer Coefficients of the Bubbles in the Liquid Radial Pulse*, Mater, Second-Union Conference of Heat Mass Transfer, Heat Massoob-Men in the Biphasic With Minsk, 1980; Vol. 5, pp 111–115.
33. Nagiyev, F. B. In *Damping of the Oscillations of Bubbles Boiling Binary Solutions,* Mater. VIII Resp. Conf. Mathematics and Mechanics: Baku, October 26–29, 1988; pp 177–178.
34. Nagyiev, F. B.; Kadyrov, B. A. Small Oscillations of the Bubbles in a Binary Mixture in the Acoustic Field. Math. *AN Az.SSR Ser. Physicotech. Mate. Sci.* **1986,** *1,* 23–26.
35. Nagiyev, F. B. In *Dynamics, Heat and Mass Transfer of Vapor-Gas Bubbles in a Two-Component Liquid,* Turkey-Azerbaijan Petrol Seminar, Ankara, Turkey, 1993; pp 32–40.
36. Nagiyev, F. B. In The Method of Creation Effective Coolness Liquids, Third Baku international Congress, Baku, Azerbaijan Republic, 1995; pp 19–22.
37. Nagiyev, F. B. The Linear Theory of Disturbances in Binary Liquids Bubble Solution. *Dep. VINITI.* **1986,** *405,* 76–79.
38. Nagiyev, F. B. Structure of Stationary Shock Waves in Boiling Binary Solutions. *Math. USSR. Fluid Dynamic.* **1989,** *1,* 81–87.
39. Lord Rayleigh, O. M. F. R. S. On the Pressure Developed in a Liquid during the Collapse of a Spherical Cavity. *Philos. Mag. Ser.* **1917,** *34* (200), 94–98.

40. Perry. R. H.; Green, D. W.; Maloney, J. O.; In *Perry's Chemical Engineers Handbook,* 7th Ed.; Perry, R. H., Green, D. W., Eds.; McGraw-Hill Professional Publishing: New York, 1997; pp 1–61.
41. Nigmatulin, R. I. *Dynamics of Multiphase Media;* Izdatel'stvo Nauka: Moscow, 1987; pp 12–14.
42. Kodura, A.; Weinerowska, K. In *The Influence of the Local Pipeline Leak on Water Hammer Properties,* Materials of the II Polish Congress of Environmental Engineering, Lublin, 2005; pp 125–133.
43. Kane, J.; Arnett, D.; Remington, B. A.; Glendinning, S. G.; Baz'an, G. Two-Dimensional versus Three-Dimensional Supernova Hydrodynamic Instability Growth. *Astrophys. J.* **2000,** *528* (2), 528–989.
44. Quick, R. S. Comparison and Limitations of Various Water Hammer Theories. *J. Hyd. Div.* **1933,** 43–45.
45. Jaeger, C. *Fluid Transients in Hydro-Electric Engineering Practice;* Blackie and Son Ltd.: Glasgow, Scotland, 1977; pp 87–88.
46. Jaime, S, A. Generalized Water Hammer Algorithm for Piping Systems with Unsteady Friction. E-Theses, 2005.
47. Fok, A.; Ashamalla, A.; Aldworth, G. In *Considerations in Optimizing Air Chamber for Pumping Plants,* Symposium on Fluid Transients and Acoustics in the Power Industry, San Francisco, USA, 1978; pp 112–114.
48. Fok, A. In *Design Charts for Surge Tanks on Pump Discharge Lines,* BHRA 3rd International Conference on Pressure Surges, Bedford, England, 1980; pp 23–34.
49. Fok, A. In *Water Hammer and its Protection in Pumping Systems,* Hydro Technical Conference, CSCE: Edmonton, 1982; pp 45–55.
50. Fok, A. A Contribution to the Analysis of Energy Losses in Transient Pipe Flow. Ph.D. Thesis, University of Ottawa, 1987.
51. Brunone, B.; Karney, B. W.; Mecarelli, M.; Ferrante, M. Velocity Profiles and Unsteady Pipe Friction in Transient Flow. *J. Water Res. Plan. Manag. ASCE.* **2000,** *126* (4), 236–244.
52. Koelle, E.; Luvizotto Jr. E., Andrade, J. P. G. In *Personality Investigation of Hydraulic Networks using MOC,* – Method of Characteristics Proceedings of the 7th International Conference on Pressure Surges and Fluid Transients, Harrogate Durham, UK, 1996; pp 1–8.
53. Filion, Y.; Karney, B. W. A In *Numerical Exploration of Transient Decay Mechanisms in Water Distribution Systems,* Proceedings of the ASCE Environmental Water Resources Institute Conference, American Society of Civil Engineers: Roanoke, Virginia, 2002; p 30.
54. Hamam, M. A.; McCorquodale, J. A. Transient Conditions in the Transition from Gravity to Surcharged Sewer Flow. *Canadian J. Civil Eng.* **1982,** *9* (2), 65–98.
55. Savic, D. A.; Walters, G. A. Genetic Algorithms Techniques for Calibrating Network Models. Report No. 95/12, Centre for Systems and Control Engineering, 1995; pp 137–146.
56. Walski, T. M.; Lutes, T. L. Hydraulic Transients Cause Low-Pressure Problems. *J. Am. Water Works Assoc.* **1994,** *75* (2), 58.
57. Chaudhry, M. H. *Applied Hydraulic Transients;* Van Nostrand Reinhold Co.: New York, 1979; pp 1322–1324.

58. Parmakian, J. *Water Hammer Analysis;* Dover Publications, Inc.: New York, NY, 1963; pp 51–58.
59. Tuckerman, D. B.; Pease, R. F. W. High Performance Heat Sinking for VLSI. *IEEE Electron Device Lett.* **1981,** *DEL-2,* 126–129.
60. Farooqui, T. A. *Evaluation of Effects of Water Reuse, on Water and Energy Efficiency of an Urban Development Area, using an Urban Metabolic Framework,* Master of Integrated Water Management Student Project Report, International Water Centre: Brisbane, Australia, 2015.
61. Ferguson, B. C.; Frantzeskaki, N.; Brown, R. R. A Strategic Program for Transitioning to a Water Sensitive City. *Landsc. Urban Plan.* **2013,** *117*, 32–45.
62. Kenway, S.; Gregory, A.; McMahon, J. Urban Water Mass Balance Analysis. *J. Ind. Ecol.* **2011,** *15* (5), 693–706.
63. Renouf, M. A.; Kenway, S. J.; Serrao-Neumann, S.; Low Choy, D. *Urban Metabolism for Planning Water Sensitive Cities. Concept for an Urban Water Metabolism Evaluation Framework.* Milestone Report. Cooperative Research Centre for Water Sensitive Cities: Melbourne: 2015.
64. Serrao-Neumann, S.; Schuch, G.; Kenway, S. J.; Low Choy, D. *Comparative Assessment of the Statutory and Non-Statutory Planning Systems: South East Queensland, Metropolitan Melbourne and Metropolitan Perth,* Cooperative Research Centre for Water Sensitive Cities: Melbourne, 2013.
65. Andrews, S.; Traynorp, P. Guidelines for Assuring Quality of Food and Water Microbiological Culture Media, August 2004.
66. Andrew, D. E.; Rice, E. W.; Clescceri, L. S. Standard Methods for the Examination of Water and Waste Water, Part 9000, 2012.
67. Instructions Procedure Rural Water and Wastewater Quality Assurance Test Results of KhorasanRazavi, 2011.
68. Abbassi, B.; Al Baz, I. Integrated Wastewater Management: A Review. Efficient Management of Wastewater – its Treatment and Reuse in Water Scarce Countries. Springer Publishing Co.: Berlin, Germany, 2008.
69. Andreasen, P. In *Chemical Stabilization. Sludge into Biosolids – Processing, Disposal and Utilization;* Spinosa, L., Vesilind, P. A., Eds.; IWA Publishing: UK, 2001; pp 392.
70. CIWEM – The Chartered Institution of Water and Environmental Management. In *Sewage Sludge: Stabilization and Disinfection – Handbooks of UK Wastewater Practice;* CIWEM Publishing: Lavenham, 1996.
71. Halalsheh, M.; Wendland, C. Integrated Anaerobic-Aerobic Treatment of Concentrated Sewage. In *Efficient Management of Wastewater – its Treatment and Reuse in Water Scarce Countries;* Baz, I. A., Otterpohl, R., Wendland, C., Eds.; Springer Publishing Co.: Berlin, Heidelberg, 2008; pp 177–186.
72. ISWA. *Handling, Treatment and Disposal of Sludge in Europe,* Situation Report 1, ISWA Working Group on Sewage Sludge and Water Works: Copenhagen, 1995.
73. Matthews, P. In *Agricultural and Other Land Uses. Sludge into Biosolids,* – Processing, Disposal and Utilization. IWA Publishing: London, 2001.
74. IWK Sustainability Report, 2012–2013.

## CHAPTER 9

# WATER DISTRIBUTION NETWORK ANALYSIS: FROM THEORY TO PRACTICE

KAVEH HARIRI ASLI[1*], SOLTAN ALI OGLI ALIYEV[1], and HOSSEIN HARIRI ASLI[2]

[1]*Department of Mathematics and Mechanics, Azerbaijan National Academy of Sciences AMEA, Baku, Azerbaijan*

[2]*Civil Engineering Department, Faculty of Engineering, University of Guilan, Rasht, Iran*

[*]*Corresponding author. E-mail: hariri_k@yahoo.com*

## CONTENTS

## ABSTRACT

Network analysis is the term used to describe the "analysis of water flows and head losses in a pressurized distribution system under a given set of demand conditions on the system."

A network is the collection of pipes, valves, booster pumps, and service reservoirs forming the water distribution system. Due to the complexity of most distribution systems, it was normal to simplify the system by considering only the key mains. As a result, this work shows the performance and possibility to include all reservoirs and mains in a distribution system, and all the various control features, with their operating constraints against unaccounted-for water.


## 9.1 INTRODUCTION

The demands and demand patterns on a network are also vital ingredients, and are made up of a number of components:

a) domestic demand,
b) metered industrial/commercial demand,
c) unaccounted-for water including leakage.

This is the process of calculating the flows and head losses in a network for a given set of demand conditions. Two types of analysis are normally used:

1. In a snapshot analysis, the flows and head losses are considered at only a single given set of demand conditions. This is frequently expressed as a single time interval dynamic or extended time.
2. In each dynamic analysis the flows and head losses are considered for a series of varying demand conditions. This is frequently a 24-h time period, and is the sort of analysis that is now most commonly used. The power and speed of computing for network analysis continue to improve.

A network model is basically an intelligent mains record drawing, allowing one to access hydraulic data as well as the position of the mains in the ground. A model represents everything we know about a particular distribution system. It will have been calibrated by the model builders to ensure that, within reason, the model gives the same flows and pressures as the real system. This is done by comparing the results from the model with huge amounts of data from field tests. It is essential to know the system of configuration on the calibration day—i.e., the day chosen as the most "typical" from the field test. The calibration process will find any significant problems with the model's representation of the distribution system, but not all of them. To calibrate a model it is necessary to get the pressures right within 1 m at virtually all points in the system at all times of the day.

Once the model is created it has to be converted to what is known as an Average Day Model. To do this, the model builder converts the demands on the model to average demands by comparing the demands for that area with the test day.[1–8]

If there is a disagreement between the computed flows and the measured flows, a number of factors can be involved. The more common are listed below:

- incorrect estimates for model demands,
- incorrect assumptions for hydraulic resistances,
- wrong pipe lengths or diameters,
- unsuspected network cross-connection,
- closed valves/opened valves,
- by-passes around PRV or meters,
- restrictions in mains,
- pressure measurement on "rider" main.

The process of model building can thus uncover many problems which may go unnoticed until a burst occurs, often wasting time and money.

Network analysis is a powerful tool for the effective management of distribution systems. Once a model exists, it allows any user to experiment with system changes before they are tried out on the ground. These could be such things as checking what reinforcements are needed to supply a new development, so that levels of service are not affected somewhere else, perhaps miles away. The model could help maximize the utilization of low cost supplies, and in pumped distribution systems, minimize the cost of pumping. It could also ensure that levels of service are achieved at customer taps by identifying areas of inadequate or excessive pressures, and areas of high leakage; corrective measures could then be simulated. It might be used for planning a trunk main shut-off, with effects over wide areas, perhaps to see how long the reservoir storage will last. It can be used to check on rehabilitation problems—re-line, renew, or up-size. It can also be used to design pressure reduction, or to alter distribution areas. As the techniques improve, it will also be used to investigate water quality problems. Network models can already tell us how old the water is throughout a system and how that changes during the day. They can also be used to tell us how different source waters blend in the system at different times of the day. All these might point to problem areas and show the results on water quality of system changes. Network models give us a better picture of the system operation, and help improve levels of service. A lot of money can be saved on capital schemes by using models to find out what size mains are really needed, or to sometimes find ways of not laying new mains at all. Network models may be useful in locating large leaks by comparing modeled pressures against actual. Large leaks cause a lowering of pressures. Network models are not perfect, but they are the only tool available to provide such detailed hydraulic information. In the past we often had to guess about the behavior of complex systems.

Distribution management is an important activity which has considerable impact on customers. The costs of distribution operations are high. It is therefore vital that management decisions are taken in a framework of knowledge and understanding of how the system operates. The development of DMAs as part of a structured operation of the distribution system allows the network to be operated in a planned way. This planned approach inevitably leads to better understanding and control of the distribution system, updated and more comprehensive records, fewer consumer complaints, and closer control of labor with more efficient and informed management. Such an approach helps to ensure that distribution managers can meet the primary objectives to the maximum benefit of the customer and the Water Supplier. District meter areas (DMAs) are the basic building blocks of a zoned distribution system. They provide a manageable unit by which the distribution customer and performance information can be linked to other activities and data systems.

Their fundamental characteristic is that their boundaries are closed except for defined, measured inputs and outputs. Ideally this should be a single metered input, but this is not always achievable in practice. DMAs in the U.K. are generally between 1000 and 5000 properties in size, and they have similar topography with limited head loss within their area.

Even it allows pressure and leakage to be managed most effectively. Larger areas are possible from a detection point of view if acoustic logging is part of the detection policy employed. The principles of DMA design and structure are very simple. Nevertheless, where possible, the design should be checked using Network Analysis to ensure that pressures are sustained at all likely demands, that no unnecessarily long water retention periods are created and that water quality variations are within an acceptable range—larger areas usually means less "dead ends." System record plans are required, preferably at a scale of 1:2500, together with property count data. This information is used, together with the local system operator's knowledge, to define the boundaries of each DMA. Other important considerations in this process are as the following:[6,9–27]

1. To cross the fewest number of distribution mains (helped by using natural boundaries such as railway lines, canals, and major roads), thereby reducing the number of meters used and the number of closed valves (which can lead to water quality problems).
2. To avoid districts with high outflows (this leads to inaccuracy in calculation of district demand as any changes in demand will be a small proportion of the total flow measured).

Having defined the limits of a DMA, it will normally be necessary to trial the area in practice. It will be necessary to ensure that all stop (stand shut, boundary) valves perform correctly, and that satisfactory flows and pressures are maintained throughout the DMA. In practice, DMAs often have to be checked very carefully during establishment. Unforeseen difficulties may be found, such as buried, or closed valves, or even unknown pipes. These problems are often discovered when the DMA is first modeled and anomalies in the model are investigated. Once satisfactorily piloted, the DMA can be fully established. This will require:

1. the installation of flow meters at all inlets and outlets;
2. the closing and marking of all boundary valves;
3. the installation of flushing, or "OXO," points;
4. the updating of plans, records, and related information systems;

The simple checklist below can be used to ensure that all of these activities are performed before a DMA is commissioned:[28–42]

### 9.1.1 DESIGN OF DMA

1000–5000 properties.
Minimum number of boundary valves.
Preferably single inlet meter.
Avoid export meters if possible.
Beware of low pressure (on peak demand).
Beware of quality problems at stop-ends.
Avoid 150 mm mechanical (Helix) meters (1 rev = 1000 L).
Typically downsize mechanical meters (not necessary for electromagnetic).
Install mechanical meters on a bypass.
Fit "out-reader" chamber for logger if meter access problems.

A methodology (1) to estimate background minimum night flows in individual DMAs, given all relevant local characteristics (mains length, number of households, and non-households, pressure), is potentially of significant value. It could indicate the night flow at which it is no longer appropriate to allocate resources to try to locate significant unreported bursts in that DMA. Such a methodology also provides an independent check on the minimum night flow achieved when a DMA is initially set up, after the "best practice" of thoroughly checking the DMA for leaks by step-testing and sounding has been carried out. The background night flow losses (when no bursts exist in a DMA) can be calculated for any DMA (given $L$ (length of mains in km), $N$ (number of properties), AZNP (average zonal night pressure in m)) from the equation:[43–58]

$$\text{NFLB (l/h)} = (C1 \times L + (C2 + C3) \times N) \times \text{PCF}, \tag{9.1}$$

Using the following values of $C1$, $C2$, and $C3$ from (Tables 9.1 and 9.2) and the pressure correction factor (PCF) from Table 9.2, based on the U.K. research of the 1980s.

**TABLE 9.1** Background Night Flow Losses.

| Background Losses Component | Units | Low | Average | High |
|---|---|---|---|---|
| C1: Dist mains | l/km/hr | 20 | 40 | 60 |
| C2: Common pipes | l/prop/hr | 1.5 | 3.0 | 4.5 |
| C3: Supply pipes | l/prop/hr | 0.5 | 1.0 | 1.5 |

**TABLE 9.2** Pressure Correction Factors.

| AZNP (metres) | 20 | 30 | 40 | 50 | 60 | 70 | 80 | 90 | 100 |
|---|---|---|---|---|---|---|---|---|---|
| PCF | .329 | .529 | .753 | 1.00 | 1.271 | 1.565 | 1.884 | 2.226 | 2.592 |

Once established, DMAs need to be maintained. For two adjacent DMAs, the opening of a single boundary stop valve is sufficient to destroy the accuracy of DMA demand monitoring. A regular regime of meter readings, boundary valve checks, and pressure monitoring must therefore be established for each DMA. For leakage control purposes, it is necessary to establish the number of domestic properties, and the demand of major industrial users within each DMA. This requires regular, usually weekly, reading of DMA meters and loggers, preferably with the input of the information into a computer analysis programmer. Careful inspection of the meter and logger readings can quickly spot any unusual results. This can be used to trigger leak detection follow-up work. Simple management procedures must be introduced to ensure that the integrity of the DMA is maintained, otherwise the cost and effort of establishment and monitoring will be wasted. The following details are worth noting for effective management:

- All boundary valves should be kept tight closed and a regular checking program should be followed.
- All boundary valves should be clearly marked and identified.
- Valves within the DMA should be fully open.
- Status quo should be re-established after bursts, rehabilitation, or other operational necessity.
- High pressure DMAs should be examined for pressure reduction.
- Logger readings of low pressure should be investigated to determine whether leakage is indicated.
- Leakage within the DMA, whether visible or not showing, should be repaired promptly.
- DMA meters should not be valve out.
- DMA meters and loggers should be operating normally.
- PRV areas should be properly isolated and operating.
- Poor quality mains should be fed forward into the capital program as candidates for renewal.
- Plans should be up to date and show new property.

The principal benefit of DMAs is that the key characteristics (e.g., demand, quality, cost) of a well-defined area of the distribution system can be closely monitored. The results of this monitoring allow management action to be prioritized and targeted on where it is most cost effective.[14,59–66] Specifically DMA's impact on:

1. leakage control,
2. pressure management and levels of service,
3. asset maintenance and renewal,
4. the monitoring and maintenance of water quality,
5. the planning and programming of repair and maintenance work.

Perhaps the most important benefit of DMAs is a little less tangible. Together with a zoned approach to distribution management, they provide a better knowledge of how the system works and how water gets to the customers in an appropriate condition. This allows the water supplier to focus attention on those activities which produce most benefits to customers—a pro-active rather than a reactive approach. For example, flow reversals and retention times can be minimized and more consistent pressures established. This results in a better knowledge of the system, improved demand management, and better and more consistent service to customers, all at a lower, long-term cost to the water supplier.

Zone metering breaks down a large supply area into several zones, typically varying between 20,000 and 50,000 properties. Again all inflows and outflows are measured continuously, including the effect of any increase or decrease in storage. Zones are too large to identify small leakages, as again these will be swamped by normal daily variations. However, they could possibly identify major leakage, especially if daily readings are collected. Zone metering may also be useful for comparing the performance of different leakage control teams, or for collecting together data for parts of this system with similar characteristics such as unit cost, age, and urban/rural character.

Within each zone, there will be several DMAs ranging in size typically from 1000 to 5000 properties. In the U.K., this would typically mean a population range of 2500–12,500 and a daily demand ranging from 0.7 to 3.5 Mld. District metering may be considered as the first level of metering which can be used for leakage detection, the previous two levels being used for performance assessment and monitoring rather than detection.

The original concept of district metering was to measure the total volume entering the DMA between the reading intervals, and hence to calculate the average daily demand. This would then be compared to previous readings, and also to the readings for all other DMAs for that period to assess climatic effects. A significant increase in demand, not generally reflected across the system, would signify a likely increase in leakage. Normally, a second cycle of readings would be taken to confirm the result before further action was taken.

This procedure suffered from a number of disadvantages:

1. It was insensitive as leakage would not be identified until it exceeded a significant proportion of the daily demand, normally at least 10%.
2. The time taken to identify the leakage and initiate further action would be two reading intervals.
3. It was not possible to differentiate between increases in leakage and increases in metered consumption, except for very large consumers whose meters may have been read as district meters.
4. Elimination of climatic factors and holiday effects was difficult, and very much a matter of judgment and experience.

Due to the large numbers of meters likely to be involved, it may not be economic for all these meters to be on telemetry, in which case data must be collected by site visit. The frequency of data collection and analysis may itself be limited by the amount of resources which can be economically justified to undertake this activity. This can be varied with the leakage growth characteristics of the area. However, district meters are now usually fitted with data loggers which will record, in addition to the total flow, the night flow over a specified period for a number of nights. This immediately achieves a better than five-fold improvement in the sensitivity of the method in the original concept, as night flows will normally be less than 20% of the average daily flow and will suffer less variation due to demand. The time taken to identify leakage is reduced to one reading interval as the night flow readings will confirm the leakage, unless it occurred at the end of the period. The effect of climatic variation is significantly reduced, although care may be needed on occasions when garden sprinklers may be left on overnight. Differentiation between leakage and metered use is easier, as any increase in metered use is less likely to take place at night, particularly at weekends. Logger manufacturers usually provide powerful software to analyze and manipulate recorded data. The equipment and economics associated with data collection are changing. Some Water Suppliers are beginning to move in favor of automated, remote, and centralized interrogation of intelligent data loggers at meters, monitoring pressure as well as flow.

The primary use of net night flow data is to provide operational data on which to decide on the need for further action. The minimum night flow (MNF) can be readily measured with reasonable accuracy for both district and waste meter areas, allowing small changes in flow volumes to be observed. Determination of the night metered consumption is more difficult. In many areas it will be negligible and can be ignored. Where it is not deemed negligible, the alternative methods available to determine it are as follows:

1. Use a percentage of average daily consumption. This is satisfactory where the total non-domestic consumption is relatively small.
2. Measure MNF immediately prior to and during a "bank holiday" period. The difference will give the night consumption of industrial users who shutdown for the holiday. Some allowance will still be required for commercial users with an element of domestic type consumption, and for industrial users with continuous processes.
3. Do a telephone survey of major consumers to determine whether there is significant night usage, e.g., replenishment of factory storage tanks. Some users may be able to supply night consumption data. In addition, on large complex sites there is a possibility of misuse of water (e.g., unauthorized use of fire mains), and it may be prudent to check such connections before embarking on leak location work.
4. Take night meter readings of the major non-domestic users. Use data loggers where the meters are logger compatible—consider changing/converting old meters on major users where this is not the case.
5. Trade effluent data may provide useful information.

It must also be remembered, however, that whilst domestic consumption is reduced to a minimum by measuring flows at night, it is not eliminated entirely. Research in the U.K. suggests an allowance of about

21/prop/hr, which includes minor undetectable leakage such as dripping taps and passing ball cocks. This consumption is included in the net night flow figure. The increasing use of domestic appliances overnight using economy electricity tariffs is also a factor which may need consideration.

Pressure is one of the most frequently measured parameters in the water industry, often being measured alongside flow. Many methods of measurement are in usage, but pressure transducers have become the most common means in distribution systems. They operate by converting fluid pressures into electrical signals.[14,59–66] Pressure measurement typically takes place for:

- general monitoring of the distribution system;
- specific monitoring at critical points (levels of service);
- particular consumer problems of inadequate pressure;
- co-ordination with particular flow tests, e.g., new housing estates, high rise flats, industrial consumers, fire fighting installations, and fire hydrants;
- network analysis calibration.

## 9.2 MATERIALS AND METHODS

Fluid mechanics is a highly visual subject, with good instrumentation and the use of dimensional analysis and modeling concepts (Chap. 5) is widespread. Thus experimentation provides a natural and easy complement to the theory. In constructing a distribution network in this book, software will apply to water distribution network calibration including:

WATER GEMS8.2 & ArcGIS9-ArcMap9.3 will assign labels and automatically demand management and resource development. When creating a schematic drawing, pipe lengths are entered manually. In a scaled drawing, pipe lengths are automatically calculated from the position of the pipes' bends and start and stop nodes on the drawing pane. In this network, the modeling of a reservoir connected to a pump simulates a connection to the main water distribution system. Simplifying the network in this way can approximate the pressures supplied to the system at the connection under a range of demands. This type of approximation is not always applicable, and care should be taken when modeling a network in this way. It is more accurate to trace the network back to the source.

## 9.3 RESULTS AND DISCUSSION

This chapter has investigated fluid condition by fluid pressure variations. The interpenetration between the mixture components could be regarded as a combination of the diffusing process and remixing process. If the former dominates the interpenetration, it is miscible; otherwise, it is immiscible. The complex microscopic interplay between the mixture components makes the simulation highly challenging. So far, there have been some dedicated reports in computer graphics dealing with immiscible mixtures, but only few works focused on miscible mixtures. Regression Equations have shown relation between two or many physical units of variables. For example, there was a relation between volume of gases and their internal temperatures. The main approach in this research was investigation of relation between: *P*-surge pressure (as a function or dependent variable with nomenclature "Y") and several factors (as independent variables with nomenclature "X") such as the following: $\rho$, density (kg/m$^3$); $C$, velocity of surge wave (m/s); $g$, acceleration of gravity (m/s$^2$); $\Delta V$, changes in velocity of water (m/s); $d$, pipe diameter (m); $T$, pipe thickness (m); $E_p$, pipe module of elasticity (kg/m$^2$); $E_w$, module of elasticity of water (kg/m$^2$); $C_1$, pipe support coefficient; $T$, time (s); $T_p$, pipe thickness (m) (for fluid at water hammer condition).[7,8]

A. Research focus attention on the steam-water flows that may occur during transients formed in pressurized-water pipeline. The method presented is applicable to liquid–solid and liquid–liquid flows as well.
B. The Laboratory has handled the development of sophisticated numerical techniques for analysis of multiphase flows and in the construction of computer codes based on these techniques (Tables 9.3 and 9.4).
C. When collisions dominate a region with many macro particles, those macro particles can be merged, and their fluid-like behavior modeled with a hydrodynamics calculation.
D. Equations were solved numerically by the method of characteristics (MOC). Transient analysis has solved transient flow in pipes for range of approximate equations to numerical solutions of the nonlinear Navier–Stokes equations.

**TABLE 9.3** Input Data Table Including of—"x" and "x²" and "y"—for Equation Finding, Transfer to Regression Software "SPSS."

$$y = a_0 + a_1 x, \qquad \text{(Line equations)}$$

| Equation | Model Summary | | | | | Parameter Estimates | | | |
|---|---|---|---|---|---|---|---|---|---|
| | *R* Square | *F* | df1 | df2 | Sig. | Constant | *b*1 | *b*2 | *b*3 |
| Linear | 0.418 | 15.831 | 1 | 22 | 0.001 | 6.062 | 0.571 | | |

$$y = a_0 + a_1 x + a_2 x^2, \qquad \text{(Second order equation)}$$

| Equation | Model Summary | | | | | Parameter Estimates | | | |
|---|---|---|---|---|---|---|---|---|---|
| | *R* Square | F | df1 | df2 | Sig. | Constant | *b*1 | *b*2 | *b*3 |
| Quadratic | 0.487 | 9.955 | 2 | 21 | 0.001 | 6.216 | −0.365 | 0.468 | |

$$y = a_0 + a_1 x + a_2 x^2, + a_3 x^3, \qquad \text{(Third order equation)}$$

| Equation | Model Summary | | | | | Parameter Estimates | | | |
|---|---|---|---|---|---|---|---|---|---|
| | *R* Square | *F* | df1 | df2 | Sig. | Constant | *b*1 | *b*2 | *b*3 |
| Cubic | 0.493 | 10.193 | 2 | 21 | 0.001 | 6.239 | 0.000 | −0.057 | 0.174 |

$$A = Ce^{kt} \text{ (Compound)}$$

| Equation | Model Summary | | | | | Parameter Estimates | | | |
|---|---|---|---|---|---|---|---|---|---|
| | *R* Square | *F* | df1 | df2 | Sig. | Constant | *b*1 | *b*2 | *b*3 |
| Compound | 0.424 | 16.207 | 1 | 22 | 0.001 | 6.076 | 1.089 | | |

**TABLE 9.3** *(Continued)*

$(dA / dT = KA$ (Growth)

| Equation | Model Summary | | | | | Parameter Estimates | | | |
|---|---|---|---|---|---|---|---|---|---|
| | ***R* Square** | ***F*** | **df1** | **df2** | **Sig.** | **Constant** | ***b*1** | ***b*2** | ***b*3** |
| Growth | 0.424 | 16.207 | 1 | 22 | 0.001 | 1.804 | 0.085 | | |

$y = ab^x$ or $\log a + x\log b = a_0 + a_1^x$, (Exponential equation)

$y = ab^x + g$, (Expression exponential equation)

| Equation | Model Summary | | | | | Parameter Estimates | | | |
|---|---|---|---|---|---|---|---|---|---|
| | ***R* Square** | ***F*** | **df1** | **df2** | **Sig.** | **Constant** | ***b*1** | ***b*2** | ***b*3** |
| Exponential | 0.424 | 16.207 | 1 | 22 | 0.001 | 6.076 | 0.085 | | |

$y = ax^b$ or $\log y = \log a + b\log x$, (**Logarithmic equation**)

$y = ax^b + g$, **(Expression logarithmic equation)**

| Equation | Model Summary | | | | | Parameter Estimates | | | |
|---|---|---|---|---|---|---|---|---|---|
| | ***R* Square** | ***F*** | **df1** | **df2** | **Sig.** | **Constant** | ***b*1** | ***b*2** | ***b*3** |
| Logistic | 0.424 | 16.207 | 1 | 22 | 0.001 | 0.165 | 0.918 | | |

**TABLE 9.4** Max. Head Compared to Equation of Regression software "SPSS."

| End Point | Max. Press. (mH) | Min. Press. (mH) | Max. Head (m) | Min. Head (m) |
|---|---|---|---|---|
| P2:J6 | 124.1 | 97.0 | 160.4 | 133.4 |
| P2:J7 | 123.7 | 96.7 | 160.0 | 133.0 |
| P3:J7 | 123.7 | 96.7 | 160.0 | 133.0 |
| P3:J8 | 121.9 | 95.1 | 158.0 | 131.2 |
| P4:J3 | 126.7 | 99.5 | 162.2 | 135.0 |
| P4:J4 | 125.1 | 97.8 | 162.3 | 135.0 |
| P5:J4 | 125.1 | 97.8 | 162.3 | 135.0 |
| P5:J26 | 125.1 | 97.8 | 162.3 | 135.0 |
| P6:J26 | 125.1 | 97.8 | 162.3 | 135.0 |
| P6:J27 | 124.1 | 97.0 | 160.5 | 133.5 |
| P7:J27 | 124.1 | 97.0 | 160.5 | 133.5 |
| P7:J6 | 124.1 | 97.0 | 160.4 | 133.4 |
| P8:J8 | 121.9 | 95.1 | 158.0 | 131.2 |
| P8:J9 | 119.9 | 93.1 | 157.9 | 131.1 |
| P9:J9 | 119.9 | 93.1 | 157.9 | 131.1 |
| P9:J10 | 117.4 | 90.8 | 155.7 | 129.2 |
| P10:J10 | 117.4 | 90.8 | 155.7 | 129.2 |
| P10:J11 | 115.8 | 89.4 | 154.4 | 128.0 |
| P11:J11 | 115.8 | 89.4 | 154.4 | 128.0 |
| P11:J12 | 112.6 | 86.4 | 152.2 | 126.0 |
| P12:J12 | 112.6 | 86.4 | 152.2 | 126.0 |
| P12:J13 | 109.1 | 83.1 | 150.1 | 124.1 |
| P13:J13 | 109.1 | 83.1 | 150.1 | 124.1 |
| P13:J14 | 107.5 | 81.5 | 149.8 | 123.8 |
| P14:J14 | 107.5 | 81.5 | 149.8 | 123.9 |
| P14:J15 | 100.9 | 60.9 | 146.1 | 106.1 |
| P15:J15 | 100.9 | 60.9 | 146.1 | 106.1 |
| P15:J16 | 102.3 | 62.3 | 145.8 | 105.8 |
| P16:J16 | 102.3 | 62.3 | 145.8 | 105.8 |
| P16:J17 | 99.3 | 59.2 | 144.3 | 104.3 |
| P17:J17 | 99.3 | 59.2 | 144.3 | 104.3 |
| P17:J18 | 101.2 | 61.1 | 144.2 | 104.1 |
| P18:J18 | 101.2 | 61.1 | 144.2 | 104.1 |
| P18:J19 | 101.8 | 61.7 | 144.1 | 104.0 |
| P19:J19 | 101.8 | 61.7 | 144.1 | 104.0 |

**TABLE 9.4** *(Continued)*

| End Point | Max. Press. (mH) | Min. Press. (mH) | Max. Head (m) | Min. Head (m) |
|---|---|---|---|---|
| P19:J20 | 99.8 | 59.8 | 144.0 | 104.0 |
| P20:J20 | 99.8 | 59.8 | 144.0 | 104.0 |
| P20:J21 | 98.3 | 58.1 | 140.9 | 100.6 |
| P21:J21 | 98.3 | 58.1 | 140.9 | 100.6 |
| P21:J22 | 94.7 | 54.4 | 139.3 | 99.0 |
| P22:J22 | 94.7 | 54.4 | 139.3 | 99.0 |
| P22:J23 | 68.4 | 28.1 | 138.3 | 98.0 |
| P23:J23 | 68.4 | 28.1 | 138.3 | 98.0 |
| P23:J24 | 56.3 | 15.9 | 138.1 | 97.7 |
| P24:J24 | 56.3 | 15.9 | 138.1 | 97.7 |
| P24:J28 | 42.4 | 0.0 | 137.6 | 95.2 |
| P25:J28 | 42.4 | 0.0 | 137.6 | 95.2 |
| P25:N1 | 16.7 | 16.7 | 112.6 | 112.6 |
| P0:J1 | 0.0 | 0.0 | 40.6 | 40.6 |
| P0:J2 | 27.5 | 2.1 | 63.0 | 37.6 |
| P1:J2 | 27.5 | 2.1 | 63.0 | 37.6 |
| P1:J3 | 27.5 | 2.1 | 63.0 | 37.6 |

Elapsed time: 12 s.

### 9.3.1 GOVERNING EQUATIONS FOR UNSTEADY (OR TRANSIENT) FLOW

Hydraulic transient flow is also known as unsteady fluid flow. During a transient analysis, the fluid and system boundaries can be either elastic or inelastic:

*Elastic theory* describes unsteady flow of a compressible liquid in an elastic system (e.g., where pipes can expand and contract). This work uses the MOC to solve virtually any hydraulic transient problems.

*Rigid-column theory* describes unsteady flow of an incompressible liquid in a rigid system. It is only applicable to slower transient phenomena. Both branches of transient theory stem from the same governing equations. Transient analysis results that are not comparable with actual system measurements are generally caused by inappropriate system data (especially boundary conditions) and inappropriate assumptions.[6,9,10]

### 9.3.2 COMPARISON OF PRESENT RESEARCH RESULTS WITH OTHER EXPERT'S RESEARCH

Comparison of present research results with other experts' research results shows similarity and advantages:

Detailed conclusions drawn on the basis of experiments and calculations for the pipeline with a local leak are similar to the results observed by Kodura and Weinerowska[42,66–73] (Fig. 9.2).

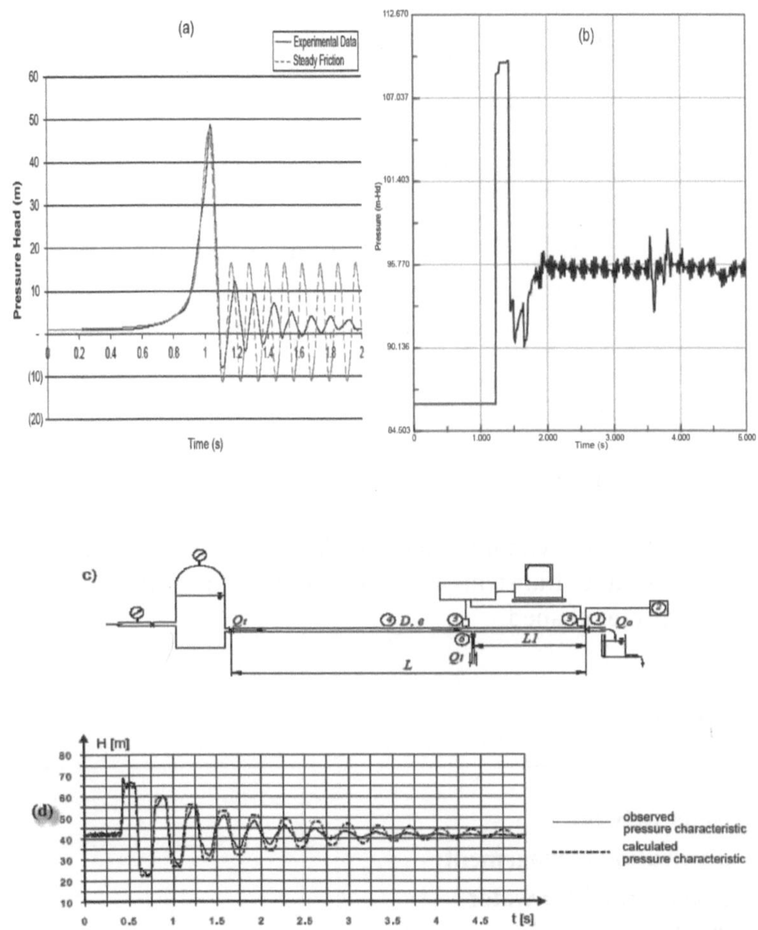


**FIGURE 9.1** (a) Pressure head histories for a single pipe system, using steady and unsteady friction.[2] (b) Existent water pipeline. (c) Pipeline with local leak. (d) Example of the measured and calculated pressure characteristics for the pipeline with local leak.

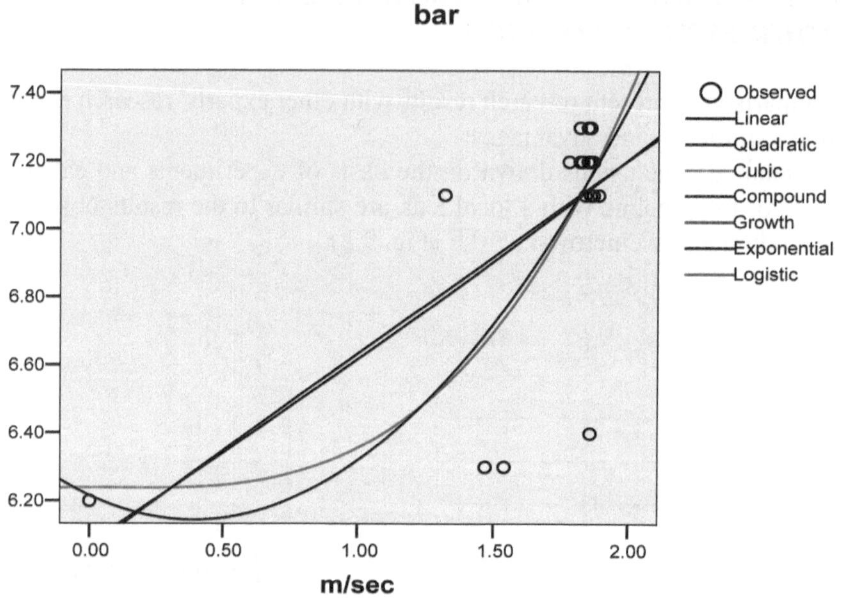


**FIGURE 9.2** "Model Summary and Parameter Estimates" for disinfection water networks and transmission lines.

## 9.4 CONCLUSION

In the computational technique presented, relationship between different variables and fluid transient as a "Dynamic" operating is presented. Results are also explained.

## KEYWORDS

- **water network**
- **distribution management**
- **unaccounted-for water**
- **district meter areas**

## REFERENCES


1. Streeter, V. L.; Wylie, E. B. *Fluid Mechanics;* McGraw-Hill Ltd.: USA, 1979, pp 492–505.
2. Leon, A. S. An Efficient Second-Order Accurate Shock-Capturing Scheme for Modeling One and Two-Phase Water Hammer Flows. Ph.D. Thesis, March 2007, 4–44.
3. Adams, T. M.; Abdel-Khalik, S. I.; Jeter, S. M.; Qureshi, Z. H. An Experimental Investigation of Single-Phase Forced Convection in Microchannels. *Int. J. Heat Mass Transf.* **1998,** *41,* 851–857.
4. Peng, X. F.; Peterson, G. P. Convective Heat Transfer and Flow Friction for Water Flow in Microchannel Structure. *Int. J. Heat Mass Transf.* **1996,** *36,* 2599–2608.
5. Mala, G.; Li, D.; Dale, J. D. Heat Transfer and Fluid Flow in Microchannels. *J. Heat Transf.* **1997,** *40,* 3079–3088.
6. Xu, B.; Ooi, K. T.; Mavriplis, C.; Zaghloul, M. E. Viscous Dissipation Effects for Liquid Flow in Microchannels. *Micorsystems.* 2002, 53–57.
7. Pickford, J. *Analysis of Surge;* Macmillian: London, 1969; pp 153–156.
8. American Society of Civil Engineers, Task Committee on Engineering Practice in the Design of Pipelines. *Pipeline Design for Water and Wastewater;* American Society of Civil Engineers: New York, 1975; pp 128.
9. Fedorov, A. G.; Viskanta, R. Three-dimensional Conjugate Heat Transfer into Microchannel Heat Sink for Electronic Packaging. *Int. J. Heat Mass Transf.* **2000,** *43,* 399–415.
10. Tuckerman, D. B. Heat Transfer Microstructures for Integrated Circuits. Ph.D. Thesis, Stanford University, 1984.
11. Harms, T. M.; Kazmierczak, M. J.; Cerner, F. M.; Holke, A.; Henderson, H. T.; Pilchowski, H. T.; Baker, K. In *Experimental Investigation of Heat Transfer and Pressure Drop through Deep Micro channels in a (100) Silicon Substrate,* Proceedings of the ASME, Heat Transfer Division, HTD 351, 1997, 347–357.
12. Holland, F. A.; Bragg, R. *Fluid Flow for Chemical Engineers;* Edward Arnold Publishers: London, 1995; pp 1–3.
13. Lee, T. S.; Pejovic, S. Air Influence on Similarity of Hydraulic Transients and Vibrations. *J. Fluids Eng.* **1996,** *118* (4), 706–709.
14. Li, J.; McCorquodale, A. Modeling Mixed Flow in Storm Sewers. *J. Hydraul. Eng.* **1999,** *125* (11), 1170–1180.
15. Minnaert, M. On Musical Air Bubbles and the Sounds of Running Water. *Phil. Mag.* **1933,** *16* (7), 235–248.
16. Moeng, C. H.; McWilliams, J. C.; Rotunno, R.; Sullivan, P. P.; Weil, J. Investigating 2D Modeling of Atmospheric Convection in the PBL. *J. Atm. Sci.* **2004,** *61,* 889–903.
17. Tuckerman, D. B.; Pease, R. F. W. High Performance Heat Dinking for VLSI. *IEEE Electron Device Lett.* **1981,** *DEL-2,* 126–129.
18. Nagiyev, F. B.; Khabeev, N. S. Bubble Dynamics of Binary Solutions. *High Temp.* **1988,** *27* (3), 528–533.
19. Shvarts, D.; Oron, D.; Kartoon, D.; Rikanati, A.; Sadot, O. Scaling Laws of Nonlinear Rayleigh-Taylor and Richtmyer-Meshkov Instabilities in Two and Three Dimensions. *C. R. Acad. Sci. Series IV.* **2000,** *719,* 719–726.

20. Cabot, W. H.; Cook, A. W.; Miller, P. L.; Laney, D. E.; Miller, M. C.; Childs, H. R. Large Eddy Simulation of Rayleigh-Taylor Instability. *Phys. Fluids.* **2005,** *17,* 91–106.
21. Cabot, W. *Phys. Fluids.* **2006,** 94–550.
22. Goncharov, V. N. Analytical Model of Nonlinear, Single-Mode, Classical Rayleigh-Taylor Instability at Arbitrary Atwood Numbers. *Phys. Rev. Lett.* **2002,** *88* (13), 10–15.
23. Ramaprabhu, P.; Andrews, M. J. Experimental Investigation of Rayleigh-Taylor Mixing at Small Atwood Numbers. *J. Fluids Mech.* **2004,** *502,* 233–271.
24. Clark, T. T. A Numerical Study of the Statistics of a Two-Dimensional Rayleigh-Taylor Mixing Layer. *Phys. Fluids.* **2003,** *15,* **2003,** 2413.
25. Cook, A. W.; Cabot, W.; Miller, P. L. The Mixing Transition in Rayleigh-Taylor Instability. *J. Fluids Mech.* **2004,** *511,* 333–362.
26. Waddell, J. T.; Niederhaus, C. E.; Jacobs, J. W. Experimental Study of Rayleigh-Taylor Instability: Low Atwood Number Liquid Systems with Single-Mode Initial Perturbations. *Phys. Fluids.* **2001,** *13,* 1263–1273.
27. Weber, S. V.; Dimonte, G.; Marinak, M. M. *Arbitrary Lagrange-Eulerian Code Simulations of Turbulent Rayleigh-Taylor Instability in Two and Three Dimensions;* Laser and Particle Beams: 21, 2003, pp 455.
28. Dimonte, G.; Youngs, D.; Dimits, A.; Weber, S.; Marinak, M. A Comparative Study of the Rayleigh-Taylor Instability using High-Resolution Three-Dimensional Numerical Simulations: The Alpha Group Collaboration. *Phys. Fluids.* **2004,** *16,* 1668.
29. Young, Y. N.; Tufo, H.; Dubey, A.; Rosner, R. On the Miscible Rayleigh-Taylor Instability: Two and Three Dimensions. *J. Fluids Mech.* **2001,** *447,* 377–408.
30. George, E.; Glimm, J. Self-Similarity of Rayleigh-Taylor Mixing Rates. *Phys. Fluids.* **2005,** *17,* 054101.
31. Oron. D.; Arazi, L.; Kartoon, D.; Rikanati, A.; Alon, U.; Shvarts, D. Dimensionality Dependence of the Rayleigh-Taylor and Richtmyer-Meshkov Instability Late-Time Scaling Laws. *Phys. Plasmas.* **2001,** *8,* 2883.
32. Nigmatulin, R. I.; Nagiyev, F. B.; Khabeev, N. S. Effective Heat Transfer Coefficients of the Bubbles in the Liquid Radial Pulse. Mater. Second-Union. Conf. Heat Mass Transfer, Heat massoob-men in the biphasic. with. Minsk, 1980, Vol. 5, 111–115.
33. Nagiyev, F. B. Damping of the Oscillations of Bubbles Boiling Binary Solutions. Mater. VIII Resp. Conf. Mathematics and Mechanics. Baku, October 26–29, 1988, 177–178.
34. Nagyiev, F. B.; Kadyrov, B. A. Small Oscillations of the Bubbles in a Binary Mixture in the Acoustic Field. Math. AN Az.SSR Ser. *Physicotech. Mate. Sci.* **1986,** 1, 23–26.
35. Nagiyev, F. B. Dynamics, Heat and Mass Transfer of Vapor-Gas Bubbles in a Two-Component Liquid. Turkey-Azerbaijan petrol semin., Ankara, Turkey, 1993, 32–40.
36. Nagiyev, F. B. The Method of Creation Effective Coolness Liquids, Third Baku international Congress. Baku, Azerbaijan Republic, 1995, 19–22.
37. Nagiyev, F. B. The Linear Theory of Disturbances in Binary Liquids Bubble Solution. *Dep. In VINITI.* **1986,** *405* (86), 76–79.
38. Nagiyev, F. B. Structure of Stationary Shock Waves in Boiling Binary Solutions. *Fluid Dyn.* **1989,** *1,* 81–87.
39. Lord Rayleigh, O. M. F. R. S. On the Pressure Developed in a Liquid during the Collapse of a Spherical Cavity. Philos. *Mag. Ser. 6.* **1917,** *34* (200), 94–98.

40. Perry. R. H.; Green, D. W.; Maloney, J. O. In *Perry's Chemical Engineers Handbook,* 7th Ed.; Perry, R. H., Green, D. W., Eds.; McGraw-Hill Professional Publishing: New York, 1997, pp 1–61.
41. Nigmatulin, R. I. *Dynamics of Multiphase Media;* Izdatel'stvo Nauka: Moscow, 1987; pp 12–14.
42. Kodura, A.; Weinerowska, K. The Influence of the Local Pipeline Leak on Water Hammer Properties, Materials of the II Polish Congress of Environmental Engineering, Lublin, 2005, 125–133.
43. Kane, J.; Arnett, D.; Remington, B. A.; Glendinning, S. G.; Baz'an, G. Two-Dimensional versus Three-Dimensional Supernova Hydrodynamic Instability Growth. *Astrophys. J.* **2000,** *528* (2), 528–989.
44. Quick, R. S. Comparison and Limitations of Various Water hammer Theories. *J. Hyd. Div.* **1933,** 43–45.
45. Jaeger, C. *Fluid Transients in Hydro-Electric Engineering Practice;* Blackie and Son Ltd.: Glasgow, 1977; pp 87–88.
46. Jaime, S. A. Generalized Water Hammer Algorithm for Piping Systems with Unsteady Friction. E-theses, 2005.
47. Fok, A.; Ashamalla, A.; Aldworth, G. Considerations in Optimizing Air Chamber for Pumping Plants, Symposium on Fluid Transients and Acoustics in the Power Industry, San Francisco, CA, 1978, 112–114.
48. Fok, A. *Design Charts for Surge Tanks on Pump Discharge Lines, BHRA 3rd Int. Conference on Pressure Surges*; Bedford: England, 1980; pp 23–34.
49. Fok, A. *Water Hammer and its Protection in Pumping Systems, Hydro Technical Conference;* CSCE: Edmonton, 1982; pp 45–55.
50. Fok, A. A Contribution to the Analysis of Energy Losses in Transient Pipe Flow. Ph.D. Thesis, University of Ottawa, 1987, 176–182.
51. Brunone, B.; Karney, B. W.; Mecarelli, M.; Ferrante, M. Velocity Profiles and Unsteady Pipe Friction in Transient Flow. *J. Water Res. Plan. Manage. ASCE.* **2000,** *126* (4), 236–244.
52. Koelle, E.; Luvizotto Jr. E.; Andrade, J. P. G. In *Personality Investigation of Hydraulic Networks using MOC–Method of Characteristics;* Proceedings of the 7th International Conference on Pressure Surges and Fluid Transients, Harrogate Durham, UK, 1–8, 1996.
53. Filion, Y.; Karney, B. W. In *A Numerical Exploration of Transient Decay Mechanisms in Water Distribution Systems,* Proceedings of the ASCE Environmental Water Resources Institute Conference, American Society of Civil Engineers, Roanoke, Virginia, 30, 2002.
54. Hamam, M. A.; McCorquodale, J. A. Transient Conditions in the Transition from Gravity to Surcharged Sewer Flow. *Canadian J. Civil Eng.* **1982,** *9* (2), 65–98.
55. Savic, D. A.; Walters, G. A. *Genetic Algorithms Techniques for Calibrating Network Models;* Report No. 95/12, Centre for Systems and Control Engineering, 137–146, 1995.
56. Walski, T. M.; Lutes, T. L. Hydraulic Transients Cause Low-Pressure Problems. *J. Am. Water Works Assoc.* **1994,** *75* (2), 58.
57. Chaudhry, M. H. *Applied Hydraulic Transients;* Van Nostrand Reinhold Co.: New York, 1979; pp 1322–1324.

58. Parmakian, J. *Water Hammer Analysis;* Dover Publications, Inc.: New York, 1963; pp 51–58.
59. Farooqui, T. A. *Evaluation of Effects of Water Reuse, on Water and Energy Efficiency of an Urban Development Area, using an Urban Metabolic Framework;* Master of Integrated Water Management Student Project Report., International Water Centre, 2015.
60. Ferguson, B. C.; Frantzeskaki, N.; Brown, R. R. A Strategic Program for Transitioning to a Water Sensitive City. *Landsc. Urban Plan.* **2013,** *117,* 32–45.
61. Kenway, S.; Gregory, A.; McMahon, J. Urban Water Mass Balance Analysis. *J. Ind. Ecol.* **2011,** *15* (5), 693–706.
62. Renouf, M. A.; Kenway, S. J.; Serrao-Neumann, S.; Low Choy, D. *Urban Metabolism for Planning Water Sensitive Cities. Concept for an Urban Water Metabolism Evaluation Framework;* Milestone Report,: Cooperative Research Centre for Water Sensitive Cities: Melbourne, 2015.
63. Serrao-Neumann, S.; Schuch, G.; Kenway, S. J.; Low Choy, D. *Comparative Assessment of the Statutory and Non-Statutory Planning Systems: South East Queensland, Metropolitan Melbourne and Metropolitan Perth;* Cooperative Research Centre for Water Sensitive Cities: Melbourne, 2113.
64. Andrews, S.; Traynor, P. *Guidelines for Assuring Quality of Food and Water Microbiological Culture Media.* August 2004.
65. Andrew D. E.; Rice, E. W.; Clesceri, L. S. Standard Methods for the Examination of Water and Wastewater. Part 9000, 2012.
66. Instructions Procedure Rural Water and Wastewater Quality Assurance Test Results of KhorasanRazavi, 2011.
67. Abbassi, B.; Al Baz, I. *Integrated Wastewater Management: A Review. Efficient Management of Wastewater–its Treatment and Reuse in Water Scarce Countries;* Springer Publishing Co.: Berlin, Heidelberg, 2008; pp 29–40.
68. Andreasen, P. In *Chemical Stabilization. Sludge into Biosolids – Processing, Disposal and Utilization.* Spinosa, L., Vesilind, P. A., Eds.; IWA Publishing: UK, 2001, pp 392.
69. Sambidge, N. E. W. Sewage Sludge: Stabilization and Disinfection. In *Handbooks of UK Wastewater Practice;* Haigh, M. D. F., Ed.; The Chartered Institution of Water and Environmental Management: London, 1996; pp 110.
70. Halalsheh, M.; Wendland, C. Integrated Anaerobic-Aerobic Treatment of Concentrated Sewage. In *Efficient Management of Wastewater – Its Treatment and Reuse in Water Scarce Countries;* Baz, I. A., Otterpohl, R., Wendland, C., Eds.; Springer Publishing Co.: Berlin, Heidelberg, 2008; pp 177–186.
71. ISWA. *Handling, Treatment and Disposal of Sludge in Europe;* Situation Report 1. ISWA Working Group on Sewage Sludge and Water Works: Copenhagen, 1995.
72. Matthews, P. Agricultural and Other Land Uses. In *Sludge into Biosolids – Processing, Disposal and Utilization;* Spinosa, 1., Vesilind, A. P., Eds.; IWA Publishing: UK, 2001; pp 43–74.
73. *IWK Sustainability Report 2012–2013;* Indah Water Konsortium Company: Kuala Lumpur, Malaysia.

## CHAPTER 10

# APPLICATION OF RENEKER'S MATHEMATICAL MODEL TO OPTIMIZE ELECTROSPINNING PROCESS

SH. MAGHSOODLOU and A. K. HAGHI*

*University of Guilan, P.O. Box 3756, Rasht, Iran*

**Corresponding author. E-mail: akhaghi@yahoo.com*

## CONTENTS

## ABSTRACT

To optimize electrospinning process, B-spline collocation methods and a new ODEs solver based on B-spline quasi-interpolation are developed. The problem consists of nonlinear ordinary differential equations. To solve the system of ODE, a 3D simulation was obtained by applying quartic B-spline collocation method. To achieve this, Reneker's model (i.e., bead-spring model) was applied and the governing equations were numerically simulated by new ODEs solver without using perturbation equations in *x* and *y* directions. Most likely, this technique can represent the results of the random perturbation. The results show that it could be possible to build mathematically a real-time simulation.

## 10.1 INTRODUCTION

The physical reasoning problem, like electrospinning phenomenon, usually requires a representational apparatus that can deal with the vast amount of physical knowledge that is used in reasoning tasks.[1] Mathematical and theoretical modeling and simulating procedure will permit to offer an in-depth prediction of electrospun fiber properties and morphology. Utilizing a model to express the effect of electrospinning parameters will assist researchers in making an easy and systematic way of presenting the influence of variables and by means of that, the process can be controlled.[2] Electrospinning, as most scientific problems, is inherently of nonlinearity, so it does not have an analytical solution and should be solved using other methods. The B-spline collocation methods and B-spline quasi-interpolation have been applied to solve many problems like this. In this chapter, cubic and quartic B-spline collocation method and cubic B-spline quasi-interpolation were used for solving this problem.[3,4] The numerical simulation of the electrospun nanofibers process was observed from momentum equation. The Reneker model, which has been chosen in this study, focuses on the whipping jet part and assumes the jet as a slender viscous object. The main novelty in this research assignment will be the application of this mathematical model for electrospinning simulation, which numerically solves the equations by B-spline quasi-interpolation ODEs solver. Additionally, perturbation function was assumed in the initial equations without using perturbation equations in *x* and *y* directions. Also, a minimum applicable time was considered for running simulation program.

## 10.2 EQUATIONS OF MATHEMATICAL MODEL

For modeling the fiber motion, bead-spring model (Reneker's model) was chosen, which describes the fiber as a chain that consisted of beads connected by springs with a viscoelastic model.[5–7] The momentum equation for the motion of the *i*-th bead can be expressed as eq 10.1. The meaning of "sign" in eq 10.1 is as follows in eq 10.2. Also, in this model, a single perturbation is added by inserting an initial bead of *i* by eq 10.3. In this formula L can be defined as eq 10.4.

$$m\frac{d^2 r_i}{dt^2}=\sum_{\substack{j=1\\ j\neq i}}^{N}\frac{e^2}{R_{ij}^2}(r_i-r_j)-e\frac{V_0}{h}\hat{k}+\frac{\pi a_{ui}^2(\overline{\sigma}_{ui}+G\ln(l_{ui}))}{l_{ui}}(r_{i+1}-r_i)-\frac{\pi a_{bi}^2(\overline{\sigma}_{bi}+G\ln(l_{bi}))}{l_{bi}}(r_i-r_{i-1})-\frac{\alpha\pi a_{a\upsilon}^2 k_i}{\sqrt{(x_i^2+y_i^2)}}sign[ix_i+jy_i] \tag{10.1}$$

$$sign(function)=\begin{cases}1 & x\succ 0\\ 0 & x=0\\ -1 & x\prec 0\end{cases} \tag{10.2}$$

$$\begin{bmatrix}x_i\\ y_i\end{bmatrix}=10^{-3}L\begin{bmatrix}\sin(\omega t)\\ \cos(\omega t)\end{bmatrix} \tag{10.3}$$

$$L=\left(\frac{4e^2}{\pi d_0^2 G}\right)^{\frac{1}{2}} \tag{10.4}$$

But in this study to reach the random perturbation, these equations cannot be considered for the simulation proposes alone. Nevertheless, in the calculation, the air drag force and gravity force are neglected. Considering the fact that both space and time are perturbations dependent. Therefore the development of whipping instability can be expected. To achieve a higher accuracy in simulation, a new solver, B-spline quasi-interpolation, by error $O(h^5)$was developed to replace common solver (Runge–Kutta),

## 10.3 NEW SOLVER: B-SPLINE QUASI-INTERPOLATION ODES SOLVER

Consider partition $\pi = \{x_0, x_1, \ldots, x_n\}$ of interval $[a,b]$ and suppose that $X_n = (x_j)^{n+d}_{j=-d}$ subject to $x_{-d} = x_{-d+1} = \ldots = x_{-1} = a,\ x_n = x_{n+1} = \ldots = x_{n+d} = b$.

According to recurrence relation of B-spline,[8] the $j$-th B-spline of degree $d$ for the knot sequence $X_n$ is denoted by $B_{Xn,j,d}$ or $B_j$ and can be defined as eqs 10.5 and 10.6. With these notations, the support of $B_{Xn,j,d}$ is Supp($B_{Xn,j,d}$) = $[x_{j-d-1}, x_j]$ and the sets = $\{B_1, B_2, \ldots, B_{n+d}\}$ form a basis over the region $a \leq x \leq b$. The univariate B-spline quasi-interpolants can be defined as operators of the form in eq 10.7.[9] For $f \in C^{d+1}(I)$ we have, $\|f - Q_d f\|_\infty = O(h^4)$. Let $f_i = f(x_j)$, $j = 0, 1, \ldots, n$, the coefficient functional for cubic QI ($d = 3$) is[9,10] in eq 10.8.

$$B_{X_{n,j,d}}(r) = \omega_{X_{n,j,d}} B_{X_{n,j,d-1}}(r) + \left(1 - \omega_{X_{n,j+1,d}}\right) B_{X_{n,j+1,d-1}}(r) \tag{10.5}$$

$$\omega_{X_{n,j,d}}(r) = \frac{r - x_j}{x_{j+d-1} - x_j}, \quad B_{X_{n,j,0}}(r) = \begin{cases} 1, & x_j \leq r < x_{j+1} \\ 0, & \rightarrow otherwise \end{cases} \tag{10.6}$$

$$Q_d f = \sum_{j=1}^{n+d} \mu_j(f) B_j \tag{10.7}$$

$$\begin{cases} \mu_1(f) = f_0 \\ \mu_2(f) = \dfrac{1}{18}(7f_0 + 18f_1 - 9f_2 + 2f_3) \\ \mu_j(f) = \dfrac{1}{6}(-f_{j-3} + 8f_{j-2} - f_{j-1}), \quad 3 \leq j \leq n+1 \\ \mu_{n+2}(f) = \dfrac{1}{18}(2f_{n-3} - 9f_{n-2} + 18f_{n-1} + 7f_n) \\ \mu_{n+3}(f) = f_n \end{cases} \tag{10.8}$$

Now, let $f$ be a given function of $x$, $y$ where $y$ is a function of $x$ and $y'$ is the first derivative with respect to $x$. Consider the first order of ordinary differential equation as follows as eq 10.9. Where the initial condition $y_0$ is a given number and the answer of $y$ is unique. Using cubic QI as an approximation of $F(x)$: $= f(x,y(x))$ in eq 10.9 and integrating in the interval $[x_i, x_{i+1}]$, $i = 0, 1, \ldots, n-1$, we have eq 10.10.

$$y'(x) = f(x, y), \quad y(0) = y_0 \tag{10.9}$$

$$\int_{x_i}^{x_{i+1}} y'(x)dx = \int_{x_i}^{x_{i+1}} \sum_{j=1}^{n+3} \mu_j(F) B_j(x) dx \tag{10.10}$$

And from B-spline properties we have eq 10.11, where $y_i = y(x_i)$. From eq 10 the following ODEs solver will be achieved, where $F_i = F(x_i)$, $i = 0, 1, \ldots, n-1$. The nonlinear system eq. 12 is formed of $n$ nonlinear equations in $n$ unknowns $y_1, y_2, \ldots, y_n$, that can be solved using trust-region-dogleg method.[11,12] Note that since we have eq 10.13, then from eq 10.7, the truncation error of this method will be $O(h^5)$.

$$y_{i+1} - y_i = \sum_{k=1}^{4} \mu_{i+k}(F) \int_{x_i}^{x_{i+1}} B_{i+k}(x)dx \tag{10.11}$$

$$\begin{cases} y_1 = y_0 + h\left(\frac{3}{8}F_0 + \frac{19}{24}F_1 - \frac{5}{24}F_2 + \frac{1}{24}F_3\right) \\ y_2 = y_1 + h\left(-\frac{7}{144}F_0 + \frac{41}{72}F_1 + \frac{1}{2}F_2 - \frac{1}{72}F_3 - \frac{1}{144}F_4\right) \\ y_{i+1} = y_i + h\left(-\frac{1}{144}F_{i-2} - \frac{1}{48}F_{i-1} + \frac{19}{36}F_i + \frac{19}{36}F_{i+1} - \frac{1}{48}F_{i+2} - \frac{1}{144}F_{i+3}\right) \quad i = 2,3,\ldots,n-3 \\ y_{n-1} = y_{n-2} + h\left(-\frac{1}{144}F_{n-4} - \frac{1}{72}F_{n-3} + \frac{1}{2}F_{n-2} + \frac{41}{72}F_{n-1} - \frac{7}{144}F_n\right) \\ y_n = y_{n-1} + h\left(\frac{1}{24}F_{n-3} - \frac{5}{24}F_{n-2} + \frac{19}{24}F_{n-1} + \frac{3}{8}F_n\right) \end{cases} \tag{10.12}$$

$$\left\| \int_{x_i}^{x_{i+1}} (F(x) - Q_3F(x))dx \right\| \le h\|F - Q_3F\| \tag{10.13}$$

## 10.4 SIMULATION ANALYSIS

According to the mathematical model described above, the time evolution of the jet whipping instability was determined by the following procedure: At $t = 0$, the initial whipping jet includes three beads, beads 1, 2, and 3. The distance was set to be small, say, $H/10{,}000$. Other initial conditions, including the stresses $\sigma_{i-1,i}$ and $\sigma_{i,i+1}$ and the initial velocity of bead $i$, $dRi/dt$, are set to be zero. For a given time, $t$, eq 10.14 was solved numerically, using B-spline quasi-interpolation algorithm. All the variables related to bead $i$, including the stresses $\sigma_{i-1,i}$ and $\sigma_{i,i+1}$, the position $r_i$, the length of the jet segment $l_{i-1,i}$, were obtained simultaneously. The new values of all the variables at time $t + \Delta t$ were calculated numerically. We denoted the last bead, pulled out of the spinneret by $i = N$. When the distance between this bead and the spinneret became long enough, say $H/5000$, a new bead $i = N + 1$ was inserted at a small distance, $H/10{,}000$, from the previous bead. Now, the positions of all beads can be traced and the path of the jet during the time can be obtained. As the jet passes through the collector, the calculation is stopped. In Table 10.1, a comparison between the present model and other works is presented. Due to the application of cubic B-spline quasi-interpolation solver for the simulation, the number of beads studied could be more, whereas the simulation time decreased significantly.

**TABLE 10.1** A Comparison between the Present Study with the Other Works.

| Researcher | Equations for Investigating Model | Method | Time | Year | Reference |
|---|---|---|---|---|---|
| Karra | Dimensionless continuity momentum | Lattice Boltzmann Method | $N = 100$<br>$\bar{t} = 0.183$<br>$\bar{t} = 0.179$ | 2007 | [6] |
| Thompson et al. | Governing quasi-one-dimensional continuity momentum and charge conservation | Lagrangian parameter | $t = 0.02$ s | 2007 | [13] |
| Zeng | Dimensionless continuity momentum | Fourth-order Runge–Kutta | $N = 100$<br>$\bar{t} = 4.471$ | 2009 | [5] |
| Lauricella | Three-dimensional continuity momentum | First-order accurate Euler | Final time: $10^{-1}$ s<br>Time step: $10^{-8}$ s | 2015 | [14] |
| | | Second-order accurate Heun (Runge–Kutta) | Final time: 2 s<br>Time step: $10^{-7}$ s | | |
| | | Fourth-order accurate Runge–Kutta | Final time: 0.5 s<br>Time step: $10^{-8}$ s | | |
| Present study | Three-dimensional continuity momentum | cubic B-spline quasi-interpolation | $N = 150$<br>$t = 0–10^{-9}$ s<br>Time step: $10^{-14}$ s | – | – |

The jet segment length increases as time develops. It demonstrates that the jet is stretched as it moves downward from the initial position to the collector. The results of the simulation of the electrospinning process can be seen in Figure 10.1.

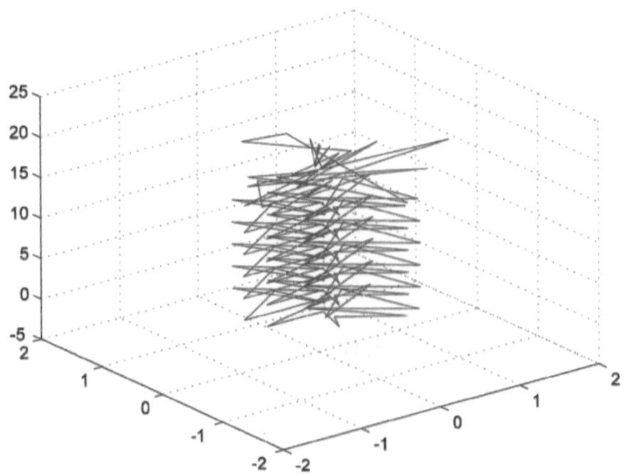


**FIGURE 10.1** The path calculated for $N = 150$ at times ranging from 0 to $10^{-9}$ s, with step time period $10^{-14}$.

Figure 10.1 illustrates the results of the model data output throughout the calculation of the jet path. As described earlier, new beads were inserted into the model when the distance between the last bead and nozzle (located at 20 cm in this case) exceeded the set value. As the beads progress downward, the perturbation added to the x and y coordinates began to grow as it fully developed into the bending instability, at which point the loops continued to grow outward as the jet moves down. It showed that the longitudinal stress caused by the external electric field acting on the charge carried by the jet stabilized the straight jet for some distance. Then a lateral perturbation grew in response to the repulsive forces between adjacent elements of the charge carried by the jet. The motion of a segment of the jet grew rapidly into an electrically driven bending instability.

## 10.5 CONCLUSION

In this chapter, Reneker's mathematical model was used to optimize electrospinning process. Meanwhile, the quartic B-spline collocation method is used to find more accurate value of $\sigma = f''(0)$. The cubic and quartic B-spline collocation methods and a new ODEs solver based on B-spline quasi-interpolation are developed and used to solve the problem. All of these methods led to a system of nonlinear equations where the unknowns are obtained by using trust-region-dogleg method. This model was employed to describe the dynamic behavior of the electrospun jet in instability part without using leading perturbation equations. The results of the bending instability phenomenon along with simulation of the model can be achieved.

## KEYWORDS

- **b-spline approximation method**
- **electrospinning process**
- **Reneker's model**
- **electrospinning simulation improvement**

## REFERENCES

1. Bhaskar, R.; Nigam, A. Qualitative Physics Using Dimensional Analysis. *Artif. Intell.* **1990,** *45* (1), 73–111.
2. Greenfeld, I.; Arinstein, A.; Fezzaa, K.; Rafailovich, M. H.; Zussman, E. *Phys. Rev.* **2011,** *84* (4), E041806.
3. Chawla, T. C.; Leaf, G.; Chen, W. A Collocation Method Using B-Splines for One-Dimensional Heat or Mass-Transfer-Controlled Moving Boundary Problems. *Nucl. Eng. Des.* **1975,** *35* (2), 163–180.
4. Saka, B.; Dag, I. A Collocation Method for the Numerical Solution of the RLW Equation Using Cubic B-spline Basis. *Arabian J. Sci. Eng.* **2005,** *30* (1), 39–50.
5. Zeng, Y.; Pei, Z.; Wang, X.; Chen, S. In *Numerical Simulation of Whipping Process in Electrospinning*, Proceedings of the 8th International Conference on Applied Computer and Applied Computational Science, Shanghai, China, 2009, pp 309–317.
6. Karra, S. Modeling Electrospinning Process and a Numerical Scheme Using Lattice Boltzmann Method to Simulate Viscoelastic Fluid Flows. MS. Thesis, Texas A&M University, May 2007.
7. Dasri, T. Mathematical Models of Bead-Spring Jets during Electrospinning for Fabrication of Nanofibers. *Walailak. J. Sci. Technol.* **2012,** *9* (4), 287–296.

8. Howarth, L. On the Solution of the Laminar Boundary Layer Equations. *Math. Phys. Eng. Sci.* **1938,** *164,* 547–579.
9. Aminikhah, H.; Jamalian, A. An Analytical Approximation for Boundary Layer Flow Convection Heat and Mass Transfer over a Flat Plate. *J. Math. Comput. Sci.* **2012,** *5* (4), 241–257.
10. Duan, Q.; Djidjeli, K.; Price, W. G.; Twizell, E. H. Weighted Rational Cubic Spline Interpolation and its Application. *J. Comput. Appl. Math.* **2000,** *117* (2), 121–135.
11. Prenter, P. M. *Splines and Variational Methods;* Dover Publications, Inc.: New York, 2008; pp 321.
12. Byrd, R. H.; Schnabel, R. B.; Shultz, G. A. A Trust Region Algorithm for Nonlinearly Constrained Optimization. *SIAM J. Numer. Anal.* **1987,** *24* (5), 1152–1170.
13. Thompson, C. J.; Chase, G. G.; Yarin, A. L.; Reneker, D. H. Effects of Parameters on Nanofiber Diameter Determined from Electrospinning Model. *Polymer.* **2007,** *48* (23), 6913–6922.
14. Lauricella, M.; Pontrelli, G.; Coluzza, I.; Pisignano, D.; Succi, S. JETSPIN: A Specific-Purpose Open-Source Software for Simulations of Nanofiber Electrospinning. *Comput. Phys. Commun.* **2015,** *197,* 227–238.

## CHAPTER 11

# MIXED-MODE DELAMINATION IN LAYERED ISOTROPIC AND LAMINATED COMPOSITE BEAM STRUCTURES

CHRISTOPHER MARTIN HARVEY*, PAUL CUNNINGHAM, and SIMON WANG

*Department of Aeronautical and Automotive Engineering, Loughborough University, Loughborough, Leicestershire, LE11 3TU, UK*

*Corresponding author. E-mail: c.m.harvey@lboro.ac.uk*

## CONTENTS

## ABSTRACT

Completely analytical theories are presented for calculating the total energy release rate (ERR) in a mixed-mode delamination in layered isotropic and laminated composite straight beam structures and for partitioning it into opening mode I and shearing mode II components. The theories are developed within the contexts of both the Euler and Timoshenko beam theories. The theories are extensively verified against numerical simulations using the finite element method. The developed theories provide a valuable means for the design of such beam structures against delamination.

## 11.1 INTRODUCTION

Layered isotropic and laminated composite straight beams are commonly used in many different engineering structures, such as aircraft, buildings, bridges, etc. Delamination is a major concern in these applications, for example, a commonly used method to repair or strengthen a concrete beam in civil engineering is to bond either a metal plate or a carbon-fiber-reinforced laminate onto it. The fracture toughness against debonding at the interface is a crucial design parameter. In general, debonding is a mixed-mode fracture, that is, it consists of both mode I opening and mode II shearing. The toughness depends on the proportions of these two individual fracture modes. Therefore, it is imperative to partition the total energy release rate (ERR) of a mixed-mode fracture into its mode I and II components which govern the fracture toughness or the fracture propagation criterion.

Some of the earliest analytical work on the topic of one-dimensional fracture—that is, fracture which propagates in one direction with mode I and mode II components only—was reported by Williams,[1] who made some significant contributions to the understanding for isotropic double cantilever beams (DCBs). A semi-analytical partition theory was given by Schapery and Davidson,[2] which was also for isotropic DCBs and based on Euler beam theory. They were not able to give Williams's[1] pair of pure modes and claimed that Euler beam theory does not provide quite enough information to obtain a decomposition of ERR into opening and shearing mode components. They therefore used the finite element method (FEM) to solve the two-dimensional continuum problem around the crack tip in order to partition the ERR. Suo and Hutchinson[3–5] used a similar approach to Schapery and Davidson,[2] but instead of using the FEM, they used integral equation methods to obtain a two-dimensional linear elasticity solution for the crack tip region. The resulting partition theory is analytical except for one parameter, which is determined numerically. Schapery and Davidson's[2] and Suo and Hutchinson's[3–5] partition theories generally give different partitions to William's[1] partition theory. Zou et al.[6] derived a completely analytical partition theory for isotropic DCBs based on Timoshenko beam theory. Bruno and Greco[7] obtained the same partition but for Euler beams instead of Timoshenko beams. Luo and Tong[8] derived the same partition theory as Bruno and Greco,[7] also for Euler beams, but by a different method. None of the work by Zou et al.,[6] Bruno and Greco,[7] and Luo and Tong[8] is in agreement with Williams's,[1] Schapery and Davidson's,[2] or Suo and Hutchinson's[3] partition theories.

Recently, based on a fundamental physical understanding and a complete mechanical representation of the problem, a powerful mathematical methodology has been created by the authors to partition the total ERR. Several challenging fracture problems have been solved analytically. The research results have been reported in a series of publications.[9–18] The authors' mixed-mode partition theory based on classical laminate theory has been shown[11,18] to agree very well with the test data obtained from a series of experimental studies by different research groups.[19–23] Also, the authors' latest work [24,25] shows that authors' mixed-mode partition theory based on first-order shear-deformable laminate theory plays a key role in the development of a local mixed-mode partition theory between two dissimilar elastic materials.

The major aim of this study is to extend the previous work[9–18] to develop analytical theories to calculate the total ERR and its mode I and II partitions for a mixed-mode delamination in layered isotropic and laminated composite straight beams under various loading conditions and boundary conditions. The work provides a valuable means for the design of such beam structures against delamination.

The structure of this chapter is as follows: Sections 11.2 and 11.3 develop analytical theories for layered isotropic beams; Section 11.4 develops analytical theories for laminated composite beams; numerical tests are presented in Section 11.5; and further discussions and conclusions are made in Section 11.6.

**Nomenclature**

| | |
|---|---|
| $a$ | length of fracture |
| $A_1$, $A_2$, $A$ | extensional stiffness of upper, lower and intact beams |
| $b$ | beam width |
| $B_1$, $B_2$, $B$ | coupling stiffness of upper, lower and intact beams |
| $D_1$, $D_2$, $D$ | bending stiffness of upper, lower and intact beams |
| $D_{op}$, $D_{sh}$ | crack tip relative opening and shearing displacements |
| $E$ | Young's modulus |
| $F_{nB}$, $F_{sB}$ | crack tip normal and shear forces |
| $F_{nB\theta 1}$ | crack tip opening force due to mode $\varphi_{\theta_1}$ |
| $F_{nBP}$ | crack tip opening force due to shearing |
| $G$, $G_I$, $F_{II}$ | total, mode I and mode II ERRs |
| $G_P$ | mode I ERR due to shearing |
| $G_{xz}$ | shear modulus |
| $G_{\theta_i}$, $G_{\beta_i}$ | ERRs due to mode $\varphi_{\theta_i}$ and mode $\varphi_{\beta_i}$ |
| $h_1$, $h_2$, $h$ | thicknesses of upper, lower and intact beams |
| $H_1$, $H_2$, $H$ | out-of-plane shearing stiffness of upper, lower and intact beams |
| $I_1$, $I_2$, $I$ | second moments of area of upper, lower and intact beams |
| $L_1$, $L_2$ | length of left and right intact parts of beam |
| $M_1$, $M_2$ | bending loads acting on upper and lower beams |
| $M_{1B_i}$, $M_{2B_i}$, $M_{B_i}$ | bending moments on upper, lower and intact beams at crack tip *i* |
| $N_1$, $N_2$ | axial loads acting on upper and lower beams |
| $N_{1B_i}$, $N_{2B_i}$, $N_{B_i}$ | axial forces on upper, lower and intact beams at crack tip *i* |
| $P_c$ | point contact force |
| $P_1$, $P_2$ | shear loads acting on upper and lower beams |
| $P_{1B_i}$, $P_{2B_i}$, $P_{B_i}$ | crack tip shear forces on upper, lower and intact beams at crack tip *i* |
| $u_1$, $u_2$, $u_3$, $u_4$ | axial displacements of the upper, lower, left and right beams |
| $w_1$, $w_2$, $w_3$, $w_4$ | deflections of the upper, lower, left and right beams |

| | |
|---|---|
| $x_{P_1}, x_{P_2}$ | distance from left crack tip to loading location on upper and lower beams |
| $\alpha_{\varphi_i}$ | mixed-mode partition coefficient for mode $\varphi_i$ |
| $\beta_i, \beta_i'$ | the two pure mode II "crack tip modes" from the *i*-th set |
| $\beta_F, \beta_F'$ | the two pure mode II "F modes" |
| $\gamma$ | thickness ratio $h_2/h_1$ |
| $\Delta G_{\varphi_i\varphi_j}$ | ERR interaction between modes $\varphi_i$ and $\varphi_j$ |
| $\theta_i, \theta_i'$ | the two pure mode I "crack tip modes" from the *i*-th set |
| $\theta_F, \theta_F'$ | the two pure mode I "F modes" |
| $\kappa^2$ | shear correction factor |
| $\varphi_{\theta_i}, \varphi_{\beta_i}$ | mode vectors for the *i*-th mode I and the *i*-th mode II |
| $\psi_1, \psi_2, \psi_3, \psi_4$ | rotations of normals to mid-surface of the upper, lower, left and right beams |

## 11.2 CLAMPED-CLAMPED ISOTROPIC BEAMS

### 11.2.1 GOVERNING EQUATIONS

A general clamped-clamped beam with a fracture is shown in Figure 11.1. Contact between the upper and lower beams is not treated initially. With reference to Figure 11.1b and using the constitutive relation from Timoshenko beam theory for isotropic materials, the following are easily derived:

$$\psi_{1,2} = \frac{1}{EI_{1,2}}\left( M_{1,2B_1}x - \frac{P_{1,2B_1}x^2}{2} - M_{1,2}\left\langle x - x_{P_{1,2}}\right\rangle + \frac{P_{1,2}}{2}\left\langle x - x_{P_{1,2}}\right\rangle^2 \right) + \psi_{1,2B_1}, \quad (11.1)$$

$$\begin{aligned} w_{1,2} = \frac{1}{EI_{1,2}}\left( \frac{M_{1,2B_1}x^2}{2} - \frac{P_{1,2B_1}x^3}{6} - \frac{M_{1,2}}{2}\left\langle x - x_{P_{1,2}}\right\rangle^2 + \frac{P_{1,2}}{6}\left\langle x - x_{P_{1,2}}\right\rangle^3 \right) \\ + \psi_{1,2B_1}x + w_{1,2B_1} + \frac{1}{bh_{1,2}G_{xz}\kappa^2}\left( P_{1,2B_1}x - P_{1,2}\left\langle x - x_{P_{1,2}}\right\rangle \right), \end{aligned} \quad (11.2)$$

$$\psi_3 = \frac{1}{EI}\left( \frac{R_A x^2}{2} - M_A x \right) \quad \text{and} \quad w_3 = \frac{1}{EI}\left( \frac{R_A x^3}{6} - \frac{M_A x^2}{2} \right) + \frac{P_{B_1}x}{bhG_{xz}\kappa^2}, \quad (11.3)$$

$$\psi_4 = \frac{1}{EI}\left( M_{B_2}x - M_{B_2}L_2 - \frac{P_{B_2}x^2}{2} + \frac{P_{B_2}L_2^{\,2}}{2} \right), \tag{11.4}$$

$$w_4 = \frac{1}{EI}\left( \frac{M_{B_2}x^2}{2} - M_{B_2}L_2x + \frac{M_{B_2}L_2^{\,2}}{2} - \frac{P_{B_2}x^3}{6} + \frac{P_{B_2}L_2^{\,2}x}{2} - \frac{P_{B_2}L_2^{\,3}}{3} \right) + \frac{P_{B_2}x}{bhG_{xz}\kappa^2}. \tag{11.5}$$

Subscripts 1 and 2 refer to the upper and lower beams, respectively. Subscripts 3 and 4 refer to the left- and right-hand intact laminates, respectively. The left crack tip is at location $B_1$. The right crack tip is at location $B_2$. In these equations, the origin of $x$ is at $B_1$ and to the right for beams 1 and 2; for beams 3 and 4 it is at the respective left-hand sides and to the right (as shown in Fig. 11.1a); $w$ is the upward deflection; the rotations, $dw/dx$ and $\psi$ are anticlockwise. The angle brackets are Macaulay brackets, denoting the ramp function. The axial displacements of the beams are

$$u_{1,2} = \left( N_{1,2B_1}x - N_{1,2}\left\langle x - x_{P_{1,2}} \right\rangle \right) / \left( bEh_{1,2} \right) + u_{1,2B_1}, \tag{11.6}$$

$$u_3 = N_{B_1}x/(bEh), \; u_4 = \left( N_{B_2}x - N_{B_2}L_2 \right)/(bEh). \tag{11.7}$$

Equilibrium can be used to describe all the forces in Figure 11.1b in terms of the six left crack tip forces $M_{1,2B_1}$, $N_{1,2B_1}$, and $P_{1,2B_1}$, and the applied loads $M_{1,2}$, $N_{1,2}$ and $P_{1,2}$.

$$R_A = -P_{B_1}, \;\; M_A = -M_{B_1} - P_{B_1}L_1, \;\; N_A = N_{B_1}, \tag{11.8}$$

$$R_C = P_{B_2}, \; M_C = P_{B_2}L_2 - M_{B_2}, \; N_C = N_{B_2}, \tag{11.9}$$

$$M_{B_{1,2}} = M_{1B_{1,2}} + M_{2B_{1,2}} + \left( h_1N_{2B_{1,2}} - h_2N_{1B_{1,2}} \right)/2, \tag{11.10}$$

$$N_{B_{1,2}} = N_{1B_{1,2}} + N_{2B_{1,2}}, \; P_{B_{1,2}} = P_{1B_{1,2}} + P_{2B_{1,2}}, \tag{11.11}$$

$$M_{1,2B_2} = M_{1,2B_1} - M_{1,2} - aP_{1,2B_1} + P_{1,2}\left( a - x_{P_{1,2}} \right), \tag{11.12}$$

$$N_{1,2B_2} = N_{1,2B_1} - N_{1,2}, \; P_{1,2B_2} = P_{1,2B_1} - P_{1,2}. \tag{11.13}$$

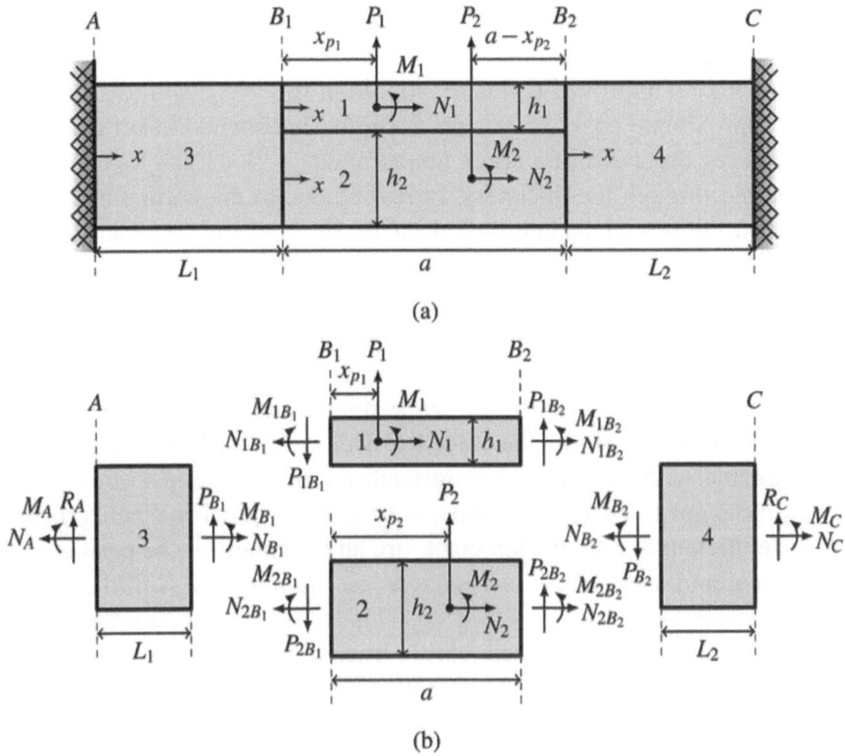


**FIGURE 11.1** A clamped-clamped beam with a fracture and its loading conditions. (a) General description. (b) Force diagram of each beam.

Equations 11.1–11.7 therefore contain 12 unknown quantities: the six left crack tip forces $M_{1,2B_1}$, $N_{1,2B_1}$, and $P_{1,2B_1}$, and the deflections, rotations and axial displacements at the left crack tip $w_{1,2B_1}$, $\psi_{1,2B_1}$, and $u_{1,2B_1}$. Twelve boundary conditions are therefore required to enforce continuity at the crack tip. There is continuity of deflection at the two crack tips.

$$w_{1B_1} = w_{2B_1} = \left(w_3\right)_{x=L_1}, \quad \left(w_1\right)_{x=a} = \left(w_2\right)_{x=a} = \left(w_4\right)_{x=0}. \tag{11.14}$$

There is also continuity of rotation at the two crack tips but this boundary condition requires special consideration. In cases where the shear modulus is finite, but the through-thickness shear effect is still small relative to bending, it is sufficient to use the following approximation:

$$\psi_{1B_1} = \psi_{2B_1} = (\psi_3)_{x=L_1}, \quad (\psi_1)_{x=a} = (\psi_2)_{x=a} = (\psi_4)_{x=0}. \tag{11.15}$$

In this work, an improved boundary condition is derived and used instead of the approximate eq 11.15. Using a single Timoshenko beam to model the intact region ahead of a crack tip, as in eqs 11.3–11.5, gives constant shear strain through the thickness. In reality, due to the normal and shear stress distributions on the interface ahead of the crack tip, the shear strain is not generally constant through the thickness. Note that the rotations $\psi_1$ and $\psi_2$ are continuous across the crack tips (although the mid-surface rotations $dw_1/dx$ and $dw_2/dx$ are not) but they are not equal. One way to represent the mechanics is to model the intact side of each crack tip using two Timoshenko beams with normal and shear stress distributions on the interface and continuous rotations across the crack tip. This would be both complex and incompatible with eqs 11.3–11.5. Instead, the method used in this work, which turns out to be very accurate, is to use a single Timoshenko beam to model the intact side of each crack tip, and account for the presence of the normal and shear stress distributions on the interface with discontinuous rotations $\psi_{1,2}$ across the crack tip. This is justified because the region affected by the crack tip in linear elastic fracture mechanics is small.

Consider the region around a general crack tip, as shown in Figure 11.2. The origin of the $\xi$ coordinate is at the crack tip B and toward the left. The deflection $w$ is upward and the rotations $dw/dx$ and $\psi$ are defined as positive in the anticlockwise direction. The interface stresses in the figure show only the sign convention rather than any representative distribution. Within the region affected by the crack tip, the through-thickness shearing equations from Timoshenko beam theory are

$$bh_{1,2}G_{xz}\kappa^2\left(dw_{1,2}/dx - \psi_{1,2}\right) = P_{1,2B} \mp b\int_0^{\xi} \sigma_n d\xi. \tag{11.16}$$

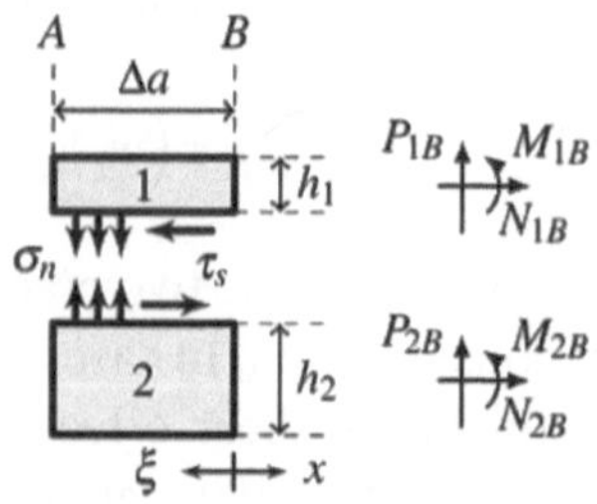


**FIGURE 11.2** Details of the crack influence region $\Delta a$ ahead of the left crack tip.

The mid-surface rotations $dw_1/dx$ and $dw_2/dx$ are discontinuous at the crack tip but for a rigid interface $(dw_1/dx)_{\xi=\delta a} = (dw_2/dx)_{\xi=\delta a}$, which are the midplane rotations of beams 1 and 2 at a very small distance $\delta a$ ahead of the crack tip B. Since the rotations $\psi_1$ and $\psi_2$ are continuous and $\delta a$ is very small; therefore, $(\psi_{1,2})_{\xi=\delta a} = \psi_{1,2B}$. Also, if the intact side of the crack tip is modeled with a single Timoshenko beam across the thickness, then $(dw_{1,2}/dx)_{\xi=\delta a} = (dw/dx)_B$. The rotation boundary conditions are therefore

$$\psi_{1,2B_1} = \left(dw_3/dx\right)_{x=L_1} - \left(P_{1,2B_1} \mp F_{nB_1}\right)/\left(bh_{1,2}G_{xz}\kappa^2\right), \quad (11.17)$$

$$\psi_{1,2B_2} = \left(dw_4/dx\right)_{x=0} - \left(P_{1,2B_2} \mp F_{nB_2}\right)/\left(bh_{1,2}G_{xz}\kappa^2\right), \quad (11.18)$$

where $F_{nB} = b\int_0^{\delta a} \sigma_n d\xi$, which is the crack tip opening force. Note that eqs 11.17 and 11.18 reduce to eq 11.15 for Euler beams, for which $bhG_{xz}\kappa^2 \to \infty$. The crack tip opening force $F_{nB}$ is known from the previously established mode partition theory for fracture in isotropic DCBs.[10,12] It is dependent on the mode partition and is a function of the crack tip forces. An expression for $F_{nB}$ is given in the following section.

There is also continuity of axial displacement at the two crack tips.

$$u_{1,2B_1} = \left(u_3\right)_{x=L_1} \mp h_{2,1}\left(\psi_3\right)_{x=L_1}/2, \; \left(u_{1,2}\right)_{x=a} = \left(u_4\right)_{x=0} \mp h_{2,1}\left(\psi_4\right)_{x=0}/2. \quad (11.19)$$

The system of 12 equations, given by eqs 11.14 and 11.17–11.19, can now be solved to give all the unknown quantities in terms of the six independent variables $M_{1,2}$, $N_{1,2}$ and $P_{1,2}$. The immediate results are not presented here for reasons of conciseness.

### *11.2.2 ERR PARTITION*

ERR can be determined by knowledge of the forces at the crack tip. If all these forces are known, then the ERRs can be found in the same way as for isotropic DCBs.[10,12] All of the crack tip forces are known from the solution of the above equations. The mode partition theory for fracture in isotropic DCBs is briefly summarized here. For full details, readers should refer to Refs. [10,12]. All expressions are for the left crack tip $B_1$. The "1" subscript is therefore dropped for convenience. It is then simple to find the ERR for the right crack tip $B_2$.

Initially the effect of through-thickness shear is ignored, as in the case of Euler beam theory where $bhG_{xz}\kappa^2 \to \infty$. It is accounted for later. The total ERR is

$$G = \frac{1}{2bE}\left[\frac{N_{1Be}^2 \gamma}{bh_1(1+\gamma)} + \frac{M_{1B}^2}{I_1} + \frac{M_{2B}^2}{I_2} - \frac{1}{I}\left(M_{1B} + M_{2B} - \frac{h_2 N_{1Be}}{2}\right)^2\right], \quad (11.20)$$

where $N_{1Be}$ is the effective axial force, defined as

$$N_{1Be} = N_{1B} - N_{2B}/\gamma. \quad (11.21)$$

Since the ERR is a function of three crack tip forces, three pure-mode vectors are required to find the partition. There are eight independent pure-mode vectors to choose from in the Euler case (two pure-mode-I modes and two pure-mode-II modes from each of the two sets). The order of the variables in the mode vectors is $\varphi = \{M_{1B}\ M_{2B}\ M_{1Be}\}^T$. It is convenient to select the following three pure modes to make the partition:

$$\phi_{\theta_1} = \{1 \quad \theta_1 \quad 0\}^{\mathrm{T}},\ \phi_{\beta_1} = \{1 \quad \beta_1 \quad 0\}^{\mathrm{T}},\ \phi_{\beta_2} = \{1 \quad 0 \quad \beta_2\}^{\mathrm{T}}, \quad (11.22)$$

where

$$\theta_1 = -\gamma^2,\ \beta_1 = \gamma^2(3+\gamma)/(1+3\gamma),\ \beta_2 = 2(3+\gamma)/\left[h_1(\gamma-1)\right]. \quad (11.23)$$

Mode $\varphi_{\theta_1}$ is a pure mode I mode, which has zero relative shearing displacement just behind the crack tip. Modes $\varphi_{\beta_1}$ and $\varphi_{\beta_2}$ are pure mode II modes, which have zero opening force ahead of the crack tip. Using these modes, the mode partition coefficients are

$$\begin{Bmatrix} \alpha_{\theta_1} \\ \alpha_{\beta_1} \\ \alpha_{\beta_2} \end{Bmatrix} = \begin{bmatrix} 1 & 1 & 0 \\ \theta_1 & \beta_1 & 0 \\ 0 & 0 & 1 \end{bmatrix}^{-1} \begin{Bmatrix} M_{1B} \\ M_{2B} \\ N_{1Be} \end{Bmatrix},\ \begin{Bmatrix} \alpha_{\theta_1} \\ \alpha_{\beta_1} \\ \alpha_{\beta_2} \end{Bmatrix} = \begin{bmatrix} 1 & 1 & 1 \\ \theta_1 & \beta_1 & 0 \\ 0 & 0 & \beta_2 \end{bmatrix}^{-1} \begin{Bmatrix} M_{1B} \\ M_{2B} \\ N_{1Be} \end{Bmatrix}, \quad (11.24)$$

for $\gamma = 1$ and $\gamma \neq 1$, respectively.

Within the context of Euler beam theory, there is interaction between the $\varphi_{\theta_1}$ mode and the $\varphi_{\beta_1}$ modes, denoted by $\Delta G_{\theta_1\beta_1}$. The mode I ERR is

$$G_I = \alpha_{\theta_1}^2 G_{\theta_1} + \alpha_{\theta_1}\alpha_{\beta_1}\Delta G_{\theta_1\beta_1} + \alpha_{\theta_1}\alpha_{\beta_2}\Delta G_{\theta_1\beta_2}, \tag{11.25}$$

where

$$G_{\theta_1} = 24\gamma/\left[Eb^2h_1^3(1+\gamma)\right], \Delta G_{\theta_1\beta_1} = 3(\gamma-1)G_{\theta_1}/(1+3\gamma),$$
$$\Delta G_{\theta_1\beta_2} = \gamma G_{\theta_1}/(1+\gamma). \tag{11.26}$$

To find the ERR using Timoshenko beam theory, the through-thickness shear effect must be considered. The interaction between the $\varphi_{\theta_1}$ mode and the $\varphi_{\beta_i}$ modes disappears, that is, $\Delta G_{\theta_1\beta_{1,2}} = 0$. There are nine independent pure modes within the context of Timoshenko beam theory (there is an extra pure mode I mode due to through-thickness shearing); however, the absence of interaction means that four of them from the second set coincide with four of them from the first set, giving five unique pure modes (three pure mode I and two pure mode II). There are also two additional contributions to the mode I ERR $G_I$: the ERR due to shearing, denoted by $G_P$; and the ERR due to interaction between the $\varphi_{\theta_1}$ mode crack tip opening force and the relative opening displacement due to shearing, and vice versa, denoted by $\Delta G_{\theta_1 P}$. The ERRs are therefore

$$G = \frac{1}{2bE}\left[\frac{N_{1Be}^2\gamma}{bh_1(1+\gamma)} + \frac{M_{1B}^2}{I_1} + \frac{M_{2B}^2}{I_2} - \frac{1}{I}\left(M_{1B} + M_{2B} - \frac{h_2 N_{1Be}}{2}\right)^2\right] + G_P + \alpha_{\theta_1}\Delta G_{\theta_1 P}, \tag{11.27}$$

$$G_I = \alpha_{\theta_1}^2 G_{\theta_1} + G_P + \alpha_{\theta_1}\Delta G_{\theta_1 P}, \tag{11.28}$$

where

$$G_P = \frac{1}{2bh_1 G_{xz}\kappa^2}\left(P_{1B}^2 + \frac{P_{2B}^2}{\gamma} - \frac{(P_{1B}+P_{2B})^2}{1+\gamma}\right),$$
$$\Delta G_{\theta_1 P} = \frac{4\sqrt{3}\,(\gamma P_{1B} - P_{2B})}{b^2h_1^2(1+\gamma)\sqrt{EG_{xz}\kappa^2}}. \tag{11.29}$$

The crack tip opening force $F_{nB}$, which is required to solve the governing equations from Timoshenko beam theory in Section 11.2.1, can now be found. Since the $\varphi_{\beta_1}$ and $\varphi_{\beta_2}$ modes are characterized by zero normal force ahead of the crack tip, only the $\varphi_{\theta_1}$ mode and the opening force due to shearing contribute to the crack tip opening force, giving

$$F_{nB} = \alpha_{\theta_1} F_{nB\theta_1} + F_{nBP}, \tag{11.30}$$

where $\alpha_{\theta_1}$ is known in terms of the crack tip forces from eq 11.24 and $F_{nB\theta_1}$ is the crack tip opening force in the $\varphi_{\theta_1}$ mode and $F_{nBP}$ is the crack tip opening force due to shearing. From Timoshenko beam theory, these quantities are

$$F_{nB\theta_1} = \left(48G_{xz}\gamma^2\kappa^2 / Eh_1^2(1+\gamma)^2\right)^{1/2}, \; F_{nBP} = \left(\gamma P_{1B} - P_{2B}\right)/(1+\gamma). \tag{11.31}$$

The governing equations in Section 11.2.1 can now be solved to find the crack tip forces and to obtain the ERR partition.

### 11.2.3 TWO SETS OF ORTHOGONAL PURE MODES

Since the crack tip forces in eq 11.24 cannot be set independently of the loads applied at $x_{P_1}$ and $x_{P_2}$, it is clearly not possible to obtain purely mode vectors $\varphi_{\beta_1}$, or $\varphi_{\beta_2}$ at a given crack tip. In this work these modes are referred to as "crack tip modes" because they relate crack tip quantities only. Some combinations of these modes, however, can give pure mode I or II fractures. These modes are now derived for the *left* crack tip for the special case when $bhG_{xz}\kappa^2 \to \infty$ (Euler beam theory) and $\gamma \neq 1$; $P_1$ and $P_2$ are applied at the same location, i.e., $x_{P_1} = x_{P_1} = x_P$ and $M_1 = M_2 = N_1 = N_2 = 0$. These fracture modes are now referred to as "F modes" and denoted by a subscript "*F*," because they relate the forces $P_1$ and $P_2$.

Because of Refs. [10,12], there are expected to be two sets of pure F modes, where the first set corresponds to zero relative shearing displacement just behind the crack tip (pure mode I) and zero opening force ahead of the crack tip (pure mode II) and the second set corresponds to zero relative opening displacement just behind the crack tip (pure mode II) and zero crack tip shearing force (pure mode I).

Mathematically, the relative shearing displacement at an infinitely small distance $\delta\alpha$ behind the crack tip, $(D_{sh})_{x=\delta\alpha}$ is expressed as

$$\begin{aligned}(D_{sh})_{x=\delta a} = (\bar{u}_1)_{x=\delta a} - (\bar{u}_2)_{x=\delta a} = (u_1)_{x=\delta a} + h_1(\psi_1)_{x=\delta a}/2 - (u_2)_{x=\delta a} \\ + h_2(\psi_2)_{x=\delta a}/2,\end{aligned} \tag{11.32}$$

where $\bar{u}$ represents the axial displacement at the interface. For the pure-mode-I mode from the first set, the relative shearing displacement at $x = \delta a$ is zero, that is $(D_{sh})_{x=\delta a} = 0$. Making the necessary substitutions and taking the limit as $\delta a \rightarrow 0$ gives

$$\begin{aligned}P_2/P_1 = \theta_F = -\left[\gamma^2\left(C_1\gamma^2 - C_2\gamma + C_3\right)\right]/\left(C_3\gamma^2 - C_2\gamma + C_1\right), \\ \phi_{\theta_F} = \{1 \quad \theta_F\}^T,\end{aligned} \tag{11.33}$$

where

$$\begin{aligned}C_1 = 2x_p\left(a - x_p\right)^2\left(a + L_1 + L_2\right)^3, C_2 = 2x_p\left(a - x_p\right)^2 \\ \left(C_7 + 3a^2\left(L_1 + L_2\right) - 2a^3\right),\end{aligned} \tag{11.34}$$

$$\begin{aligned}C_3 = 2a^3x_p^3 - ax_p^2\left(C_7 + 6a^2L_2 + 4a^3\right) + 2a^2x_p \\ \left(C_7 - 3aL_1L_2 + 3a^2L_2 + a^3\right) - aC_8,\end{aligned} \tag{11.35}$$

$$C_7 = \left(L_1 + L_2\right)^3 + 3a\left(L_1 + L_2\right)^2, \; C_8 = a^2L_1^2\left(L_1 + 3L_2 + 3a\right), \tag{11.36}$$

and $\varphi_F$ represents mode vector format $\{P_1 \;\; P_2\}^T$. The orthogonal condition to the zero relative shearing displacement condition is zero opening force ahead of the crack tip. The mode corresponding to this condition could be derived by applying orthogonality through eq 11.20; however, it is more convenient in this instance to simply enforce $F_{nB} = 0$. Noting that $F_{nB\theta_1}$ is infinite in Euler beam theory and that therefore the finite $F_{nBP}$ in eq 11.30 is negligible, the requirement for $F_{nB} = 0$ is simply $\alpha_{\theta_1} = 0$, which from eq 11.24 gives

$$\begin{aligned}P_2/P_1 = \beta_F = \gamma^2\left(C_1\gamma^3 + C_4\gamma^2 + C_5\gamma + C_6\right)/\left(C_6\gamma^3 + C_5\gamma^2 + C_4\gamma + C_1\right), \\ \phi_{\beta_F} = \{1 \quad \beta_F\}^T,\end{aligned} \tag{11.37}$$

where

$$C_4 = 2x_p\left(a - x_p\right)^2\left(2C_7 + 6a^2\left(L_1 + L_2\right) + 5a^3\right), \tag{11.38}$$

$$\begin{aligned} C_5 = {} & x_p^3\left(2C_7 + 6a^2\left(L_1 + L_2\right) + 14a^3\right) - ax_p^2\left(C_7 + 12a^2L_1 - 6a^2L_2 + 28a^3\right) \\ & - 2a^2x_p\left(2C_7 - 3a^2L_1 + 6a^2L_2 - 7a^3 - 9aL_1L_2\right) + 3aC_8, \end{aligned} \tag{11.39}$$

$$\begin{aligned} C_6 = {} & 6a^3x_p^3 - 3ax_p^2\left(C_7 + 6a^2L_2 + 4a^3\right) + 6a^2x_p \\ & \left(C_7 - 3aL_1L_2 + 3a^2L_2 + a^3\right) - 3aC_8. \end{aligned} \tag{11.40}$$

Now considering the second set of pure modes, the pure mode I mode is given by zero shear force at the crack tip. The shear force at the crack tip is

$$F_{sB} = \alpha_{\theta_1}F_{sB\theta_1} + \alpha_{\beta_1}F_{sB\beta_1} + \alpha_{\beta_2}F_{sB\beta_2}. \tag{11.41}$$

The mode partition coefficients are known from eq 11.24 and the crack tip shear forces relating to each mode vector are known from Refs. [10,12].

$$\begin{aligned} F_{sB\theta_1} = 6\gamma(1-\gamma)/\left[h_1(1+\gamma)^2\right], \; F_{sB\beta_1} = 6\gamma/\left[h_1(1+3\gamma)\right], \\ F_{sB\beta_2} = 2\gamma^2/\left[h_1\left(\gamma^2 - 1\right)\right]. \end{aligned} \tag{11.42}$$

Making these substitutions and setting $F_{sB} = 0$ gives

$$P_2/P_1 = \theta_F' = -1, \quad \phi_{\theta_F'} = \{1 \quad \theta_F'\}^T. \tag{11.43}$$

Finally, for the pure mode II mode from the second set, the relative opening displacement at $x = \delta a$ is zero $(D_{op})_{x=\delta a} = 0$.

$$\left(D_{op}\right)_{x=\delta a} = \left(w_1\right)_{x=\delta a} - \left(w_2\right)_{x=\delta a}. \tag{11.44}$$

Making the necessary substitutions and taking the limit as $\delta a \rightarrow 0$ gives

$$P_2/P_1 = \beta_F' = \gamma^3, \quad \phi_{\beta_F'} = \{1 \quad \beta_F'\}^T. \tag{11.45}$$

That $\theta_F'$ and $\beta_F'$, relating $P_2$ to $P_1$, are the same as $\theta_I'$ and $\beta_I'$, which relate $M_{2B}$ to $M_{1B}$,[10,12] should be no surprise. The axial forces $N_{1B}$ and $N_{2B}$, induced at the crack tip by $P_1$ and $P_2$, clearly have no effect on the opening displacement. Therefore the condition $(D_{op})_{x=\delta a} = 0$ is unaffected by $N_{1B}$ and $N_{2B}$, and $\beta'_F = \beta'_I$. Also, if $P_1$ and $P_2$ are equal and opposite $P_2/P_1 = -1$, then regardless of how beams 1 and 2 deflect, beams 3 and 4 do not deflect and the crack tip rotations are zero. Therefore $N_{2B}$ and $N_{2B}$ are both zero and the two crack tip bending moments are in the ratio of $\theta'_I$. If $P_1$ and $P_2$ are applied in different locations then this would not be observed because each load would have a different moment arm around the crack tip.

If the above procedure is repeated for Timoshenko beams, it is found that the $\varphi_{\theta'_F}$ mode changes to coincide with the $\varphi_{\theta_F}$ mode, which is different to that obtained from Euler beam theory due to the static indeterminacy of clamped-clamped beams (the relative shearing displacement is otherwise not affected by through-thickness shear). Similarly the $\varphi_{\beta'_F}$ mode coincides with the $\varphi_{\beta_F}$ mode, which is also different to that obtained from Euler beam theory. For cases when the through-thickness shear effect is not excessively large, eqs 11.33 and 11.37 are good approximations. The expressions for the Timoshenko "F modes" are not as simple as those for the Euler "F modes," so are not presented here in general form. They are, however, easily derived for specific cases with numerical—rather than algebraic—quantities.

### 11.2.4 CONTACTING FRACTURES

For some values of $P_1$ and $P_2$, the beams either side of the fracture will come into contact. This raises two questions: where is the point of first contact, and what happens after contact?

To find the point of first contact $x_c$ using Euler beam theory, two conditions must be satisfied: first, the relative opening displacement at this point must be zero; second, the relative opening displacement must be a minimum at $x = x_c$, which implies that it is the point of first contact.

$$\left(D_{op}\right)_{x=x_c} = 0,\ \left(\partial D_{op}/\partial x\right)_{x=x_c} = 0. \tag{11.46}$$

Solving eqs 11.46 simultaneously for $P_2/P_1$ and $x_c$, and ignoring the obvious and unavailing solutions for the crack tips, gives

$$P_2/P_1 = \gamma^3 = \beta_F, \tag{11.47}$$

for all values of $x$. This implies simultaneous contact everywhere along the fracture for this value of $P_2/P_1$.

If $P_2/P_1$ is increased beyond $\beta'_F$, the contact can either be at a point or distributed. Obviously the solution must not allow interpenetration between the upper and lower beams anywhere. In addition, since linear elastic mechanics is being used, there can only be one valid solution. Therefore point contact at $x = x_p$, which is a reasonable assumption, will be considered and shown to satisfy the requirements, thus demonstrating that it is the correct solution.

Say that two loads $P_{1c}$ and $P_{2c}$ are applied to the beam at $x = x_p$ and that they cause point contact at this same location. Call the point contact force $P_c$. It acts to prevent non-physical intersection. The net shear loads $P_1$ and $P_2$ acting on the beams are therefore

$$P_1 = P_{1c} + P_c, \quad = P_2 = P_{2c} + P_c. \tag{11.48}$$

Note that $P_1$ and $P_2$ in eq 11.48 are the same quantities that appear in all the equations thus far. The final condition that must be satisfied is

$$(D_{op})_{x=xp} = 0 \tag{11.49}$$

Solving eqs 11.48 and 11.49 for $P_1$, $P_2$ and $P_c$ gives

$$\begin{gathered} P_1 = (P_{1c} + P_{2c})/(1+\gamma^3), \; P_2 = \gamma^3 (P_{1c} + P_{2c})/(1+\gamma^3), \\ P_c = (P_{2c} - \gamma^3 P_{1c})/(1+\gamma^3). \end{gathered} \tag{11.50}$$

Substituting eq 11.50 into $D_{op} = w_1 - w_2$ reveals that $D_{op} = 0$ for all values of $x$. Therefore the requirements for physical contact behavior are satisfied by this solution and the ERR partition is then found in the usual way.

Using Timoshenko beam theory, at $P_2/P_1 = \beta_F = \beta'_F$ there is both zero opening force ahead of the crack tip and zero relative opening displacement just behind. Therefore crack tip running contact occurs at $P_2/P_1 = \beta_F$ and a pure mode II fracture is obtained. Since there is running contact, if the loading ratio $P_2/P_1$ is increased further then the crack tip remains closed as the contacting region grows.

## 11.3 SIMPLY SUPPORTED ISOTROPIC BEAMS

The theory presented in Section 11.2 is easily modified for simply supported isotropic beams. In this section, the modified theory is briefly summarized. For this new case, eqs 11.51–11.53 replace eqs 11.3–11.5 and 11.7.

$$\psi_3 = \frac{1}{EI_3}\left(\frac{R_A x^2}{2} - M_A x\right) + \psi_A,\ w_3 = \frac{1}{EI_3}\left(\frac{R_A x^3}{6} - \frac{M_A x^2}{2}\right) + \psi_A x + \frac{P_{B_1} x}{bhG_{xz}\kappa^2}, \tag{11.51}$$

$$\psi_4 = \frac{1}{EI_4}\left(M_{B_2} x - \frac{P_{B_2} x^2}{6}\right) + \psi_{B_2}, \tag{11.52}$$

$$w_4 = \frac{1}{EI_4}\left(\frac{M_{B_2} x^2}{2} - \frac{M_{B_2} L_2^2}{2} - \frac{P_{B_2} x^3}{6} + \frac{P_{B_2} L_2^3}{6}\right) + \psi_{B_2}(x - L_2) + \frac{P_{B_2}(x - L_2)}{bhG_{xz}\kappa^2}. \tag{11.53}$$

Since zero rotation is no longer enforced at the supports, two additional boundary conditions are required at these locations. For simple supports these are

$$M_A = 0 = -P_{B_1} L_1 - M_{B_1},\ \ M_C = 0 = P_{B_2} L_2 - M_{B_2}. \tag{11.54}$$

This new system of equations is now easily solved for the *left* crack tip for the special case when $bhG_x\kappa^2 \to \infty$ (Euler beam theory). All the unknown quantities are then known in terms of the independent variables $P_1$ and $P_2$. Again, the immediate results are not presented here for reasons of conciseness. If the analysis is carried out for when $\gamma \neq 1$, $P_1$ and $P_2$ are applied at the same location, i.e., $x_{P_1} = x_{P_2} = x_P$ and $M_1 = M_2 = N_1 = N_2 = 0$, the four pure mode relationships for both the left and right crack tips are

$$P_2/P_1 = \theta_F = \gamma^2\left(C_1\gamma^2 - C_1\gamma - C_2\right)/\left(C_2\gamma^2 + C_1\gamma - C_1\right), \tag{11.55}$$

$$P_2/P_1 = \beta_F = \gamma^2\left(C_1\gamma^3 + 2C_1\gamma^2 + C_3\gamma + C_4\right)/\left(C_4\gamma^3 + C_3\gamma^2 + 2C_1\gamma + C_1\right), \tag{11.56}$$

$$P_2/P_1 = \theta_F' - 1,\quad P_2/P_1 = \beta_F' = \gamma^3,\ P_2/P_1 = \beta_F' = \gamma^3, \tag{11.57}$$

where

$$C_1 = 2x_p\left(a - x_p\right)^2\left(a + L_1 + L_2\right),\ C_2 = C_5x_p^2 + ax_p\left(a^2 - 2C_5\right) + a^3L_1, \quad (11.58)$$

$$C_3 = 2x_p^3\left(a + L_1 + L_2\right) - C_5x_p^2 + ax_p\left(3a^2 - 4C_5\right) + 3a^3L_1, \quad (11.59)$$

$$C_4 = -3C_5x_p^2 + ax_p\left(6C_5 - 3a^2\right) - 3a^3L_1,\ C_5 = a\left(a + L_1 + L_2\right). \quad (11.60)$$

The contact behavior is found to be identical to the clamped-clamped case, i.e., contact at $P_2 / P_1 = \gamma^3$ for all values of $x$ between beams 1 and 2 and point contact at $x = x_p$ afterward.

If the above procedure is repeated for Timoshenko beams then as before, it is found that the $\varphi_{\theta'_F}$ mode coincides with the $\varphi_{\theta_F}$ mode; the $\varphi_{\beta'_F}$ mode coincides with the $\varphi_{\beta_F}$ mode; and the $\varphi_{\beta_F}$ and $\varphi_{\beta_F}$ modes are different to those obtained from Euler beam theory. However, for cases when the through-thickness shear effect is not excessively large, eqs 11.55 and 11.56 are good approximations. The expressions for the Timoshenko "F modes" are not as simple as those for the Euler "F modes," so are not presented here in general form. They are, however, easily derived for specific cases with numerical quantities.

## 11.4 CLAMPED-CLAMPED LAMINATED COMPOSITE BEAMS

### *11.4.1 GOVERNING EQUATIONS*

A general clamped-clamped laminated composite beam with a delamination now receives the same analysis. Contact between the upper and lower sub-laminates is not treated initially. The extensional, coupling, bending and shearing stiffness are denoted by *A*, *B*, *D*, and *H*, respectively. Note that these quantities take different values under the plane-strain assumption from those from under the plane-stress assumption, however, they make no difference to the following development. Subscripts 1 and 2 are used to indicate the upper and lower sub-laminates, respectively. No subscript is used for the intact part of the laminate $A_1$ is therefore the extensional stiffness of the upper sub-laminate and $A$ is the extensional stiffness of the intact laminate, etc. With reference to Figure 11.1 and using the constitutive relation from classical laminate theory,

$$\begin{Bmatrix} N_{1,2}(x)/b \\ -M_{1,2}(x)/b \\ P_{1,2}(x)/b \end{Bmatrix} = \begin{bmatrix} A_{1,2} & B_{1,2} & 0 \\ B_{1,2} & D_{1,2} & 0 \\ 0 & 0 & H_{1,2} \end{bmatrix} \begin{Bmatrix} du_{1,2}/dx \\ -d\,\psi_{1,2}/dx \\ dw_{1,2}/dx - \psi_{1,2} \end{Bmatrix}, \tag{11.61}$$

$$\begin{Bmatrix} N_{3,4}(x)/b \\ -M_{3,4}(x)/b \\ P_{3,4}(x)/b \end{Bmatrix} = \begin{bmatrix} A & B & 0 \\ B & D & 0 \\ 0 & 0 & H \end{bmatrix} \begin{Bmatrix} du_{3,4}/dx \\ -d\,\psi_{3,4}/dx \\ dw_{3,4}/dx - \psi_{3,4} \end{Bmatrix}, \tag{11.62}$$

where

$$N_{1,2}(x) = N_{1,2B_1} - N_{1,2}\left\langle x - x_{P_{1,2}} \right\rangle^0, \quad P_{1,2}(x) = P_{1,2B_1} - P_{1,2}\left\langle x - x_{P_{1,2}} \right\rangle^0, \tag{11.63}$$

$$M_{1,2}(x) = M_{1,2B_1} - P_{1,2B_1}x - M_{1,2}\left\langle x - x_{P_{1,2}} \right\rangle^0 + P_{1,2}\left\langle x - x_{P_{1,2}} \right\rangle, \tag{11.64}$$

$$N_{3,4}(x) = N_{B_{1,2}}, \quad P_{3,4}(x) = P_{B_{1,2}}, \tag{11.65}$$

$$M_3(x) = M_{B_1} - P_{B_1}(x - L_1), \quad M_4(x) = M_{B_2} - P_{B_2}x. \tag{11.66}$$

As shown in Figure 11.1a, the origin of $x$ in these equations depends on the beam in question: it is at location $B_1$ and to the right for beams 1 and 2; for beams 3 and 4 it is at the respective left-hand sides and to the right; Positive deflection, $w$ is always upward; the rotations $dw/dx$ and $\psi$ are positive in the anticlockwise direction. From eqs 11.61–11.66, the following are easily derived:

$$\psi_{1,2} = \frac{B_{1,2}\left(N_{1,2B_1}x - N_{1,2}\left\langle x - x_{P_{1,2}} \right\rangle\right)}{b\left(A_{1,2}D_{1,2} - B_{1,2}^2\right)} + \psi_{1,2B_1} + \frac{A_{1,2}\left(2M_{1,2B_1}x - 2M_{1,2}\left\langle x - x_{P_{1,2}} \right\rangle + P_{1,2}\left\langle x - x_{P_{1,2}} \right\rangle^2 - P_{1,2B_1}x^2\right)}{2b\left(A_{1,2}D_{1,2} - B_{1,2}^2\right)}, \tag{11.67}$$

$$w_{1,2} = \frac{B_{1,2}\left(N_{1,2B_1}x^2 - N_{1,2}\left\langle x - x_{P_{1,2}} \right\rangle^2\right)}{2b\left(A_{1,2}D_{1,2} - B_{1,2}^2\right)} + \psi_{1,2B_1}x + w_{1,2B_1} + \frac{P_{1,2B_1}x - P_{1,2}\left\langle x - x_{P_{1,2}} \right\rangle}{bH_{1,2}} + \frac{A_{1,2}\left(3M_{1,2B_1}x^2 - 3M_{1,2}\left\langle x - x_{P_{1,2}} \right\rangle^2 + P_{1,2}\left\langle x - x_{P_{1,2}} \right\rangle^3 - P_{1,2B_1}x^3\right)}{6b\left(A_{1,2}D_{1,2} - B_{1,2}^2\right)}, \tag{11.68}$$

$$\psi_3 = \frac{2BN_{B_1}x + 2AM_{B_1}x + 2AL_1P_{B_1}x - AP_{B_1}x^2}{2b\left(AD - B^2\right)}, \tag{11.69}$$

$$w_3 = \frac{3BN_{B_1}x^2 + 3AM_{B_1}x^2 + 3AL_1P_{B_1}x^2 - AP_{B_1}x^3}{6b\left(AD - B^2\right)} + \frac{P_{B_1}x}{bH}, \tag{11.70}$$

$$\psi_4 = \frac{\left(2BN_{B_2} + 2AM_{B_2}\right)\left(x - L_2\right) - AP_{B_2}\left(x^2 - L_2^2\right)}{2b\left(AD - B^2\right)}, \tag{11.71}$$

$$w_4 = \frac{\left(3BN_{B_2} + 3AM_{B_2}\right)\left(x^2 - 2L_2x + L_2^2\right) - AP_{B_2}\left(x^3 - 3L_2^2x + 2L_2^2\right)}{6b\left(AD - B^2\right)} + \frac{P_{B_2}\left(x - L_2\right)}{bH}, \tag{11.72}$$

$$u_{1,2} = \frac{D_{1,2}\left(N_{1,2B_1}x - N_{1,2}\left\langle x - x_{P_{1,2}}\right\rangle\right)}{b\left(A_{1,2}D_{1,2} - B_{1,2}^2\right)} + u_{1,2B_1}$$
$$+\frac{B_{1,2}\left(2M_{1,2B_1}x - 2M_{1,2}\left\langle x - x_{P_{1,2}}\right\rangle + P_{1,2}\left\langle x - x_{P_{1,2}}\right\rangle^2 - P_{1,2B_1}x^2\right)}{2b\left(A_{1,2}D_{1,2} - B_{1,2}^2\right)}, \tag{11.73}$$

$$u_3 = \frac{2DN_{B_1}x + 2BM_{B_1}x + 2BL_1P_{B_1}x - BP_{B_1}x^2}{2b\left(AD - B^2\right)}, \tag{11.74}$$

$$u_3 = \frac{2DN_{B_1}x + 2BM_{B_1}x + 2BL_1P_{B_1}x - BP_{B_1}x^2}{2b\left(AD - B^2\right)}, \tag{11.75}$$

As before, there are 12 unknown quantities: the six left crack tip forces $M_{1,2B_1}$, and $w_{1,2B_1}$, and the deflections, rotations and axial displacements at the left crack tip $w_{1,2B_1}$, $\psi_{1,2B_1}$, and $u_{1,2B_1}$. Equations 11.14 and 11.19 are still applicable. The continuity of rotation at the two crack tips is treated in the same way as in Section 11.2.1 and the following boundary conditions are obtained:

$$u_4 = \frac{2DN_{B_2}\left(x - L_2\right) + 2BM_{B_2}\left(x - L_2\right) - BP_{B_2}\left(x^2 - L_2^2\right)}{2b\left(AD - B^2\right)}. \tag{11.76}$$

$$\psi_{1,2B_1} = \left(dw_3/dx\right)_{x=L_1} - \left(P_{1,2B_1} \mp F_{nB_1}\right)/\left(bH_{1,2}\right), \tag{11.77}$$

Note that eqs 11.76 and 11.77 reduce to eq 11.15 for Euler beams, for which $bH \to \infty$. The crack tip opening force $F_{nB}$ is known from the previously established mode partition theory for one-dimensional fracture in laminated composite DCBs.[10,13] It is dependent on the mode partition and is a function of the crack tip forces. An expression for $F_{nB}$ is given in the following section.

The algebraic solution for the general case is extensive. The solution for the much simpler symmetric case with Euler beams is instead given for reasons of practicality. From symmetry we have

$$\psi_{1,2B_2} = \left(dw_4/dx\right)_{x=0} - \left(P_{1,2B_2} \mp F_{nB_2}\right)/\left(bH_{1,2}\right). \tag{11.78}$$

Symmetry provides two additional boundary conditions, which simplify the calculations. These are zero axial displacement and zero rotation at the mid-span.

$$\left(u_1\right)_{x=a/2} = \left(u_2\right)_{x=a/2}, \quad \left(\psi_1\right)_{x=a/2} = \left(\psi_2\right)_{x=a/2}. \tag{11.79}$$

The resulting crack tip forces are

$$\begin{aligned}
M_{1,2B_1} &= C_1 P_{1,2}\Big[2aA_1A_2L^2\left(h_1+h_2\right)^2 - 8A_{1,2}L^2\left(2D_{1,2}L - aD_{2,1} \pm ah_1B_{2,1} \pm ah_2B_{2,1}\right) \\
&+ 4A_{2,1}L^2\left(-4D_{1,2}L + 2aD_{2,1} \pm ah_1B_{1,2} \pm ah_2B_{1,2} - 2h_1B_{1,2}L - 2h_2B_{1,2}L\right) \\
&+ aAL\left(ah_1^2A_2 + ah_2^2A_1 \mp 4ah_{1,2}B_{2,1} \pm 2ah_{2,1}B_{1,2} \mp 4h_{2,1}B_{1,2}L - 8D_{1,2}L + 4aD_{2,1}\right) \\
&+ 4aBL\left(-aB_{1,2} + 2B_{1,2}L - 2aB_{2,1} + ah_1A_2 - ah_2A_1\right) + 4a^2DL\left(A_1 + A_2\right) \\
&+ 8L^2\left(2B_{1,2}^2L + 2B_1B_2L - aB_{2,1}^2 - aB_1B_2\right) - 2a^3B^2 + 2a^3AD\Big] \\
&- 2C_1LP_{2,1}\left(a+2L\right)\big[-4B_{1,2}L\left(B_1+B_2\right) + 4D_{1,2}L\left(A_1+A_2\right) \\
&+ aA\left(2D_{1,2} \pm h_{2,1}B_{1,2}\right) - 2aBB_{1,2} \pm 2A_{2,1}B_{1,2}L\left(h_1+h_2\right)\big],
\end{aligned} \tag{11.80}$$

$$\begin{aligned}
N_{1,2B_1} = \pm 2C_1L(a+2L)(P_1+P_2)\big[&2A_1A_2L\left(h_1+h_2\right) + ah_{2,1}AA_{1,2} \\
&+4L\left(A_2B_1 - A_1B_2\right) \pm 2a\left(AB_{1,2} - A_{1,2}B\right)\big],
\end{aligned} \tag{11.81}$$

where

$$\begin{aligned} C_1 = &\left[64A_1L^2\left(D_1 + D_2 - h_1B_2 - h_2B_2\right) + 64A_2L^2\left(D_1 + D_2 + h_1B_1 + h_2B_1\right)\right. \\ &+ 32aL\left(A_1D + AD_1 + A_2D + AD_2 - h_2A_1B + h_1A_2B + h_2AB_1 - h_1AB_2\right) \\ &+ 16A_1A_2L^2\left(h_1 + h_2\right)^2 + 8aAL\left(h_2^2A_1 + h_1^2A_2\right) - 64L^2\left(B_1 + B_2\right)^2 \\ &\left. -64aBL\left(B_1 + B_2\right) + 16a^2\left(AD - B^2\right)\right]^{-1}. \end{aligned} \quad (11.82)$$

## 11.4.2 ERR PARTITION

All of the crack tip forces are known from the solution of the above equations. The ERRs can therefore be found. The mode partition theory for one-dimensional fracture in laminated composite DCBs is briefly summarized here. For full details, readers should refer to Refs.[10,13]. All expressions are for the left crack tip $B_1$. The "1" subscript is therefore dropped for convenience. It is then simple to find the ERR for the right crack tip $B_2$.

Initially the effect of through-thickness shear is ignored, as in the case of Euler beam theory where $bH \to \infty$. It is accounted for later. The total ERR is

$$G = \frac{1}{2b^2}\left(\frac{M_{1B}^2}{D_1^*} + \frac{M_{2B}^2}{D_2^*} - \frac{M_B^2}{D^*} + \frac{N_{1B}^2}{A_1^*} + \frac{N_{2B}^2}{A_2^*} - \frac{N_B^2}{A^*} - \frac{2B_1M_{1B}N_{1B}}{B_1^*} - \frac{2B_2M_{2B}N_{2B}}{B_2^*} + \frac{2BM_BN_B}{B^*}\right), \quad (11.83)$$

where

$$A_i^* = A_i - B_i^2/D_i,\ B_i^* = B_i^2 - A_iD_i,\ D_i^* = D_i - B_i^2/A_i. \quad (11.84)$$

The range of subscript $i$ is 1–2, which again refers to the upper and lower sub-laminates, respectively. For the intact part of the laminate, no subscript is used. Other terms in eq 11.83 are

$$N_B = N_{1B} + N_{2B},\ M_B = M_{1B} + M_{2B} + \frac{1}{2}\left(h_1N_{2B} - h_2N_{1B}\right). \quad (11.85)$$

Since the ERR is a function of four crack tip forces, four pure modes are required to find the partition. There are 12 pure modes to choose from in the Euler case (six pure mode I and six pure mode II). The order of the variables in the mode vectors is $\phi = \{M_{1B}\ \ M_{2B}\ \ N_{1B}\ \ N_{2B}\}^T$. It is convenient to select the following four modes to make the partition:

$$\phi_{\theta_1} = \{1 \quad \theta_1 \quad 0 \quad 0\}^{\mathrm{T}}, \quad \phi_{\beta_1} = \{1 \quad \beta_1 \quad 0 \quad 0\}^{T}$$

$$\phi_{\beta_2} = \{1 \quad 0 \quad \beta_2 \quad 0\}^{T}, \quad \phi_{\beta_3} = \{1 \quad 0 \quad 0 \quad \beta_3\}^{T}, \tag{11.86}$$

where

$$\theta_1 = \left(B_2^2 - A_2 D_2\right)\left(B_1 + h_1 A_1/2\right) \Big/ \left[\left(B_1^2 - A_1 D_1\right)\left(B_2 - h_2 A_2/2\right)\right], \tag{11.87}$$

$$\beta_1 = -D_2^*\left(D_1^* + D_1^*\theta_1 - D^*\right) \Big/ \left[D_1^*\left(D_2^* + D_2^*\theta_1 - D^*\theta_1\right)\right], \tag{11.88}$$

$$= \frac{\theta_2\left[h_2/\left(2D^*\right) - B_1/B_1^* + B/B^*\right] + 1/D_1^* - 1/D^*}{\theta_2\left[Bh_2/B^* - 1/A_1^* + 1/A^* + h_2^2/\left(4D^*\right)\right] - h_2/\left(2D^*\right) + B_1/B_1^* - B/B^*}, \tag{11.89}$$

$$\beta_3 = \frac{\theta_3\left[h_1/\left(2D^*\right) - B/B^*\right] - 1/D_1^* + 1/D^*}{\theta_3\left[Bh_1/B^* + 1/A_2^* - 1/A^* - h_1^2/\left(4D^*\right)\right] - h_1/\left(2D^*\right) + B/B^*}, \tag{11.90}$$

$$\begin{Bmatrix} \alpha_{\theta_1} \\ \alpha_{\beta_1} \\ \alpha_{\beta_2} \\ \alpha_{\beta_3} \end{Bmatrix} = \begin{Bmatrix} \left(M_{2B}\beta_2 + N_{1B}\beta_1 - M_{1B}\beta_1\beta_2\right)/\beta_2\left(\theta_1 - \beta_1\right) + N_{2B}\beta_1/\left[\beta_3\left(\theta_1 - \beta_1\right)\right] \\ \left(M_{1B}\theta_1\beta_2 - M_{2B}\beta_2 - N_{1B}\theta_1\right)/\left[\beta_2\left(\theta_1 - \beta_1\right)\right] - N_{2B}\theta_1/\left[\beta_3\left(\theta_1 - \beta_1\right)\right] \\ N_{1B}/\beta_2 \\ N_{2B}/\beta_3 \end{Bmatrix}. \tag{11.91}$$

Within the context of Euler beam theory, which has interaction between the $\varphi_{\theta_1}$ mode and the $\varphi_{\beta_1}$ modes, the mode I ERR is

$$G_I = \alpha_{\theta_1}{}^2 G_{\theta_1} + \alpha_{\theta_1}\alpha_{\beta_1}\Delta G_{\theta_1\beta_1} + \alpha_{\theta_1}\alpha_{\beta_2}\Delta G_{\theta_1\beta_2} + \alpha_{\theta_1}\alpha_{\beta_3}\Delta G_{\theta_1\beta_3}, \tag{11.92}$$

where

$$G_{\theta_1} = \frac{1}{2b^2}\left[\frac{1}{D_1^*} + \frac{\theta_1^2}{D_2^*} - \frac{\left(1 + \theta_1\right)^2}{D^*}\right], \quad \Delta G_{\theta_1\beta_1} = \frac{F_{nB\theta_1}\,\delta a}{4b^2}\left(\frac{1}{D_1^*} - \frac{\beta_1}{D_2^*}\right), \tag{11.93}$$

$$\Delta G_{\theta_1\beta_2} = \frac{F_{nB\theta_1}\,\delta a}{4b^2}\left(\frac{1}{D_1^*} - \frac{B_1\beta_2}{B_1^*}\right), \quad \Delta G_{\theta_1\beta_3} = \frac{F_{nB\theta_1}\,\delta a}{4b^2}\left(\frac{1}{D_1^*} + \frac{B_2\beta_3}{B_2^*}\right), \tag{11.94}$$

$$F_{nB\theta_1}\,\delta a = \frac{2\left(D_1^* D^* \theta_1^2 + D_2^* D^* - D_1^* D_2^* - 2D_1^* D_2^* \theta_1 - D_1^* D_2^* \theta_1^2\right)}{D^*\left(D_2^* - D_1^* \theta_1\right)}. \quad (11.95)$$

There are 13 pure modes within the context of Timoshenko beam theory, however, the absence of interaction means that six of them from the first set coincide with six of them from the second set, giving seven unique pure modes (four pure mode I and three pure mode II). There are also two additional contributions to the mode I ERR $G_I$: the ERR due to shearing, denoted by $\varphi_{\theta_1}$; and the ERR due to interaction between the $\varphi_{\theta_1}$ mode crack tip opening force and the relative opening displacement due to shearing, and vice versa, denoted by $\Delta G_{\theta_1 P}$. The ERRs are therefore

$$G = \frac{1}{2b^2}\left(\frac{M_{1B}^2}{D_1^*} + \frac{M_{2B}^2}{D_2^*} - \frac{M_B^2}{D^*} + \frac{N_{1B}^2}{A_1^*} + \frac{N_{2B}^2}{A_2^*} - \frac{N_B^2}{A^*} - \frac{2B_1 M_{1B} N_{1B}}{B_1^*} - \frac{2B_2 M_{2B} N_{2B}}{B_2^*} + \frac{2BM_B N_B}{B^*}\right) + G_P + \alpha_{\theta_1}\Delta G_{\theta_1 P}, \quad (11.96)$$

$$G_I = \alpha_{\theta_1}^2 G_{\theta_1} + G_P + \alpha_{\theta_1}\Delta G_{\theta_1 P}, \quad (11.97)$$

where

$$G_P = \left(H_1 P_{2B} - H_2 P_{1B}\right)^2 / \left[2b^2 H_1 H_2 \left(H_1 + H_2\right)\right], \quad (11.98)$$

$$\Delta G_{\theta_1 P} = \frac{1}{b^2}\left(\frac{P_{1B}}{H_1} - \frac{P_{2B}}{H_2}\right)\left[\frac{H_1 H_2}{2(H_1 + H_2)}\left(\frac{1}{D_1^*} + \frac{\theta_1^2}{D_2^*} - \frac{(1+\theta_1)^2}{D^*}\right)\right]^{1/2}. \quad (11.99)$$

The crack tip opening force $F_{nB}$, which is required to solve the governing equations from Timoshenko beam theory in Section 11.4.1, can now be derived. Since the $\varphi_{\beta_1}$ modes are characterized by zero opening force ahead of the crack tip, the crack tip opening force $F_{nB}$ is given by

$$F_{nB} = \alpha_{\theta_1} F_{nB\theta_1} + F_{nBP}, \quad (11.100)$$

where $\alpha_{\theta_1}$ is known in terms of the crack tip forces from eq 11.91, $F_{nB\theta_1}$ is the crack tip opening force in mode $\varphi_{\theta_1}$ and $F_{nBP}$ is the crack tip opening force due to shearing. From Timoshenko beam theory, they are

$$F_{nB\theta_1} = \left[ \frac{H_1 H_2}{2(H_1 + H_2)} \left( \frac{1}{D_1^*} + \frac{\theta_1^2}{D_2^*} - \frac{(1+\theta_1)^2}{D^*} \right) \right]^{1/2}, \quad F_{nBP} = \frac{H_2 P_{1B} - H_1 P_{2B}}{H_1 + H_2}. \quad (11.101)$$

The governing equations in Section 11.4.1 can now be solved to find the crack tip forces and obtain the ERR partition.

### 11.4.3 TWO SETS OF ORTHOGONAL PURE MODES

For the symmetric case with Euler beams, for which the crack tip forces are given by eqs 11.80 and 11.81, the F modes arising from the displacement conditions (i.e., zero relative shearing when $P_2/P_1 = \theta_F$, and zero relative opening displacement when $P_2/P_1 = \beta_F'$) can be presented algebraically. By substituting the displacements and crack tip forces for this symmetric case into eqs 11.32 and 11.44 and equating them to zero, the following F modes are obtained:

$$P_2/P_1 = \theta_F = -B_2^*(2B_1 + h_1 A_1) / \left[ B_1^*(2B_2 - h_2 A_2) \right], \quad \phi_{\theta_F} = \{1 \quad \theta_F\}^T, \quad (11.102)$$

$$P_2/P_1 = \beta_F' = D_2^*/D_1^*, \quad \phi_{\beta_F'} = \{1 \quad \beta_F'\}^T. \quad (11.103)$$

The F modes arising from the zero crack tip opening force when condition is too extensive to be presented here algebraically. However, for specific cases, a numerical value for $\beta_F$ can be calculated by enforcing orthogonality with $\theta_F$. The ERR can be written as

$$G = \{P_1 \quad P_2\}[C]\{P_1 \quad P_2\}^T, \quad (11.104)$$

where $[C]$ is found by examining coefficients of $P_1$ and $P_2$ in eq 11.83 when eqs 11.80 and 11.81 have been substituted in. Therefore $\beta_F$ can be found by solving

$$0 = \{1 \quad \beta_F\}[C]\{1 \quad \theta_F\}^T. \quad (11.105)$$

Similarly $\theta_F'$ can be found by solving

$$0 = \{1 \quad \theta_F'\} [C] \{1 \quad \beta_F'\}^T, \quad (11.106)$$

which always gives

$$\theta'_F = -1. \tag{11.107}$$

If the above procedure is repeated for Timoshenko beams then as before, it is found that the $\varphi_{\theta'_F}$ mode changes to coincide with the $\varphi_{\theta_F}$ mode, which is different to that obtained from Euler beam theory. Similarly the $\varphi_{\beta'_F}$ mode coincides with the $\varphi_{\beta_F}$ mode, which is also different to that obtained from Euler beam theory. The expressions for the Timoshenko "F modes" are long and complex in their general form, so are not presented here. They are, however, easily derived for specific cases with numerical quantities. Furthermore, when the through-thickness shear effect is not excessively large, eqs 11.102 and 11.105 are good approximations.

### *11.4.4 CONTACTING FRACTURES*

To find the point of first contact $x_c$ using Euler beam theory, again the two conditions given by eq 11.46 must be satisfied. Solving these equations simultaneously for $P_2/P_1$ and $x_c$ and ignoring the obvious and unavailing solutions for the crack tips, gives

$$P_2/P_1 = D_2^*/D_1^* = \beta'_F, \tag{11.108}$$

for all values of $x$. This implies simultaneous contact everywhere along the fracture for this value of $P_2/P_1$.

If $P_2/P_1$ is increased beyond $\beta'_F$, the contact can either be at a point or distributed. In the same way as before for the isotropic case, point contact at $x = x_p$ is assumed, which is a reasonable assumption, and shown to satisfy the requirement that it prevents intersection between the upper and lower sub-laminates for all values of $x$.

Two loads $P_{1c}$ and $P_{2c}$ are applied to the beam at $x = x_p$ and they cause point contact at this same location. The point contact force $P_c$ acts to prevent non-physical interpenetration. The net shear loads $P_1$ and $P_2$ acting on the beams are given by eq 11.48. Equation 11.49 is the equation that must be satisfied to prevent intersection at $x = x_p$. Solving eqs 11.48 and 11.49 for $P_1$, $P_2$ and $P_c$ gives

$$P_1 = A_2 B_1^* \left(P_{1c} + P_{2c}\right) / \left(A_2 B_1^2 + A_1 B_2^2 - A_1 A_2 D_1 - A_1 A_2 D_2\right), \tag{11.109}$$

$$P_2 = A_1 B_2^* \left(P_{1c} + P_{2c}\right) / \left(A_2 B_1^2 + A_1 B_2^2 - A_1 A_2 D_1 - A_1 A_2 D_2\right), \tag{11.110}$$

$$P_c = \left[\left(A_1 A_2 D_2 - A_1 B_2^2\right) P_{1c} - \left(A_1 A_2 D_1 - A_2 B_1^2\right) P_{2c}\right] / \left(A_2 B_1^2 + A_1 B_2^2 - A_1 A_2 D_1 - A_1 A_2 D_2\right). \tag{11.111}$$

Substituting eqs 11.109–11.111 into $D_{op} = w_1 - w_2$ reveals that $D_{op} = 0$ for all values of $x$. Therefore the requirements for physical contact behavior are satisfied by this solution, demonstrating that it is the correct one.

Using Timoshenko beam theory, at $P_2/P_1 = \beta_F = \beta'_F$ there is both zero opening force beyond the crack tip and zero relative opening displacement just behind. Therefore crack tip running contact occurs at $P_2/P_1 = \beta_F$ and a pure mode II fracture is obtained. Since there is running contact, if the loading ratio $P_2/P_1$ is increased further then the crack tip remains closed as the contacting region grows.

## 11.5 NUMERICAL INVESTIGATIONS

To verify the theory, a finite element method (FEM) simulation capability was developed based on the Euler and Timoshenko beam theories and 2D elasticity. Normal and shear point interface springs with the very high stiffness of $10^{14}$ N/m were used to model perfectly bonded plies.[14,26–29] Through convergence studies this value was found to be large enough to approach the behavior of a rigid interface, but not so high as to introduce excessive numerical error. The ERR partition was calculated using the virtual crack closure technique in conjunction with these interface springs.[14,26–29] A contact algorithm was also implemented to deal with any possible contact in loading.

Two clamped-clamped beam cases were investigated. The first case is an asymmetric, isotropic one, the data for which is given in Table 11.1. The second case is a symmetric laminated composite one. It has a quasi-isotropic lay-up with 16 plies. There is a delamination between the fourth and fifth plies, which gives a thickness ratio of $\gamma = 3$. The data for this case is given in Table 11.2. The material properties are for a T300/976 graphite/epoxy ply.[28]

One set of simulations, which used linear Timoshenko beam elements, is compared against the Euler beam theory. Very large out-of-plane shear moduli $G_{xz} = G_{13} = G_{23} = 10^{16}$N/m were used to simulate Euler beam theory. As is the case for the spring stiffness, convergence studies were carried out and this value for $G_x$ was found to be large enough to approach the behavior of Euler beams, but not so high as to introduce excessive numerical error. Two layers of elements were used to represent the beams with one on either side of the fracture. The elements were distributed uniformly. To avoid shear locking, reduced integration was applied. Use of linear Timoshenko beams correctly enforces continuity along the interface ahead of the crack tip.

Another set of simulations, which was the same as the first set but which instead used the normal out-of-plane shear moduli (those given in Tables 11.1 and 11.2) and a shear correction factor of $\kappa^2 = 5/6$, is compared against the Timoshenko beam theory.

**TABLE 11.1** Data for Numerical Simulations of a Clamped-Clamped Isotropic Beam.

| | |
|---|---|
| Elastic modulus, $E$ | 70 GPa |
| Shear modulus, $G_{xz}$ | 26 GPa |
| Poisson's ratio, $v_{xz}$ | 0.35 |
| Beam thicknesses, $h_1$ and $h_2$ | 1 and 2 mm |
| Intact lengths of beam, $L_1$ and $L_2$ | 10 and 25 mm |
| Length of fracture, $b$ | 65 mm |
| Width of beam, $b$ | 10 mm |
| Loading location, $x_p$ | 20 mm |
| Euler pure modes $\theta_F$, $\beta_F$, $\theta'_F$, and $\beta'_F$ | −3.92, 2.81, −1, and 8 |
| Timoshenko pure modes $\theta_F$ and $\beta_F$ | −3.84 and 2.75 |

The final simulations used four-node quadrilateral (QUAD4) finite elements with the normal out-of-plane shear moduli. Layers of QUAD4 elements model the sub-laminates and they are also joined with very high stiffness normal and shear interface springs. In the composite case, a layer of QUAD4 elements was used for each individual ply. This was found to be necessary to obtain converged results. In the isotropic case, two and four layers of QUAD4 elements were needed in the top and bottom beams, respectively, for sufficient convergence. The elements were distributed uniformly along the length and thickness. The results from these simulations are compared against the Euler and Timoshenko theories and an averaged partition rule.

**TABLE 11.2** Data for Numerical Simulations of a Clamped-Clamped Laminated Composite Beam.

| | |
|---|---|
| Ply longitudinal modulus, $E_{11}$ | 139.3 GPa |
| Ply transverse modulus, $E_{22}$ | 9.72 GPa |
| Out-of-plane modulus, $E_{33}$ | 9.72 GPa |
| In-plane shear modulus, $G_{12}$ | 5.58 GPa |
| Out-of-plane shear moduli, $G_{13}$ | 5.58 GPa |
| Out-of-plane shear moduli, $G_{23}$ | 3.45 GPa |
| In-plane Poisson's ratio, $v_{12}$ | 0.29 |
| Out-of-plane Poisson's ratio, $v_{13}$ | 0.29 |
| Out-of-plane Poisson's ratio, $v_{23}$ | 0.4 |
| Ply thickness, $t_p$ | 0.125 mm |
| Sub-laminate lay-up 1 (top) | $90/-45/0/45$ |
| Sub-laminate lay-up 2 (bottom) | $(45/0/-45/90)_2/90/-45/0/45$ |
| Laminate thicknesses, $h_2$ and $h_2$ | 0.5 and 1.5 mm |
| Intact lengths of beam, $L_1 = L_2$ | 25 mm |
| Length of fracture, $a$ | 50 mm |
| Width of beam, $b$ | 10 mm |
| Loading location, $x_p$ | 25 mm |
| Euler pure modes $\theta_F$, $\beta_F$, $\theta_F'$, and $\beta_F'$ | −26.45, 4.98, −1, and 66.90 |
| Timoshenko pure modes $\theta_F$ and $\beta_F$ | −23.20 and 4.74 |

The following sections present the results from these three sets of simulations for the two different cases. The only applied loads are $P_1$ and $P_2$; $P_1$ was held constant at 1 N and $P_2$ was varied.

### 11.5.1 TESTS WITH CLAMPED-CLAMPED ISOTROPIC BEAMS

Results from the various analytical theories and numerical simulations of the isotropic clamped-clamped beam are presented in Tables 11.3 and 11.4 and Figure 11.3. Plane stress is assumed in all analytical and numerical calculations. The ERR partition for the left crack tip is given. In Figure 11.3 and for every figure in this section, unfilled data markers indicate results from simulations with contact modeling and filled markers indicate results from simulations without. The results from the simulations using Timoshenko beam elements and the very large shear modulus are

**TABLE 11.3** Comparison between Various Theories for Clamped-Clamped Isotropic Beam ERR Partitions with Varying $P_2$ and $P_1 = 1$N.

| $P_2$ (N) | $G_I/G$ (%) | | | | | | |
|---|---|---|---|---|---|---|---|
| | **Analytical Euler** | **Numerical Euler (100 × 2 Timo. beams)** | **Analytical Timo.** | **Numerical Timo. (800 × 2 Timo. beams)** | **Numerical Timo. (200 × 2 Timo. beams)** | **Averaged Analytical (Euler & Timo.)** | **2D FEM (400 × 6 QUAD4s)** |
| −10 | 70.30 | 70.30 | 88.54 | 87.39 | 84.07 | 79.94 | 80.23 |
| −8 | 77.28 | 77.27 | 92.34 | 91.38 | 88.62 | 85.15 | 85.05 |
| −6 | 86.94 | 86.93 | 96.73 | 96.09 | 94.27 | 91.92 | 91.28 |
| −4 | 99.46 | 99.45 | 99.97 | 99.92 | 99.77 | 99.49 | 98.13 |
| −2 | 107.63 | 107.62 | 92.34 | 93.60 | 96.40 | 99.58 | 97.69 |
| 0 | 76.34 | 76.34 | 48.13 | 51.43 | 56.70 | 61.84 | 60.69 |
| 2 | 13.35 | 13.36 | 2.92 | 4.24 | 6.15 | 7.78 | 8.07 |
| 4 | −7.25 | −7.23 | 4.36 | 3.05 | 0.99 | −0.87 | 0.91 |
| 6 | −5.45 | −5.44 | 16.42 | 14.24 | 10.38 | 6.64 | 9.10 |
| 8 | 0.00 | 0.01 | 26.16 | 23.76 | 19.15 | 14.47 | 17.14 |
| 10 | 5.01 | 5.02 | 33.09 | 30.65 | 25.70 | 20.54 | 23.25 |

**TABLE 11.4** Comparison between Various Theories for Clamped-Clamped Isotropic Beam Contact Behavior with Varying $P_2$ and $P_1$ = 1N.

| | First Contact | | After First Contact | |
|---|---|---|---|---|
| | $P_2$ (N) | $G_I/G$ (%) | $P_2$ (N) | $G_I/G$ (%) |
| Analytical Euler | 8 | 0 | 10 | 0 |
| Numerical Euler (100 × 2 Timo. beams) | 7.99 | 0 | 0 | 0 |
| Analytical Timo. | 2.75 | 0 | 10 | 0 |
| Numerical Timo. (800 × 2 Timo. beams) | 3.06 | 0 | 10 | 0 |
| Numerical Timo. (200 × 2 Timo. beams) | 3.67 | 0 | 10 | 0 |
| Averaged analytical (Euler & Timo.) | 4.33 | 0 | 10 | 0 |
| 2D FEM (400 × 6 QUAD4s) | 3.52 | 0 | 10 | 0 |

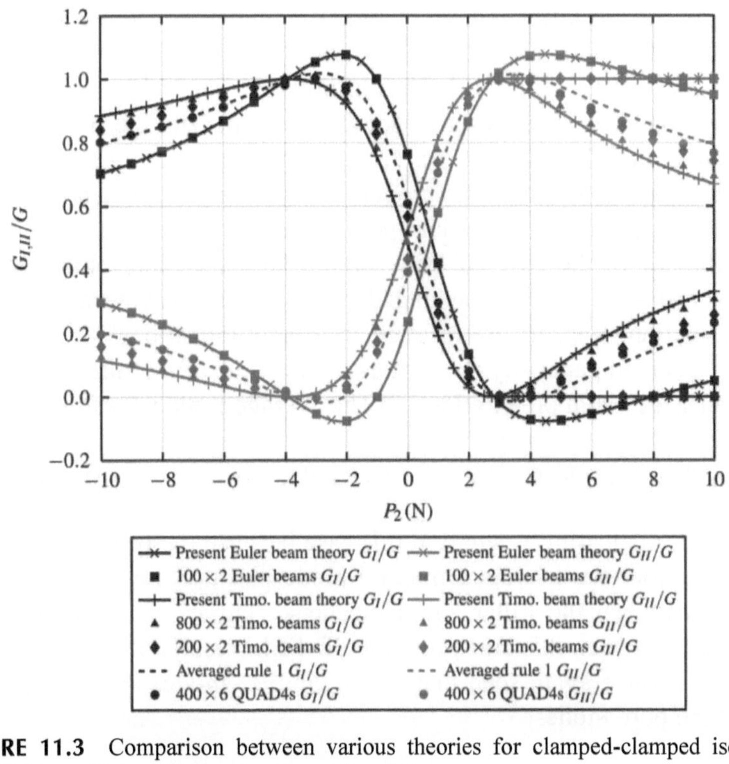


**FIGURE 11.3** Comparison between various theories for clamped-clamped isotropic beam ERR partitions with varying $P_2$ and $P_1$ = 1N.

compared against the Euler beam partition theory. Excellent agreement is seen between the two sets of data. The two sets of pure modes are plainly visible where $G_I / G = 0$ and $G_I / G = 1$. The two methods are in agreement that point contact at the loading location and at the crack tips contact occur simultaneously at $P_2 = 8$ N; and that after first contact, both crack tips remain closed and the fracture is pure mode II.

The Timoshenko beam partition theory is compared with results from numerical simulations with the normal shear modulus. As expected, the $\varphi_{\theta'_F}$ and $\varphi_{\beta'_F}$ modes coincide with the $\varphi_{\theta_F}$ and $\varphi_{\beta_F}$ modes, respectively. The numerical results with 800×2 elements very closely follow the analytical values. The results with 200×2 elements are in less good agreement. This demonstrates that the element size $\delta a$ needs to be very small otherwise $F_{nB\theta_1}\delta a$ is not negligible and a second set of pure modes is generated numerically. This is consistent with the discussion and observations in previous work.[10–13] As expected, crack tip running contact begins at the $\varphi_{\beta_F}$ mode. Crack tip running contact necessarily gives $G_I / G = 0$. Beyond the $\varphi_{\beta_F}$ mode, the crack tips remain closed. The numerical simulations model this contact behavior very closely.

An "averaged partition rule" has been tested in previous work[10–13] and has been found to generally give good agreement with the fracture mode partition from 2D elasticity for: (1) all thickness ratios, (2) all loading conditions, and (3) all material properties, including laminated composite. Particularly regarding material properties, there is some complex mechanical behavior in the case of even simple laminates like biomaterials.[20] However, despite this, the averaged rule can still provide a reasonable approximation. Readers are referred to these publications[10–13] for further details. Detailed papers by the authors on the topic of fractures on bimaterial interfaces are in preparation.[24,25] The averaged rule is as follows:

$$G_I = \alpha_{\theta_1}{}^2 G_{\theta_1} + \alpha_{\theta_1}\alpha_{\beta_1}\Delta G_{\theta_1\beta_1}/2 + \alpha_{\theta_1}\alpha_{\beta_2}\Delta G_{\theta_1\beta_2}/2 + G_P + \alpha_{\theta_1}\Delta G_{\theta_1 P}. \quad (11.112)$$

The effect of shearing is small in this case because the beam is relatively thin. Therefore the averaged fracture mode partition lies approximately midway between the Euler and Timoshenko curves. There is excellent agreement between this curve and the 2D FEM results. In addition to the above, it is once again seen that the $\varphi_{\theta_F}$ and $\varphi_{\beta_F}$ modes are still approximately the pure modes.

### 11.5.2 TESTS WITH CLAMPED-CLAMPED LAMINATED COMPOSITE BEAMS

The data are now presented for the clamped-clamped laminated composite beam. The plane-strain assumption was used in all these analytical and numerical calculations. Under this assumption, $A = A_{11}$, $B = B_{11}$, $= D = D_{11}$, and $H = A_5$.

Since many of the observations are the same as for the isotropic case, they are not repeated. New observations are simply added. Tables 11.5 and 11.6 and Figure 11.4 present results from the various analytical theories and numerical simulations of the laminated composite clamped-clamped beam are presented in. There is excellent agreement between the Euler beam partition theory and the Euler numerical results. There is also excellent agreement between the Timoshenko beam theory and the Timoshenko numerical results.

In this composite case there is a much larger difference between the $\varphi_{\theta_F}$ and $\varphi_{\theta'_F}$ modes than what was seen for the isotropic case (compare Tables 11.1 and 11.2). Having an Euler curve with substantially different $\varphi_{\theta_F}$ and $\varphi_{\theta'_F}$ modes makes it substantially different to the Timoshenko curve. This large difference might therefore have strained the accuracy of the average partition approximation. Despite this possibility, the agreement observed between the averaged partition and the 2D FEM is excellent for the whole the range of $P_2$ simulated.

## 11.6 FURTHER DISCUSSION AND CONCLUSIONS

Analytical theories have been developed for mixed-mode delamination in layered isotropic and laminated composite straight beam structures. Unlike the theories developed elsewhere[9–18] for mixed-mode cracks in layered isotropic and laminated composite DCBs, in these beam structures the internal forces at the crack tips are generally complex functions of remotely applied loads. It is not generally possible to obtain pure "crack tip modes," i.e., modes which relate crack tip quantities, because these quantities cannot be set independently of each other. Instead some combinations of these modes can give pure mode I or II fractures.

This work mainly focused on the most common practical cases of layered isotropic and laminated composite straight beam structures with

**TABLE 11.5** Comparison between Various Theories for Clamped-Clamped Laminated Composite Beam ERR Partitions with Varying $P_2$ and $P_1$ = 1N.

| $P_2$ (N) | $G_I/G$ (%) | | | | | | |
|---|---|---|---|---|---|---|---|
| | **Analytical Euler** | **Numerical Euler (100 × 2 Timo. beams)** | **Analytical Timo.** | **Numerical Timo. (800 × 2 Timo. beams)** | **Numerical Timo. (200 × 2 Timo. beams)** | **Averaged Analytical (Euler & Timo.)** | **2D FEM (200 × 16 QUAD4s)** |
| −10 | 145.33 | 145.33 | 87.57 | 89.20 | 98.46 | 113.51 | 101.99 |
| −8 | 147.63 | 147.62 | 79.89 | 82.13 | 93.05 | 111.45 | 100.40 |
| −6 | 145.01 | 145.00 | 68.83 | 71.93 | 84.24 | 105.37 | 95.65 |
| −4 | 134.61 | 134.60 | 54.04 | 58.15 | 71.19 | 93.47 | 86.17 |
| −2 | 114.10 | 114.09 | 36.51 | 41.43 | 53.97 | 74.79 | 70.87 |
| 0 | 83.82 | 83.81 | 19.27 | 24.13 | 34.51 | 50.90 | 50.73 |
| 2 | 48.24 | 48.24 | 6.41 | 9.90 | 16.58 | 26.46 | 29.71 |
| 4 | 14.30 | 14.30 | 0.47 | 1.65 | 3.83 | 6.72 | 12.78 |
| 6 | −12.60 | −12.61 | 0.98 | −0.24 | −2.39 | −5.75 | 2.50 |
| 8 | −30.77 | −30.77 | 5.66 | 2.59 | −3.12 | −11.51 | −1.66 |
| 10 | −41.30 | −41.29 | 12.21 | 7.99 | −0.36 | −12.56 | −1.69 |

**TABLE 11.6** Comparison between Various Theories for Clamped-Clamped Laminated Composite Beam Contact Behavior with Varying $P_2$ and $P_1$ = 1N.

| | First Contact | | After First Contact | |
|---|---|---|---|---|
| | $P_2$ (N) | $G_I/G$ (%) | $P_2$ (N) | $G_I/G$ (%) |
| Analytical Euler | 66.90 | 0 | 100 | 0 |
| Numerical Euler (100×2 Timo. beams) | 66.77 | 0 | 100 | 0 |
| Analytical Timo. | 4.74 | 0 | 100 | 0 |
| Numerical Timo. (800×2 Timo. beams) | 6.38 | 0 | 100 | 0 |
| Numerical Timo. (200×2 Timo. beams) | 10.18 | 0 | 100 | 0 |
| Averaged Analytical (Euler & Timo.) | 18.23 | 0 | 100 | 0 |
| 2D FEM (200×16 QUAD4s) | 11.60 | 0 | 100 | 0 |

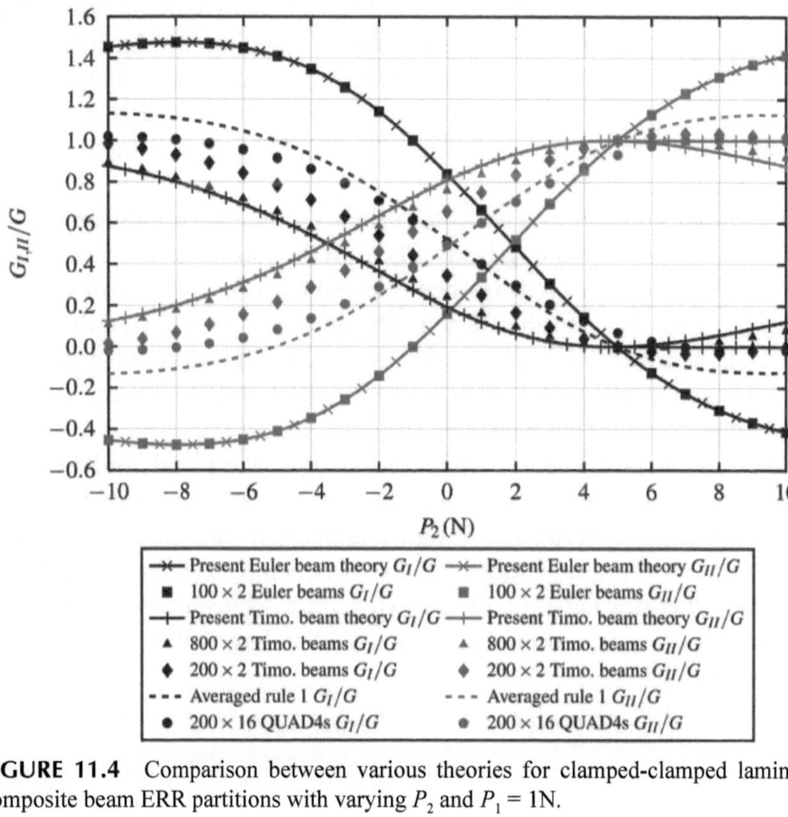


**FIGURE 11.4** Comparison between various theories for clamped-clamped laminated composite beam ERR partitions with varying $P_2$ and $P_1$ = 1N.

shear forces applied at an arbitrary location in the delaminated region. For these beams, the "F modes" have been derived for each crack tip. The F modes give the ratios required between applied shear forces $P_1$ and $P_2$, to give pure fractures modes.

The theories have been developed based on the Euler and Timoshenko isotropic and laminated composite beam theories. Both theories have their own orthogonal $\varphi_{\theta_F}$ and $\varphi_{\beta_F}$ pure modes which are called the first set. They correspond to zero relative shearing displacement just behind the crack tip and zero crack tip opening force, respectively. For the statically indeterminate beam structures examined in this chapter, the first set of pure modes from Euler beam theory is generally different in value to the first set from Timoshenko beam theory. However, when the through-thickness effect is small, the Euler pure modes may be a close approximation to the Timoshenko pure modes.

In Euler beam theory, there is a second set of orthogonal pure modes $\varphi_{\theta'_F}$ and $\varphi_{\beta'_F}$, which are different to the first set. They correspond to zero crack tip shearing force and zero relative opening displacement just behind the crack tip, respectively. Within the context of Timoshenko beam theory, the $\varphi_{\theta'_F}$ and $\varphi_{\beta'_F}$ modes coincide with the $\varphi_{\theta_F}$ and $\varphi_{\theta_F}$ modes. Therefore the $\varphi_{\theta_F}$ and $\varphi_{\beta_F}$ modes form a complete basis for mixed-mode partitions.

The Euler and Timoshenko beam theory mode partitions agree very well with the corresponding beam FEM predictions. The averaged partition approximation, which has been described in previous work by the authors,[10–13] has been further tested. The approximation generally agrees very well with the 2D FEM results, even when the difference between the Euler and Timoshenko curves is substantial and the accuracy of the approximation might have become strained.

The developed theories will be a valuable analytical tool in many applications for example for analytical researchers to develop fracture propagation criteria; for design engineers to design high integrity structures and for numerical analysts to benchmark their simulations, etc. These theories have also been extended to isotropic and composite axisymmetric plates, curved beams and shells; they will be reported in a future paper.

## KEYWORDS

- **energy release rate**
- **fracture mechanics**
- **mixed-mode delamination**
- **straight beam structures**
- **Euler beam theory**
- **Timoshenko beam theory**
- **double cantilever beams**

## REFERENCES

1. Williams, J. G. On the Calculation of Energy Release Rates for Cracked Laminates. *Int. J. Fract. Mech.* **1988,** *36,* 101–119.
2. Schapery, R. A.; Davidson, B. D. Prediction of Energy Release Rate for Mixed-Mode Delamination Using Classical Plate Theory. *Appl. Mech. Rev.* **1990,** *43,* S281–S287.
3. Suo, Z. Delamination Specimens for Orthotropic Materials. *J. Appl. Mech.* **1990,** *56,* 627–634.
4. Suo, Z.; Hutchinson, J. W. Interface Crack between Two Elastic Layers. *Int. J. Fract. Mech.* **1990,** *43,* 1–18.
5. Hutchinson, J. W.; Suo, Z. Mixed Mode Cracking in Layered Materials. *Adv. Appl. Mech.* **1992,** *29,* 63–191.
6. Zou, Z.; Reid, S. R.; Li, S.; Soden, P. D. General Expressions for Energy Release Rates for Delamination in Composite Laminates. *Proc. R. Soc. A.* **2002,** *458,* 645–667.
7. Bruno, D.; Greco, F. Mixed Mode Delamination in Plates: A Refined Approach. *Int. J. Solids Struct.* **2001,** *38,* 9149–9177.
8. Luo, Q.; Tong, L. Calculation of Energy Release Rates for Cohesive and Interlaminar Delamination Based on the Classical Beam-Adhesive Model. *J. Compos. Mater.* **2009,** *43,* 331–348.
9. Wang, S.; Harvey, C. M. Mixed Mode Partition in One-Dimensional Fracture. *J. Key Eng. Mater.* **2011,** *462–463,* 616–621. Also, a Plenary Lecture in the 8th International Conference on Fracture and Strength of Solids (FEOFS 2010), 7–9th June, 2010, Kuala Lumpur, Malaysia.
10. Wang, S.; Harvey, C. M. A Theory of One-Dimensional Fracture. *Compos. Struct.* **2012,** *94,* 758–767. Also, a Plenary Lecture in the 16th International Conference on Composite Structures (ICCS16), 28–30th June, 2011, Porto, Portugal.

11. Harvey, C. M.; Wang, S. Experimental Assessment Of Mixed-Mode Partition Theories. *Compos. Struct.* **2012,** *94,* 2057–2067.
12. Wang, S.; Harvey, C. M. Mixed Mode Partition Theories for One Dimensional Fracture. *Eng. Fract. Mech.* **2012,** *79,* 329–352.
13. Harvey, C. M.; Wang, S. Mixed-Mode Partition Theories for One-Dimensional Delamination in Laminated Composite Beams. *Eng. Fract. Mech.* **2012,** *96,* 737–759.
14. Harvey, C. M. Mixed-Mode Partition Theories for One-Dimensional Fracture. Ph.D. Dissertation, Loughborough University, UK, 2012.
15. Wang, S.; Guan, L. On Fracture Mode Partition Theories. *Comp. Mater. Sci.* **2012,** *52,* 240–245.
16. Wang, S.; Harvey, C. M.; Guan, L. Partition of Mixed Modes in Layered Isotropic Double Cantilever Beams with Non-Rigid Cohesive Interfaces. *Eng. Fract. Mech.* **2013,** *111,* 1–25.
17. Harvey, C. M.; Wood, J. D.; Wang, S.; Watson, A. A Novel Method for the Partition of Mixed-Mode Fractures in 2D Elastic Laminated Unidirectional Composite Beams. *Compos. Struct.* **2014,** *116,* 589–594.
18. Harvey, C. M.; Eplett, M. R.; Wang, S. Experimental Assessment of Mixed-Mode Partition Theories for Generally Laminated Composite Beams. *Compos. Struct.* **2015,** *124,* 10–18.
19. Charalambides, M.; Kinloch, A. J.; Wang, W.; Williams, J. G. On the Analysis of Mixed-Mode Failure. *Int. J. Fract. Mech.* **1992,** *54,* 269–291.
20. Hashemi, S.; Kinloch, A. J.; Williams, G. Mixed-Mode Fracture in Fiber-Polymer Composite Laminates. In *Composite materials: Fatigue and fracture, ASTM STP 1110;* O'Brien, T. K., Ed.; American Society for Testing and Materials: Philadelphia, PA, 1991; Vol. 3, pp 143–168.
21. Davidson, B. D.; Gharibian, S. J.; Yu, L. Evaluation of Energy Release Rate-Based Approaches for Predicting Delamination Growth in Laminated Composites. *Int. J. Fract. Mech.* **2000,** *105,* 343–365.
22. Davidson, B. D.; Bialaszewski, R. D.; Sainath, S. S. A Non-Classical, Energy Release Rate Based Approach for Predicting Delamination Growth in Graphite Reinforced Laminated Polymeric Composites. *Compos. Sci. Technol.* **2006,** *66,* 1479–1496.
23. Conroy, M.; Sørensen, B. F.; Ivankovic, A. In *Combined Numerical and Experimental Investigation of Mode-Mixity in Beam Like Geometries*, Proceeding of the 7th International Conference on Fracture of Polymers, Composites and Adhesives, Les Diablerets, Switzerland, September 14–18, 2014.
24. Harvey, C. M.; Wood, J. D.; Wang, S. Brittle Interfacial Cracking between Two Dissimilar Elastic Layers. Part 1—Analytical Development. *Compos. Struct.* **2015,** *134,* 1076–1086.
25. Harvey, C. M.; Wood, J. D.; Wang, S. Brittle Interfacial Cracking between Two Dissimilar Elastic Layers. Part 2—Numerical Validation. *Compos. Struct.* **2015,** *134,* 1087–1094.
26. Harvey, C. M.; Wang, S. Modelling of Delamination Propagation in Composite Laminated Beam Structures. *AIP Conf. Proc.* **2012,** *1504,* 1146–1149.

27. Harvey, C. M.; Wang, S. Numerical and Analytical Study of Delamination in Composite Laminates. *Int. J. Eng. Syst. Model. Sim.* **2012,** *4,* 120–137.
28. Wang, S.; Zhang, Y. Buckling, Post-Buckling and Delamination Propagation in Debonded Composite Laminates. Part 2: Numerical applications. *Compos. Struct.* **2009,** *88,* 131–146.
29. Kutlu, Z.; Chang, F. K. Composite Panels Containing Multiple Through-The-Width Delaminations and Subjected to Compression. Part II: Experiments and Verification. *Compos. Struct.* **1995,** *31,* 297–314.

CHAPTER 12

# CRACK PARAMETER IDENTIFICATION IN ELASTIC PLATE STRUCTURES BASED ON REMOTE STRAIN FIELDS AND SOLUTION OF INVERSE PROBLEMS

RAMDANE BOUKELLIF* and ANDREAS RICOEUR

*Institute of Mechanics, University of Kassel, Kassel 34125, Germany*

**Corresponding author. E-mail: ramdane.boukellif@uni-kassel.de*

## CONTENTS

## ABSTRACT

In this work, a concept for the detection of cracks in plate structures is presented. This method is based on strains measured at different locations on the surface of a structure and the application of the dislocation technique. Solving the inverse problem with the particle swarm optimization (PSO) algorithm, this allows the identification of crack position parameters, such as length, location, angles, and the calculation of stress intensity factors (SIF). Numerical as well as physical experiments are performed using pre-cracked plates made of the aluminum alloy Al-7075. The parameter identification shows satisfactory results.

## 12.1 INTRODUCTION

Engineering structures are in general exposed to cyclic or stochastic mechanical loading. Exhibiting incipient cracks, particularly light-weight shell and plate structures suffer from fatigue crack growth, limiting the life time of the structure and supplying the risk of a fatal failure. Due to the uncertainty of loading boundary conditions and the geometrical complexity of many engineering structures, numerical predictions of fatigue crack growth rates and residual strength are not reliable. Most experimental monitoring techniques, nowadays, are based on the principle of wave scattering at the free surfaces of cracks. Many of them are working well, supplying information about the position of cracks. One disadvantage is that those methods do not yield any information on the loading of the crack tip. Goal of our work is the development of a monitoring concept supplying both the information on the actual crack position and the stress intensity factors.

## 12.2 PRINCIPLES OF THE DISLOCATION TECHNIQUE

The dislocation method was proposed elsewhere.[1] An infinite plate containing a crack with a length $2a$ under remote stresses $\sigma_{yy}^{\infty}$ is considered as an easy example. Applying the superposition principle, the total stress field $\sigma_{ij}^{*}(x,y)$ is as follows:

$$\sigma_{ij}^{*}(x,y) = \sigma_{ij}^{A}(x,y) + \sigma_{ij}^{D}(x,y), \tag{12.1}$$

where $\sigma_{ij}^{A}(x,y)(ij = xx, yy, xy)$ is the stress field induced in the plate without a crack. The stresses induced at a point $(x, y)$ due to a single dislocation located at the origin of the coordinate system with Burgers vector $\vec{b}$ and its components $b_x$ and $b_y$ may be found from the corresponding Airy stress functions.

Assuming pure mode-I loading, the normal stresses arising along the crack faces due to a single dislocation are given by setting $b_x = 0$ and $y = 0$ in the Airy stress functions so that

$$\sigma_{yy}^{D}(x, y=0) = \frac{2\mu}{\pi(\kappa+1)}\frac{db_y(\xi)}{x-\xi} = \frac{2\mu}{\pi(\kappa+1)}\frac{B_y(\xi)}{x-\xi}d\xi, \tag{12.2}$$

where $B_y(\xi)$ is the density of the dislocation at the source point $\xi$.

In case that the crack faces are traction free, i.e., $\sigma_{yy}^{*} = 0$, the stresses due to a continuous distribution of dislocations along the crack line are then given by:

$$\sigma_{yy}^{A}(x) = \sigma_{yy}^{\infty}(x) = -\frac{2\mu}{\pi(\kappa+1)}\int_{-a}^{+a}\frac{B_y(\xi)}{x-\xi}d\xi. \tag{12.3}$$

The singular integral eq 12.3 has to be normalized and the dislocation density is rewritten as $B(s) = w(s)\phi(s)$, where $\phi(s)$ is a bounded unknown function and $w(s)$, with $s = \xi/a$, characterizes the fundamental solution. Using the Gauss–Chebyshev numerical quadrature, eq 12.3 can be written as follows:

$$-\frac{(\kappa+1)}{2\mu}\sigma_{yy}^{\infty}(t_k) = \frac{1}{N}\sum_{i=1}^{N}\frac{\phi_y(s_i)}{t_k - s_i} \qquad k = 1,\ldots,N-1, \tag{12.4}$$

where $s_i$ and $t_k$ are, respectively, discrete integration and collocation points and $N$ is typically chosen between 50 and 80.

The solution for the total stresses $\sigma_{ij}^{*}(x,y)(ij = xx, yy, xy)$ is obtained as follows:

$$\sigma_{xx}^{*}(x,y) = \frac{2\mu}{(\kappa+1)}\frac{1}{N}\sum_{i=1}^{N}\phi_y(s_i)\left[\frac{(x/a - s_i)\left[(x/a - s_i)^2 - (y/a)^2\right]}{\left[(x/a - s_i)^2 + (y/a)^2\right]^2}\right], \tag{12.5}$$

$$\sigma_{yy}^{*}(x,y)=\frac{2\mu}{(\kappa+1)}\frac{1}{N}\sum_{i=1}^{N}\phi_y(s_i)\left[\frac{(x/a-s_i)\left[(x/a-s_i)^2+3(y/a)^2\right]}{\left[(x/a-s_i)^2+(y/a)^2\right]^2}\right]+\sigma_{yy}^{\infty}, \quad (12.6)$$

$$\sigma_{xy}^{*}(x,y)=\frac{2\mu}{(\kappa+1)}\frac{1}{N}\sum_{i=1}^{N}\phi_y(s_i)\left[\frac{y/a\left[(x/a-s_i)^2-(y/a)^2\right]}{\left[(x/a-s_i)^2+(y/a)^2\right]^2}\right]. \quad (12.7)$$

The solution of the total strains is calculated as follows:

$$\varepsilon_{xx}^{*}=\frac{1}{E}\left(\sigma_{xx}^{*}-\nu\sigma_{yy}^{*}\right),\ \varepsilon_{yy}^{*}=\frac{1}{E}\left(\sigma_{yy}^{*}-\nu\sigma_{xx}^{*}\right),\ \gamma_{xy}^{*}=\frac{2(1+\nu)}{E}\sigma_{xy}^{*}, \quad (12.8)$$

where $E$ is Young's modulus and $\nu$ is Poisson's ratio.

## 12.3 SOLVING THE INVERSE PROBLEM USING THE PSO

The solution of the inverse problem for a cracked structure is carried out using the PSO algorithm[2] minimizing a *fitness* function, defined by a square sum of residuals between measured and computed strains $\overline{\varepsilon}_{ij}$ and $\varepsilon_{ij}$ at positions $P_m$:

$$fitness=\sum_{m=1}^{M}\sum_{ij=xx,yy,xy}\left\{\overline{\varepsilon}_{ij}(P_m)-\varepsilon_{ij}(P_m)\right\}^2. \quad (12.9)$$

## 12.4 VERIFICATION OF THE CONCEPT

First verifications of the monitoring concept have been carried out numerically as shown in Figure 12.1 (left). Here, $\sigma_{yy}^{\infty}$ and $a$ are used as testing parameters for numerical calculations. The formulation of the direct problem is more complex than depicted above, however, following the same principles. The table in Figure 12.1 (right) shows the given problem and the obtained numerical results.

Besides numerical simulations, hardware experiments have been performed. Here, $\overline{\sigma}_{xx}$, $a$ and $K_I$ are used as testing parameters. A fatigue edge crack located in the middle of an Al-7075-plate driven by a cyclic load is considered. The strain $\bar{\varepsilon}_{ij}(P_m)$ is measured using strain gauges at points $P_m$ ($m = 1,\ldots,7$), the crack length ranges from $a = 16$ mm to $a = 39.5$ mm. and the loading stresses vary from $\bar{\sigma}_{xx} = 2$ MPa to $\bar{\sigma}_{xx} = 24$ MPa. Figure 12.2 shows the obtained experimental results.[3]

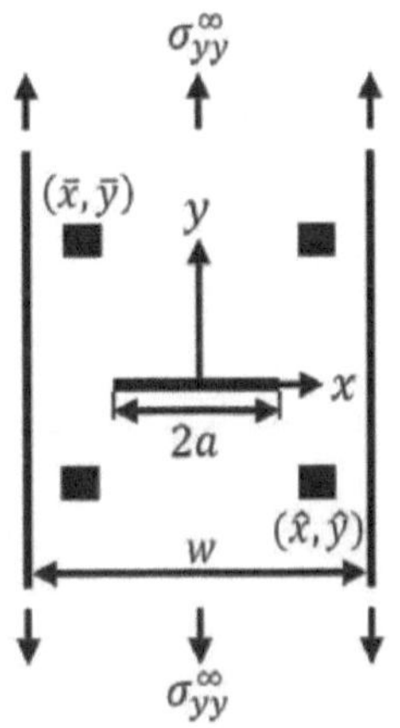


| Given Problem | Obtained Results |
|---|---|
| $(\bar{x},\bar{y}) = (-4.6, 4.6\text{mm})$<br>$(\hat{x},\hat{y}) = (4.6, 4.6\text{mm})$<br>$w = 20$ mm | |
| $\sigma_{yy}^{\infty} = 50\,\text{MPa}$ | $\sigma_{yy}^{\infty} = 49.988\,\text{MPa}$ |
| $a = 1$ mm | $a = 1.0$ mm |
| $K_I = 88.6227\,\text{MPa}\sqrt{\text{mm}}$ | $K_I = 88.601\,\text{MPa}\sqrt{\text{mm}}$ |

**FIGURE 12.1** (Left): crack in a semi-infinite plate with two vertical boundaries; right: results of the inverse problem solution based on the dislocation technique; locations of strain monitoring are indicated by dark squares at $(\bar{x},\bar{y})$ and $(\hat{x},\hat{y})$.

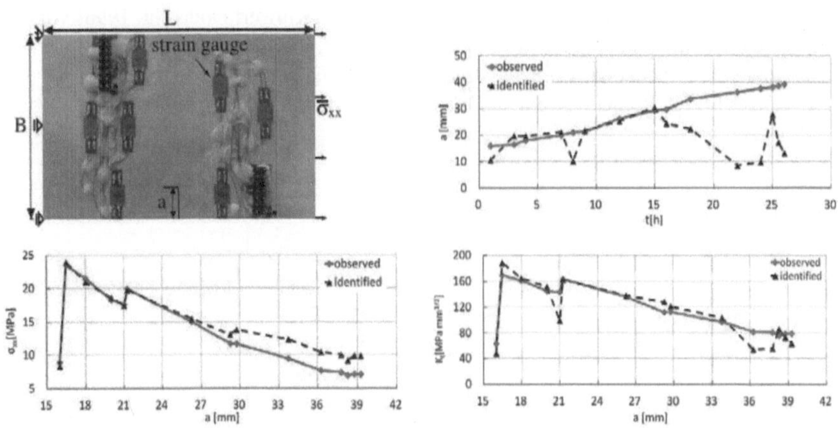


**FIGURE 12.2** Plate with edge crack and strain gauges ($B = 70$ mm, $L = 150$ mm). The plots show the identified and observed results of the crack length $a$, stress intensity factor and load $\bar{\sigma}_{xx}$.

## 12.5 CONCLUSIONS

The concept of distributed dislocations is used for monitoring fatigue crack growth and stress intensity factors in plate structures. The method is verified numerically assuming an internal crack in an elastic sheet with two boundaries and homogeneous external loads and experimentally for an edge crack exposed to external cyclic loads. The solution of the inverse problem is in a good agreement with the model data within a wide range of parameters.[4]

## KEYWORDS

- **monitoring fatigue cracks**
- **stress intensity factors**
- **inverse problem**
- **dislocation method**

## REFERENCES

1. Hills, D. A. *Solid Mechanics and its Applications;* Kluwer: Dodrecht, 1996; Vol. 44.
2. Chen, D. H.; Nisitani, H. *Eng. Fract. Mech.* **1993,** *45,* 671–685.
3. Boukellif, R.; Ricoeur, A. *Proc. Appl. Math. Mech.* **2012,** *12,* 141–142.
4. Boukellif, R.; Ricoeur, A. *Int. J. Solids Struct.* **2014,** *51,* 2123–2132.

## CHAPTER 13

# FRACTURE ENERGY OF STEEL FIBER-REINFORCED HIGH-STRENGTH CONCRETE BEAMS

SRIMAN NARAYAN, H. N.[1*], MURALIDHARA, S.[1], and B. K. RAGHU PRASAD[2]

[1]*Department of Civil Engineering, BMS College of Engineering, Bangalore, India.*

[2]*Department of Civil Engineering, Indian Institute of Science, Bangalore, India.*

[*]*Corresponding author. E-mail: sriman.hn@gmail.com*

## CONTENTS

## ABSTRACT

The chapter presents some results of an experimental study on the fracture behavior of steel fiber-reinforced high strength concrete (SFRHSC) beams. The fracture energy and ductility factors of SFRHSC beams are obtained by carrying out three point bend tests on notched beams. Presence of steel fibers is known to enhance the ductility of concrete. Instead of the very small softening branch of the load–deflection curve observed in plain high strength concrete (HSC) beams without fibers, larger softening branch of the load–deflection curve occurs in SFRHC. The fracture energy and ductility factors are found to be high in SFRHSC. The ductility decreases as the size of the specimen increases which is an indication of reduction of energy absorption. The compressive strength of SFRHSC is observed to be about 95 MPa with a tensile strength of about 7.2 MPa. The energy absorption is found to be very high in SFRHSC in comparison with HSC. The softening branch of the load–deflection curve in the post peak behavior, which indicates ductility, is found to be comparatively longer in the case of SFRHSC.

## 13.1 INTRODUCTION

Fracture energy is an important material property for design of concrete structures under "Fracture Mechanics." Under the name fracture mechanics, a study which combines the mechanics of cracked bodies and mechanical properties is made. Fracture mechanics deals with fracture phenomena. There have been several disasters in the past which have forced a reasonable significant progress in the understanding and estimation of fracture and fatigue.

Steel fiber-reinforced high strength concrete (SFRHSC) is being investigated as an advanced cementitious material. SFRHSC offers good potential in innovative structural design. An optimized material composition results in high compressive strength. Multi micro cracking provides significant ductility which permits high tensile forces to be sustained by structural elements in bending. Since the material is highly resistant to environmental degradation, offshore and ocean structures can be constructed.

Fracture mechanics is a rapidly developing field that has great potential for application to concrete structural design. The stress strain curve is linearly up to the maximum stress in ideally brittle material. At the maximum stress, an initial crack propagates leading to failure. Tensile stress elongation curve is linear for which linear elastic fracture mechanics (LEFM) is valid. While, for a quasi-brittle material like concrete, non-linearity exists substantially before the maximum stress is reached. There has not been a clear understanding of the mechanism of deformation beyond the proportional limit. Randomly distributed micro cracks are formed initially, and at some point before the peak stress, micro cracks begin to localize into macro cracks that propagate leading to a failure. Under the steady-state propagation of these cracks, strain softening is observed.

High strength concrete (HSC) does not develop significant combined cracks until about 90% of peak stress. Addition of fibers to HSC provides better strengthening and stiffening and tends to result in a strain-softening behavior. Reinforcement in the form of short discrete fibers act effectively as rigid inclusions in the concrete matrix and enhances the bridging mechanism.

The fibers intersecting with the crack ensure stress transfer over the crack. The maximum bridging stress that can be reached depends on the number of fibers, the elasticity of the fibers, shape, and its inclination with respect to the crack face, the pullout behavior, and the aspect ratio.

HSC is more brittle than normal strength concrete. It is ideal to add fibers to make HSC more ductile. With the fibers of different materials and different types being available, the choice of the right type of fibers that will improve the ductility and the energy absorption is required. Steel fibers are chosen for this study owing to its highly contributive mechanical properties.

Because of concrete being high strength, the interface between the course aggregate and the matrix will be very strong. The interface between fibers and surrounding matrix may not allow the fibers to slip and there could be a certain snapping after the yield strength is reached. However, the sudden snapping would jeopardize the energy absorption capacity and should hence be avoided. Tests are being conducted on SFRHSC to obtain fracture energy and process zone experimentally.

RILEM FMC 50 (1985) has proposed the methodology to evaluate the fracture energy under TPB tests on concrete beams using displacement controlled equipment. The energy required to produce a unit crack surface is termed as Fracture energy. This is obtained by the area under pre peak and post peak load displacement curves.

Fiber-reinforced concrete is a composite material. Fibers present in the concrete improve the tensile strength and energy absorption capacity of concrete. Explosive type of failures occurs when the peak stress is reached in HSC due to increase in brittleness. The failure will be sudden and catastrophic in structures subjected to blast, earthquake, and suddenly applied loads. The fracture behavior of concrete is greatly influenced by the fracture process zone (FPZ). FPZ is defined as the zone in which the material undergoes softening damage FPZ in concrete is quite small. Such materials are now commonly called quasi-brittle material. HSC does not develop significant combined cracks until about 90% of peak stress. Addition of fibers to HSC tends to result in strain-softening branch of the load–deflection curve. Strain softening of the load–deflection curve is observed under steady-state propagation of crack. Often the crack arrest mechanism in the matrix produces a pseudo-debonding crack parallel to the fiber at some distance away from the actual interface. Steel fibers provide better strengthening and stiffening. The fibers act to enhance the bridging mechanism in SFRHSC. An attempt has been made in the present investigation to evaluate the fracture energy of SFRHSC beams and its comparison with HSC beams without fibers. Attempt has also been made to determine the ductility of SFRHSC and its comparison with HSC without fibers.

## 13.2 EXPERIMENTAL PROGRAM

Ordinary Portland cement of 53 grade conforming to IS: 8112-1989 was used. Fine aggregate was natural river sand passing through 2.36 mm sieve with specific gravity of 2.604. Crushed white granite coarse aggregate of 20 mm down and 12 mm down sizes were used in 50:50 ratios. Water cement ratio was standardized at 0.37 with 130 L/m$^3$. Super plasticizer BASF GLENIUM B230 was used at 3.8 L/m$^3$. Hooked end Dramix 8060 Steel fibers with aspect ratio of 80 of 0.75 mm diameter and 60 mm length were used for SFRHSC beams at 25 kg/m$^3$ of the concrete. Modulus of elasticity of the steel fibers was 210 GPa. The constituents of the concrete mix in kg/m$^3$ are as shown in Table 13.1.

Specimens of 150, 200, and 250 mm depths with 100 mm constant widths are cast. Span to depth ratio of 4 is adopted. Three samples with fiber and without fiber are cast for each variant summing up to 18 beams. Notches of 0.25 d were made using diamond blade on a high speed circular cutting machine. The specimens were subjected to three point bend tests and tested in closed loop strain controlled equipment. The 28 day compressive strength was found to be 95 MPa for HSC and 96 MPa for SFRHSC tested on 150 mm cubes. Tensile strength was ascertained by split tensile test and was found to be 7.2 and 7.5 MPa for SFRHSC. Load point deflections and Crack mouth opening displacement (CMOD) are measured.

**TABLE 13.1** Constituent Materials Used per m$^3$.

| | Cement (kg) | GGBS (kg) | Alccofine (kg) | Fine aggregate (kg) | Coarse Aggregate in (kg) | Steel Fibers (kg) |
|---|---|---|---|---|---|---|
| HSC | 350 | 120 | 80 | 550 | 1120 | 0 |
| SFRHSC | 350 | 120 | 80 | 550 | 1120 | 25 |

## 13.3 FRACTURE ENERGY $G_F$

The equation adopted for evaluating the fracture energy is

$$G_F = (W_O + mg\delta_0)/A_{lig} \text{ [N/m ]}$$

where W is the area under the load deflection curve, m is weight of beam, g = acceleration due to gravity in m/s$^2$, $\delta_o$ = deformation at the final failure of the beam, $A_{lig}$= area of the ligament.

## 13.4 DUCTILITY FACTOR

**Ductility** is defined as the ability to possess nonlinear deformations under loading. The ductility of concrete beam with initial notches may be defined as the ratio of CMOD at the failure load to the CMOD at the yield.

$$\text{Ductility factor} = \frac{CMOD_u}{CMOD_y}.$$

## 13.5 RESULTS AND DISCUSSION

SFRHSC displayed slightly higher peak load than HSC. But the post peak response of SFRHSC was observed to be much longer and prolonged than that of HSC. The fracture energy in SFRHSC 150 mm deep beam is

1118 N/m. The fracture energy of HSC of the same size beam is 147 N/m. The fracture energy in SFRHSC is increased by 7.5 times in comparison to HSC with the addition of 25 kg/m$^3$ of steel fibers. The fracture energy in 200 and 250 mm deep SFRHSC beams are 646 and 272 N/m, respectively, in comparison to that in HSC which are 43 and 39 N/m. This indicates the huge amount of energy absorption taking place in SFRHSC. The results are reported in Table 13.2. There is a decrease in fracture energy with the increase in the size of SFRHSC beams. This trend with HSC of 95 MPa concrete follows the earlier findings for concrete of 55 MPa (7). In the pre peak region, the initial load–CMOD curve is linear almost up to the peak load. Figure 13.1 shows the variation of fracture energy in HSC and SFRHSC beams of different depths.

**TABLE 13.2**

| MIX | Depth of Specimen in mm | Fracture Energy N-m/m$^2$ | CMOD$_y$ mm | CMOD$_u$ mm | Ductility Factor |
|---|---|---|---|---|---|
| HSC | 150 | 147 | 0.02243 | 0.5598 | 25 |
| SFRHSC | 150 | 1118 | 0.0438 | 3.606 | 82 |
| HSC | 200 | 111 | 0.0496 | 0.5346 | 11 |
| SFRHSC | 200 | 646 | 0.0799 | 3.394 | 43 |
| HSC | 250 | 99 | 0.0551 | 0.5004 | 9 |
| SFRHSC | 250 | 272 | 0.15 | 5.724 | 38 |

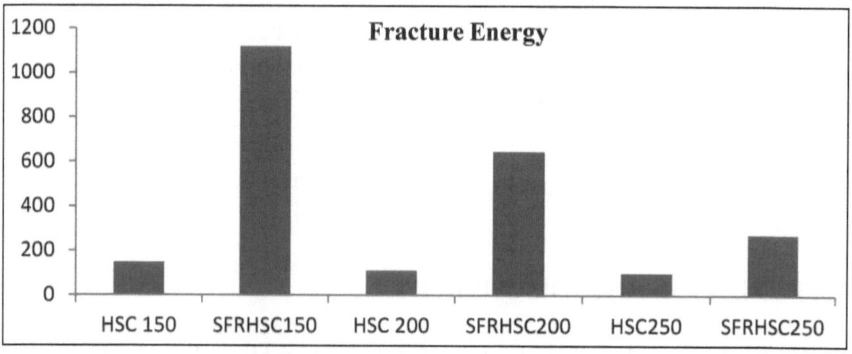


**FIGURE 13.1** Variation of fracture energy in HSC and SFRHSC.

The post peak response is gradual with the softening branch of the load–CMOD curve increasing. The CMOD at failure in 150 mm deep HSC is 0.6 mm while in SFRHSC, it is found to be 3.606 mm which is almost six times that of HSC. The same trend follows in the 200 and 250 mm deep beams. Ductility factor as defined by the relation mentioned above is 39 and 9 in SFRHSC and HSC, respectively, in 250 mm deep beams. The ductility of SFRHSC is thus about four times that of HSC in 250 mm deep beams with the addition of 25 kg/m$^3$ fibers. The average increase in ductility in SFRHSC is about 220–330% over HSC. Ductility is seen to be decreasing with the increase in the size. This strongly indicates the size effect in SFRHSC beams. The failures were very gradual in SFRHSC unlike that in HSC where it was a sudden failure. Figure 13.2 shows the vast improvement in ductility of SFRHSC in comparison to HSC.

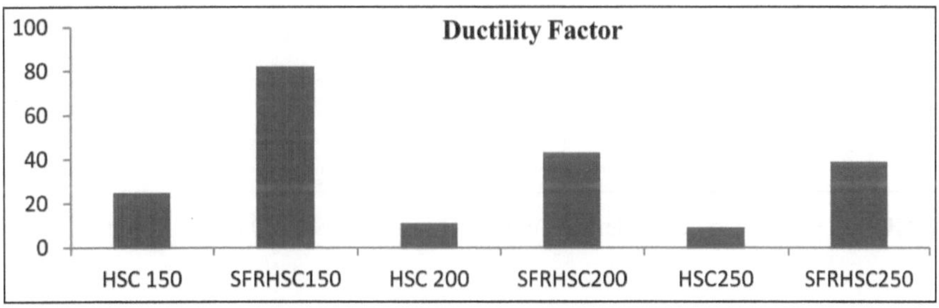


**FIGURE 13.2** Variation of ductility factor in HSC and SFRHSC.

The load–deflection curve for 150, 250, and 250 mm deep HSC and SFRCHC beams are shown in Figures 13.3–13.5, respectively.

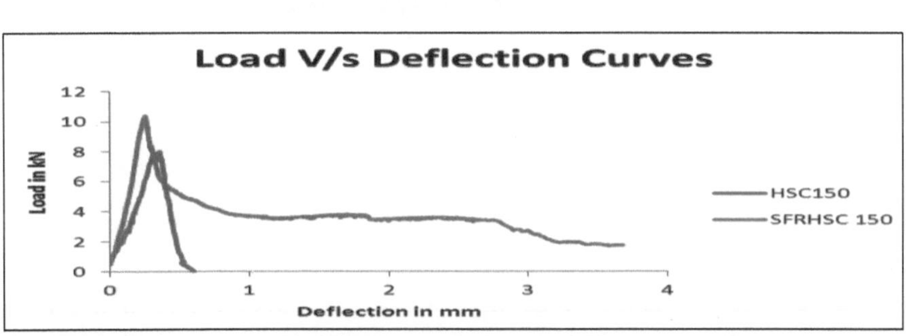


**FIGURE 13.3** Load versus deflection in HSC and SFRHSC 150 mm deep beams.

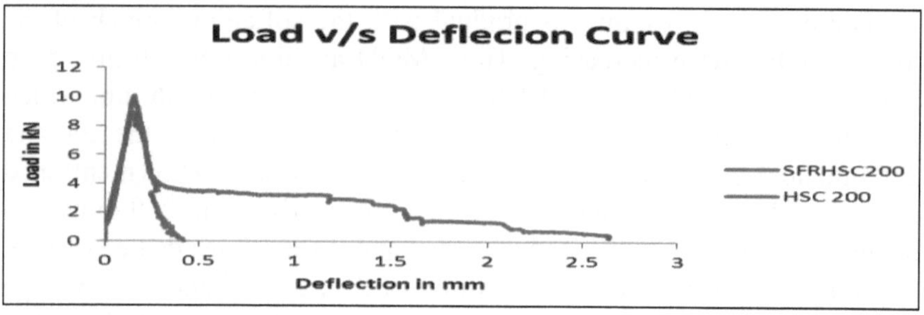


**FIGURE 13.4** Load versus deflection in HSC and SFRHSC 200 mm deep beams.

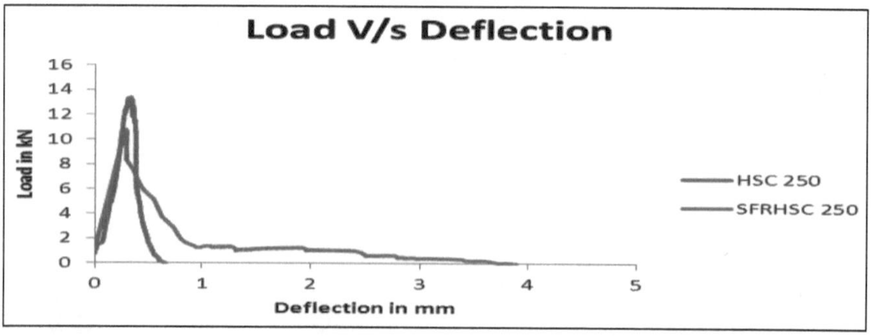


**FIGURE 13.5** Load versus deflection in HSC and SFRHSC 250 mm deep beams.

The load–CMOD curve for 150, 250, and 250 mm deep HSC and SFRCHC beams are shown in Figures 13.6–13.8, respectively.

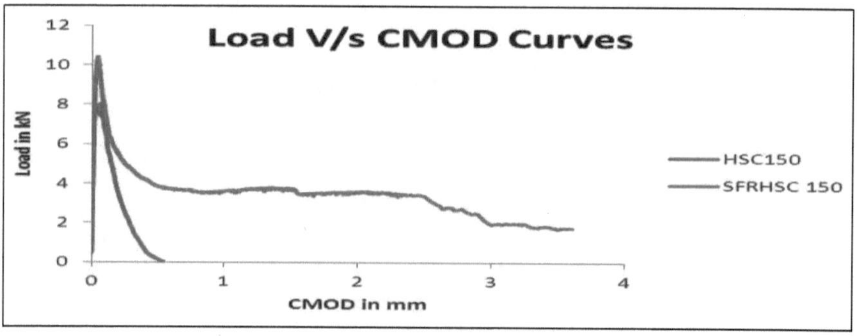


**FIGURE 13.6** Load versus CMOD in HSC and SFRHSC 150 mm deep beams.

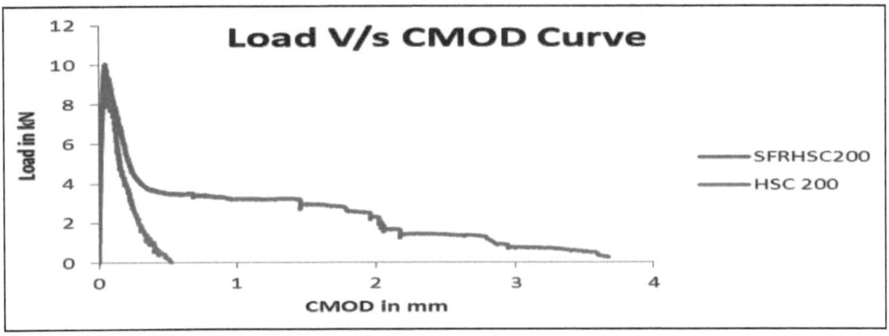


**FIGURE 13.7** Load versus CMOD in HSC and SFRHSC 200 mm deep beams.

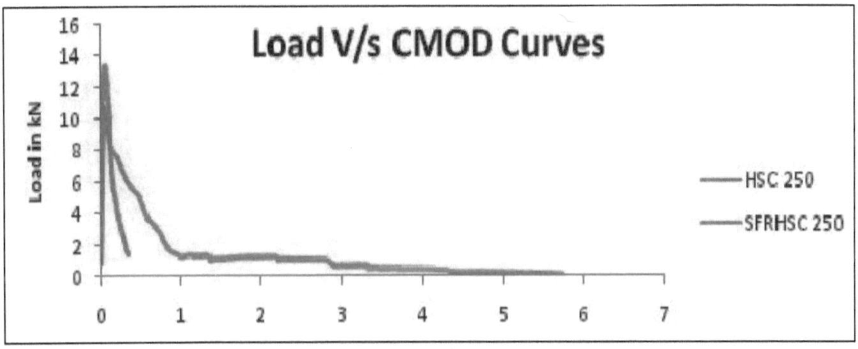


**FIGURE 13.8** Load versus CMOD in HSC and SFRHSC 250 mm deep beams.

The load–deflection curves and load–CMOD curves for 150, 250, and 250 mm deep HSC and SFRCHC beams are shown in Figures 13.9 and 13.10, respectively,

Figure 13.9 shows the variation of load with Deflection in HSC and SFRHSC in 150, 200, and 250 mm deep beams. The post peak tension deflection in SFRHSC is large, the length of the descending curve being very large in comparison to HSC.

Figure 13.10 shows the variation of load with CMOD in HSC and SFRHSC in 150 mm deep beams. The post peak softening response in SFRHSC is gradual with large plastic deformation being observed. In SFRHSC, the length of the descending curve is very large in comparison to HSC. The CMOD value is very large in SFRHSC than that of HSC.

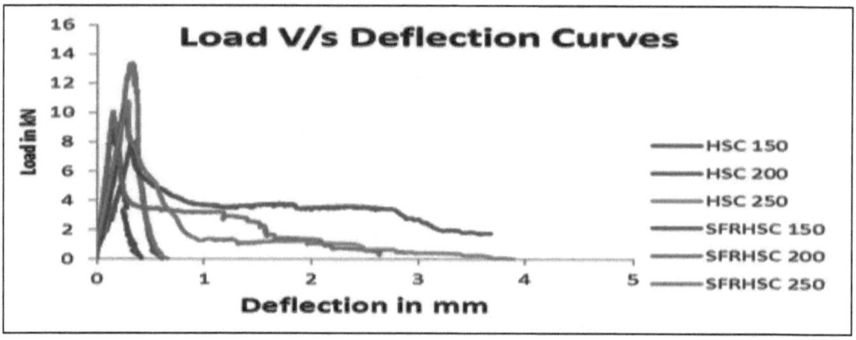


**FIGURE 13.9** Load versus deflection in HSC and SFRHSC 150, 250, and 250 mm deep beams.

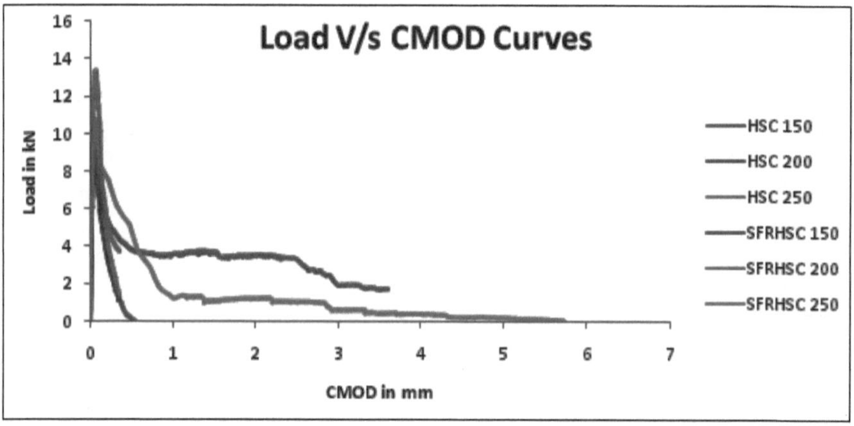


**FIGURE 13.10** Load versus CMOD in HSC SFRHSC 150, 250, and 250 mm deep beams.

## 13.6 CONCLUSIONS

Based on the experimental investigations on the effect of addition of steel fibers on the fracture characteristics of HSC such as fracture energy, ductility, the following conclusions are made:

1. Fracture energy in HSC decreases with the increase in specimen size despite increase in peak load in SFRHSC which can be attributed to the size effect.

2. A marginal improvement in peak load is observed in SFRHSC in comparison to HSC.
3. Addition of 25 $kg/m^3$ steel fibers to HSC has increased the fracture energy of SFRHSC to around 5–7 times that of HSC.
4. SFRHSC is more ductile than HSC which is indicated by the post peak softening branch of the load vs. CMOD curve which is predominantly extended.
5. The failure mode of structure changes from catastrophic to gradual with the addition of fibers.
6. SFRHSC could be a suitable substitute to HSC where ductile structures are required like in earthquake, blast, and radiation resistant structures.

## ACKNOWLEDGMENTS

The first author sincerely thanks M/s. Bekaert for providing steel fibers and M/s. BASF for providing super plasticizer for the experiments.

## KEYWORDS

- **steel fibers**
- **fracture energy**
- **ductility factors**
- **strain softening**

## REFERENCES

1. Ya-Cheng, K.; Kuang-Chih, P.; HSUAN-Chih, Y. In *An Investigation on Toughness of Steel Fiber Reinforced Heavy Concret,* 17th International Conference on structural Mechanics in Reactor Technology, Prague, Czech Republic, paper #H01-6, August 17 –22, 2003, pp 1–8.
2. Karihaloo, B. L.; Xiao, Q. Z. Size Effect in the Strength of Concrete Structures. *Sadhana.* **2002,** *27* (4), 449–459.

3. Dattatreya, J. K.; Harish, K. V.; Neelamegam, M. Investigation of the Flexural Toughness and Fracture Energy of High and Ultra High Performance Fiber Reinforced Concretes, Seventh India Rilem Symposium, on FRC, 2008; pp 231–242.
4. Jin-Keun Kim and Seong-Tae Yi, Application of Size Effect to Compressive Strength of Concrete Members. *Sadhana.* **2002,** *27* (4), 467–484.
5. Appa Rao, G.; Raghu Prasad, B. K. Fracture Energy of FRC. *J. Struct. Eng.* **2005,** *31* (4), 249–255.
6. Barros, J. A. O.; SenaCriz, J. Fracture Energy of Steel FRC. *IasJornadas de Estruturas de Betao, NovosComposit.* **1999,** 151–170.
7. Nguyen Van Chanh, *Steel Fiber Reinforced Concrete;* Ho Chi Mink City University of Technology: Vietnam, 2004; pp 108–116.
8. Zongjin, L.; Xianyu, J.; Chen L. Fracture Toughness and Microstructure of Concrete at Early Ages. Advances in Civil Engineering Materials, 50year Research Anniversary, Nanjing, China, 2008; pp 139–149.
9. Wittmann, F. H. Crack Formation and Fracture Energy of Normal and High Strength Concrete. *Sadhana.* **2002,** 27 (4), 412–423.
10. Amanda Bordelon, "Manual for Three-Point Bending Single-Edge Notched Fracture Test with Plain or Fiber Reinforced Concrete, pp 1–14, University of Illinois, Dec 07.
11. Meda, A.; Pizzari, G. A.; Sorelli, L. *Fracture Properties of Concrete Reinforced with Hybrid Fibers;* People and Practice: Dundee, UK, 2003.
12. RILEM Draft Recommendation: Fracture 50—FMC Committee Fracture Mechanics of Concrete. *Mat. Str.* **1985,** *18* (206), 285–290.
13. Janson. A.; Lifgren. I; Gylluft. K. Material Testing and Structural Analysis of FRC Beams-A Fracture Mechanics Approach, 7th Rilemsymp, FRC: pp 85–103.
14. Anderson, T. L. *Fracture Mechanics, Fundamentals and Applications;* CRC Press Publications: London, 1994.
15. Jan G. M. vanMier, *Fracture Process of Concrete;* CRC Press Publications: London, 1997.

CHAPTER 14

# STRUCTURAL CHARACTERIZATION AND MECHANICAL BEHAVIOR OF CARBON NANOTUBE-REINFORCED ALUMINUM MATRIX COMPOSITE —A REVIEW

SHAHRUKH SHAMIM*, GAURAV SHARMA, C. SASIKUMAR, and VIKRAM SINGH RAGHUVANSHI

*Department of Materials Science and Metallurgical Engineering, MANIT, Bhopal, Madhya Pradesh, India*

**Corresponding author. E-mail: shahrukh.manitb@gmail.com*

## CONTENTS

## ABSTRACT

Preliminary investigation shows that some nanocomposites with special filler material may achieve superlative mechanical properties like remarkable improvement in Fracture toughness, Vibration damping, etc. Recently nano-sized particles are superseding the micro-sized particles because of their improved physical and mechanical properties. Particulate-reinforced nanocomposites have been considerably employed in the automotive and aerospace industry for their potential to withstand high temperature and pressure. This chapter reviews relevant literature which deals with various processing method, mechanical properties, effect of carbon nanotube's morphology on the properties and structure of aluminum matrix composites. Among the varieties of metallurgical processing techniques are available, powder metallurgy technique is generally accepted process route due to its simplicity, adaptability, and uniform distribution. This review will provide the reader with an overview of aluminum nanocomposite manufacturing methods along with its mechanical properties and structural characterization.


## 14.1 INTRODUCTION

The necessity for high strength and lightweight materials has been acknowledged since the development of the airplane and vehicles. The incompetence of alloys and metals in providing both strength and stiffness to a structure has led to the development of various composites distinctly metal matrix composites (MMCs).[1] MMCs today are extensively used in automobile and aerospace applications.[2–3] Conventional MMCs reinforced with ceramic particles possess agreeable combination of strength, stiffness, wear, and creep-resistant properties over monolithic alloys.[4] Particulate-reinforced MMCs have received increasingly attention because of their lower cost, ease of fabrication, and near-isotropic properties.[5] Fabrication of carbon fibers of high strength in the 1960s and 1970s made them the first choice for the manufacture of advanced composites for use in rocket nozzle exit cones, missile nose tips, and re-entry heat shields, etc. Since 1970, carbon fiber-reinforced composites have been extensively used in a wide range of applications such as space structures, aircraft brakes, military and commercial planes, and structural reinforcement.[1] The discovery of carbon nanotubes (CNT) by Iijima[6] revolutionized the field of research in carbon fibers. CNTs have inspired extensive interests in current Nano science and nanotechnology since their discovery in 1991. Since the last decade, a number of investigations have been carried out using CNT as reinforcement in different materials, namely polymer, ceramic, and metals. A major hindrance to the successful reinforcement of metallic matrices with CNT has been the poor distribution/dispersion and agglomeration of CNTs within the metallic matrix and strong interfacial bonding between the CNTs and the matrix to ensure load transfer.[7] CNTs have recently emerged as effectual reinforcements for the aluminum (Al) matrix.[8–10] Al composites have been paid great attention due to their use in automobile and aerospace industries because of their light weight, high strength and good corrosion resistance. The development of bulk fabrication method for Al–CNTs composite was started on the basis of a mechanical milling approach.[11–12]

CNT-reinforced metal matrix can be prepared by a variety of processing techniques. Figure 14.1 shows the various processes that have been adopted for synthesis of CNT-reinforced MMC. Powder metallurgy (PM) is the most popular and widely used process technique. The main conventional blending method for preparing the CNT/metal composite powders is high-energy ball milling.[13–16] The matrix powder and the CNTs are subjected together to the impact and friction effects of the milling media. Under the harsh milling conditions, the structure and integrity of CNTs may be damaged or even destroyed.[17–18]

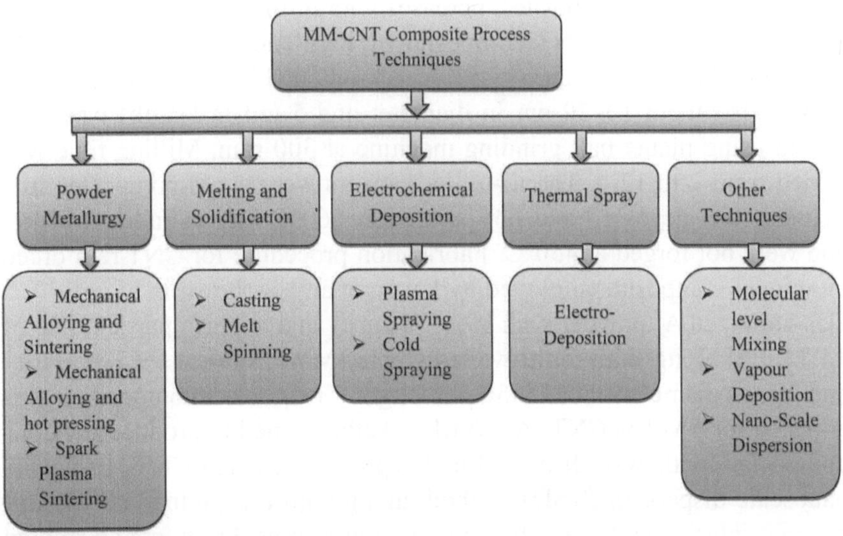


**FIGURE 14.1** Various processing routes for metal matrix-CNTs.

## 14.2 ALUMINUM NANOCOMPOSITE

Aluminum matrix based MMCs have been used as structural material in the aerospace, automotive and railway sectors. Aluminum MMCs are composed of an aluminum matrix and a reinforcement, or filler material, which exhibits excellent mechanical performance. Most of the studies on Al–CNT composites have been carried out using the PM method. The basic step of manufacturing Al–CNT composite is mixing CNTs with metal powder by mechanical alloying or grinding, followed by consolidation by compaction and sintering. The composite compacts were subjected to post-sintering process like extrusion, etc. Main focus is on obtaining high strength and reinforcement by homogeneous dispersion of CNT in the metal matrix.

Most critical issues in processing of CNT-reinforced aluminum composites are (1) dispersion of CNTs and (2) interfacial bond strength between CNT and the Al matrix. Aluminum nanocomposite reinforced with CNTs can be prepared by mechanical alloying and sintering technique. Choi et al.[19] prepared aluminum nanocomposite by dispersing multiwall CNTs (4.5 vol.%) with aluminum powder (99.5% in purity and 150 μm in average diameter) in planetary ball mill. Stearic acid (1 wt.%) was added to prevent agglomeration of the powder. Samples were prepared with varying milling speed and milling time to verify the effect of milling conditions. Density of 3.5 vol.% CNT composites increases with milling time and speed. Liu et al.[20] fabricated Al–CNT composite by high-energy ball mill and PM technique. Pure Al powder (99.5% purity, about 13 μm in diameter) and MWCNTs (about 10–20 nm in diameter and 5 μm in length) was ball-milled using planet ball grinding machine at 300 rpm. Milling time was varied from 2 to 12 h. The as-milled powers were could-compacted in a cylinder die, degassed and hot-pressed at 560°C into cylindrical billets and were hot forged at 450°C. Fabrication procedure for CNT-reinforced aluminum composite employed by Kwon et al.[21] is shown in Figure 14.2. Gas-atomized Al powder with 99.85% purity and average particle size of 14.82 μm) along with multi-walled CNTs (MWCNTs) of 99.5 % purity and 20 nm diameter with 15–50 μm length. The powder composition was adjusted for 5 vol.% CNT–Al powder. Author aimed to produce Al–CNT composites with well-dispersed and regularly oriented CNTs using the nanoscale dispersion (NSD) method, along with a combination of spark plasma sintering (SPS) and hot extrusion processes. Fre´ty et al.[22] studied the influence of the CNT concentration on the composite Young's modulus and yield strength, which leads to an estimation of the nanotubes Young's modulus and aspect ratio using analytical models using $N_2$-atomized AA5083 (Al–Mg alloy) powder and 0.2 wt.% multiwall CNTs. Liao et al.[23] studied the mechanical mixing methods, viz. high energy and low energy ball millings, and compared them to a novel polyester binder-assisted (PBA) mixing method. The Al–CNTs mixture was subsequently consolidated by PM technique. Small addition of CNTs (0.5 wt.%) in pure Al metal powder, evidently improved the tensile strength and hardness of the composite by comparing with the pure matrix. For achieving high dispersion of MWCNTs in Aluminum matrix, Singhal et al.[24] fabricated Al-matrix composites reinforced with amino-functionalized multiwalled CNTs (fCNTs). Using this method fCNTs (1.5 wt.%) were dispersed in Al powder by high energy ball milling. Al–fCNTs composites (1.5 wt.%) were fabricated by the consolidation of powders followed by sintering. Characterization of these composites revealed the formation of a thin transition layer of $Al_4C_3$ between fCNTs and Al matrix which is responsible for the load transfer from AL matrix to fCNTs. Jeyasimman et al.[25] fabricated the composite by blending microcrystalline Al 6061 metal powder with 97% pure MWCNT with various weight percentages in a two station planetary ball mill at 280 rpm for 2 h. Consolidation by compaction is performed using a double action compaction die. MWCNTs were added during the last 2 h of milling resulting in higher hardness values.

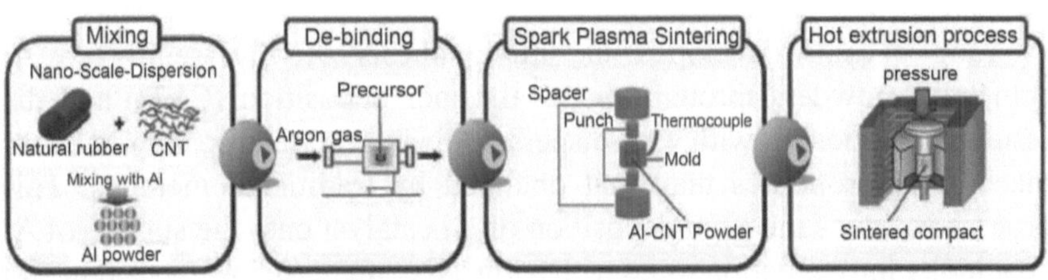


**FIGURE 14.2** Fabrication technique for Al–CNT composite.[21] (Reprinted from Hansang, K.; Mehdi, E.; Kenta, T.; Takamichi, M.; Akira, K. Combination of Hot Extrusion and Spark Plasma Sintering for Producing Carbon Nanotube Reinforced Aluminium Matrix Composites. Carbon., 47, 570–577. © 2009, with permission from Elsevier.)

Li et al.[26] used a strategy called flake powder metallurgy (flake PM) to achieve a uniform distribution of CNTs in CNT/Al composites. The process involved processing of spherical Al particles by ball milling into two-dimension (2-D) nanoflakes. Surface of Al particles were modified with a polyvinyl alcohol (PVA) hydrosol to make surface more compatible with carboxyl functionalized CNTs. Multiwall CNTs (50 nm in diameter, 2.5 μm in length) functionalized with carboxyl groups were used. It was found out that structural integrity of the CNTs was well maintained. Flake PM allows fabrication of a strong and ductile CNT/Al composite with much higher tensile strength. The microstructure of the composites prepared is shown in Figure 14.3.

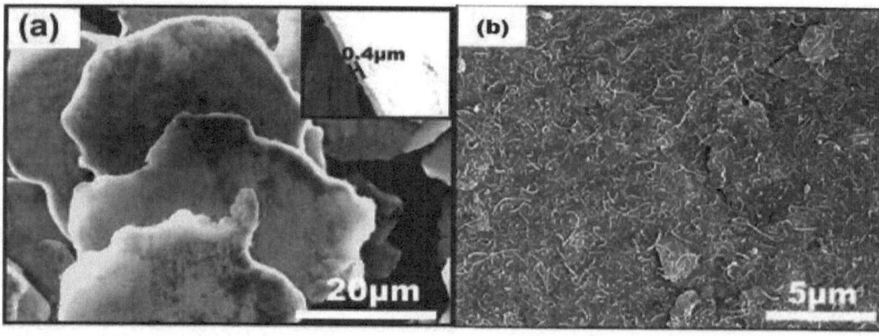


**FIGURE 14.3** (a) FESEM of as-prepared nanoflakes, inset shows the relevant thickness; (b) FESEM of CNT/Al composite powders with homogeneously distributed CNTs after PVA removing.[26] (Reprinted from Lin, J.; Zhiqiang, L.; Genlian, F.; Linlin, C.; Di, Z. The Use of Flake Powder Metallurgy to Produce Carbon Nanotube (CNT)/Aluminium Composites with a Homogenous CNT Distribution. Carbon. 50, 1993–1998. © 2012, with permission from Elsevier.)

Yang et al.[27] developed in situ synthesis of CNT-reinforced Al composite powders through chemical vapor deposition (CVD) and the resultant composites with well-dispersed CNTs in Al matrix showed better mechanical properties than that obtained by traditional methods. This process involves the even deposition of Ni catalyst onto the surface of Al powder by impregnation route with a low Ni content (0.5 wt.%) and in situ synthesis of CNTs in Al powder. The density of the composites increases with 0.5 wt.% CNT content, while it decreases with further increase of the CNT content. The compressive yield stress and elastic modulus of 1.5 wt.%-CNT/Al composites synthesized by hot extrusion are 2.2 and 3.0 times as large as that of the pure Al matrix. Cong et al.[28] developed single-walled carbon nanotube (SWNT)-reinforced nanocrystalline Al composite (SWNT/nano-Al). Nano-Al particles used were produced by an active $H_2$ plasma evaporation method in a chamber with a mixed gas of 60 pct $H_2$

and 40 pct Ar from commercial Al with a purity of 99.85%. For homogeneous dispersion of SWCNT, nano-Al particles mixed with 5.0 wt.% SWNTs were soaked in alcohol and stirred ultrasonically for 30 min and then dried. Ma et al.[29] reported a manufacturing route, combining friction stir processing and subsequent rolling processing in order to fabricate 1.5–4.5 vol.% CNT-reinforced 2009 Al composites. Powders of 2009 Al alloy is mixed with CNTs in a bi-axis rotary mixer and were cold-compacted in a cylinder die. To completely disperse the CNTs into the aluminum matrix, overlapping friction stir process is employed. Fabrication route involved in manufacturing the composite is shown in Figure 14.4.

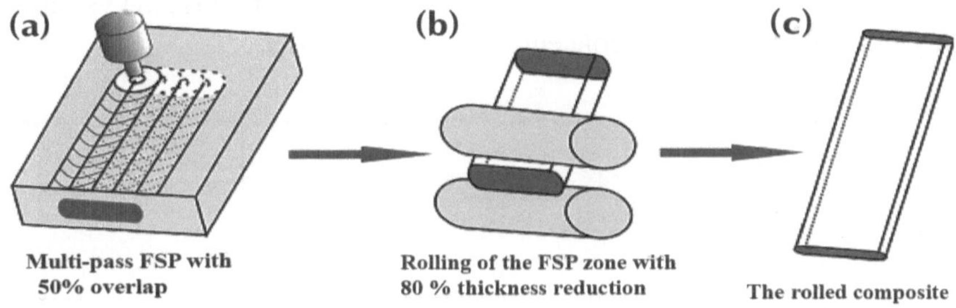


**FIGURE 14.4** Schematic of the CNT/2009Al composite fabrication flow.[29] (Reprinted from Liu, Z. Y.; Xiao, B. L.; Wang, W. G. Ma, Z. Y. Developing High-Performance Aluminium Matrix Composites with Directionally Aligned Carbon Nanotubes by Combining Friction Stir Processing and Subsequent Rolling. Carbon. 62, 35–42. © 2013, with permission from Elsevier.)

FSP-rolled composites with directionally aligned CNTs exhibited much higher strength, ductility, and modulus values as compared to FSP composites with randomly distributed CNTs. Kurita et al.[30] prepared CNT–Al matrix composite powders with uniform multi-walled CNT dispersion by using hetero-agglomeration principles, and fabricated fully dense 1.0 vol.% CNT–Al matrix composites by using SPS and a subsequent hot extrusion process. Their study suggested that the tensile improvement achieved in extruded/SPS composites is directly caused by an effective load transfer at the CNT–Al interface, because no evidence of Al work hardening or the formation of interfacial $Al_4C_3$ crystals was detected. Bakshi et al.[31] showed the macro- and nano-scale mechanical and wear properties of CNT-reinforced Al–Si composite coatings prepared by plasma spraying. The differences in mechanical and wear properties at nano and macro scales are explained in terms of the bimodal CNT dispersion (well dispersed and clusters) in Al–Si matrix, fraction and carbide formation and CNT cluster size. They[31] concluded that there could be a critical CNT cluster size fraction which would lead to the best macroscale wear resistance of these coatings. P′erez-Bustamante et al.[32] developed novel Al-based nanocomposites by mechanical milling followed by pressure-less sintering at 823 K under vacuum. MWCNTs showed high mechanical and chemical stability. The yield stress ($\sigma_y$), maximum strength ($\sigma_{max}$), and hardness values obtained for the nanocomposites were considerably higher than those for pure Al (not milled, and milled and sintered). They reported that hardness values increases as the milling time and MWCNT concentration is increased. Figure 14.5 reports the variation of yield strength with the concentration of MWCNTs.

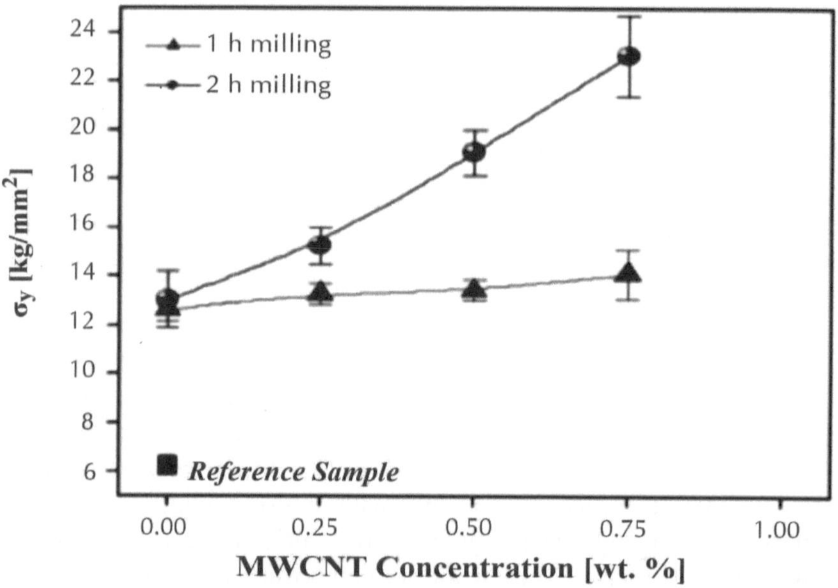


**FIGURE 14.5** Yield strength of Al-nanocomposite as a function of MWCNT concentration.[32] (Reprinted from Pérez-Bustamante, R.; Estrada-Guel, I.; Antúnez-Flores, W.; Miki-Yoshida, M.;Ferreira, P. J.; Martínez-Sánchez, R. Novel Al-Matrix Nanocomposites Reinforced with Multi-Walled Carbon Nanotubes. J. Alloys Compound. 450, 323–326. © 2008, with permission from Elsevier.)

Maqbool et al.[33] fabricated the nanocomposite with gas-atomized Al powder (99.84% pure, with 10 μm average particle size) and multi-walled

CNTs (purity > 90%, diameter 10–30 nm). For the uniform distribution of CNTs, electrolessplating, and ultrasonic assisted mixing with $N_2$ purging were utilized. In order to enhance the interface bonding of Al matrix and CNTs, a Cu/CNT precursor was prepared by using an electrolessplating process and then mixed by ball milling with Al powder. Due to strong interfacial bonding of Cu-coated CNTs with Al matrix, micro-hardness of hot rolled Cu-coated CNT/Al composites is more effective at 80% reduction in thickness. Laha et al.[34] employed composite fabrication route using thermal spray forming. Gas-atomized and spherical hypereutectic Al–23 wt.% Si alloy powder, 15–45 μm, was selected as the matrix. For the reinforcement in the synthesis of the MMC, 10 wt.% MWCNTs (95% purity, 40–70 nm diameter, 0.5–2.0 μm length) were used. Synthesis of the composite was carried out by two different thermal spray methods, namely, plasma spray forming (PSF) and HVOF spraying followed by Al–Si powder and MWCNTs blended and mixed in a ball mill for 48 h in order to achieve homogeneous mixing. Improvement in the wettability and the interfacial adhesion between the MWCNT reinforcement and the Al–Si matrix is reported due to the formation of b-SiC. Javadi et al.[35] successfully fabricated the MWCNT-reinforced Al composite by mechanical alloying route. MWCNT was synthesized via CVD. The Al powders (purity 99%, 200 mesh) were irregular in shape with a flake morphology. MWCNT-Al powder mixtures were prepared by a mixing process that involved an intensive sonication in ethanol. Mechanical alloying was carried out for 24 h. Finally, the ball-milled powder mixtures were compacted in a cylindrical-diameter compaction dye. Figure 14.6 shows the microstructure of Al-MWCNT composite. The MWCNTs were homogeneously dispersed and rarely tangled together after sonication.

Lipecka et al.[36] developed the composite by mechanical alloying route. Two 7XXX aluminum alloy (S790) containing 3% of CNTs were used. Al powder and CNTs were first high energy milled, axial cold compacted to form a billet and finally extruded into a rod-shape at 370°C. The samples were annealed at various temperatures ranging from 100 to 550°C for 10 min in order to evaluate their thermal stability and mechanical property. Balani et al.[37] successfully grown in situ CNTs onto $Al_2O_3$ particles via the catalytic chemical vapor deposition (CCVD) route (Fig. 14.7). Consequently, plasma spraying of the $Al_2O_3$ is 0.5 wt.% in situ grown CNT powder induced splat-reinforcement by CNT bridging. It was concluded that enhancement of mechanical properties is attributed to the presence of strong chemical bonding between CNTs and $Al_2O_3$.

**FIGURE 14.6** FE-SEM image of the Al–CNT composite after sonication.[35] (Reprinted from Javadi, A. H.; Sh. Mirdamadi, Faghihisani, M. A.; Shakhesi, S.; Soltani, R. Fabrication of Well-Dispersed, Multiwalled Carbon Nanotubes-Reinforced Aluminium Matrix Composites. New Carbon Mater. 27 (3), 161–165. © 2012, with permission from Elsevier.)

**FIGURE 14.7** SEM images of $Al_2O_3$ particles (a) and (b) in situ CNTs grown over $Al_2O_3$ particles.[37] (Reprinted from Balani, K.; Zhang, T.; Karakoti, A.; Li, W. Z.; Seal, S.; Agarwal, A. In Situ Carbon Nanotube Reinforcements in a Plasma-Sprayed Aluminium Oxide Nanocomposite Coating. Acta Materialia. 56, 571–579. © 2008, with permission from Elsevier.)

## 14.3 MECHANICAL PROPERTIES OF AL–CNT COMPOSITES

Bradbury et al.[38] showed that up to 9 wt.% of MW-CNTs can be incorporated into an Al matrix by planetary milling. The hardness reaches a maximum at HV equal to 151 for 6 wt.% MW-CNTs. As far as the authors know, this is among the highest wt.% of MW-CNTs and hardness values for pure Al matrix using milling to disperse the MW-CNTs. Some CNTs were observed linking the crack surfaces together, signifying a load transfer mechanism and a good interfacial bonding between the matrix and the reinforcement materials. A maximum of 262% increase in tensile strength is reported with 3%-CNT concentration. Researchers have tried to incorporate 0.5–4.5 wt.% CNT into Al matrix by PM routes[19–20,23–26,32,35,36,38] and Figure 14.8a shows a comparative analysis of variation in tensile strength with CNTs concentration. On the other hand, Javadi et al.[35] reported the macro hardness of sintered MWCNT-Al composite was about 48 HV, whereas that of pure Al was 22 HV. Figure 14.8b describes the effect of MWCNTs concentration on Hardness of Al–CNT composites. Li et al.[26] fabricated 0.5 vol.% CNT-reinforced Al composites which showed Young's modulus of 75 GPa and tensile stress of 312 MPa. While, a 2 vol.% CNT-reinforced Al composite exhibited Young's modulus of 89 GPa and tensile stress of 435 MPa, which is almost two times higher than that of unreinforced Al matrix. Singhal et al.[24] reported an average compressive strength of 270 ± 20 MPa for Al composites loaded with 1.5 wt.% fCNTs. The average compressive strength of these composites was found to be around 250 MPa. Microhardness of the composite were found to be Hv = 100 ± 10 kg/mm$^2$. Ma et al.[29] fabricated 3 vol.% CNT/aluminum composites based on Friction Stir Processing and hot plastic deformation route which exhibited an ultimate tensile strength of 600 MPa and elongation of 10%. Bakshi et al.[31] conducted micro- and macro-scale testing on Al–Si–5% CNT composite. Micro/nano-indentation testing indicated an increase in the elastic modulus by 19 and 39% and an increase in the yield strength by 17.5 and 27% by the addition of 5 and 10 wt.% CNTs, respectively. Macro-scale compression tests indicated no improvement in the elastic modulus but an increases in the compressive yield strength by 27 and 77%, respectively, by addition of 5 and 10 wt.% CNTs. Fractured surface of Al-Si-5% CNTs and 10% CNTs is shown in Figure 14.9. Maqbool et al.[33] reported that in 1.0 wt.% uncoated and Cu-coated CNT/Al composites, compared to pure Al, the microhardness increased by 44 and 103%, respectively. Similarly in 1.0 wt.% uncoated CNT/Al composite,

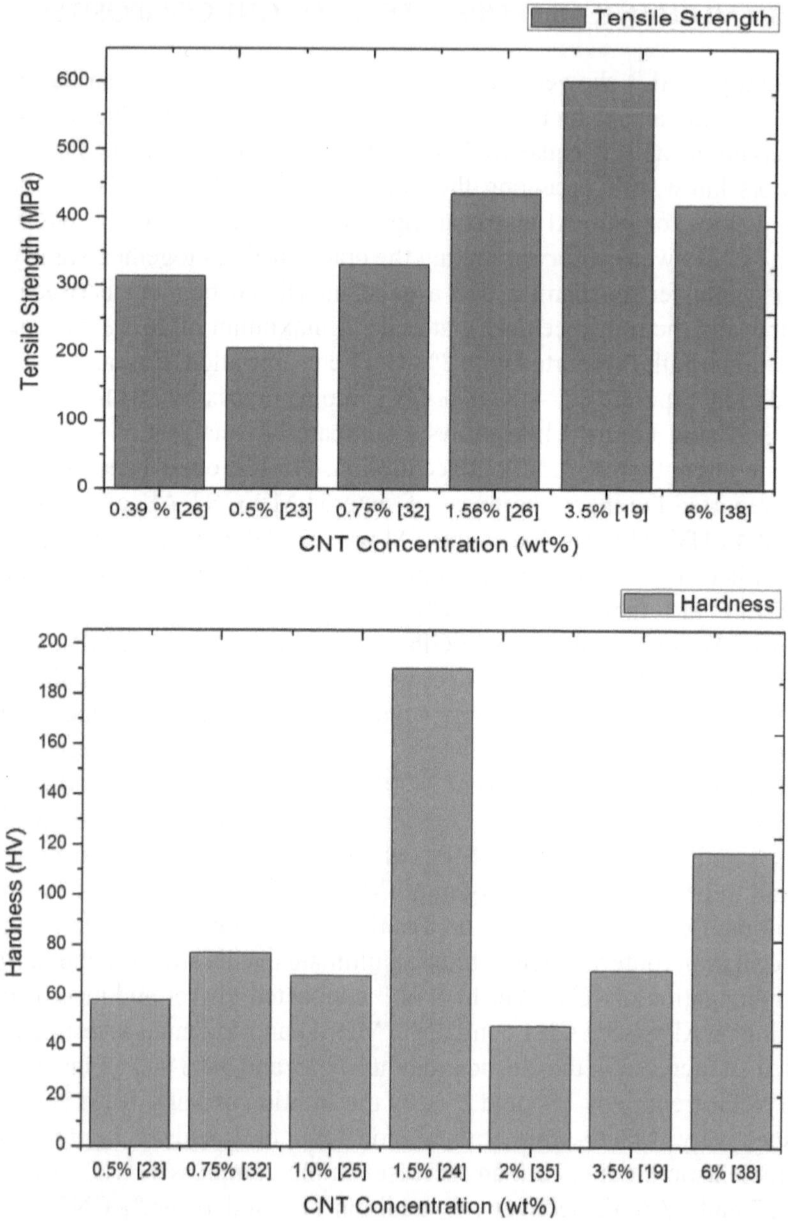


**FIGURE 14.8** Comparative assessment of (a) variation in tensile strength with CNT concentration and (b) effect of CNT concentration on hardness of Al/CNTs composite reported by researchers.

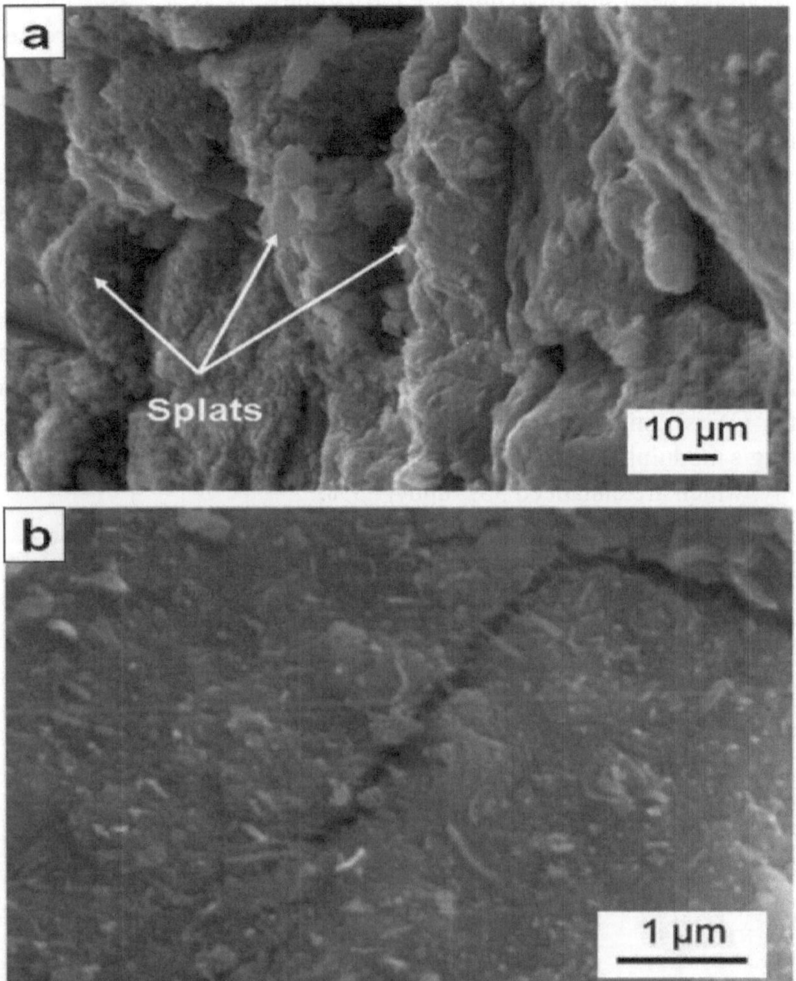


**FIGURE 14.9** Fractured surface of (a) Al-Si-5 wt.% CNT and (b) Al-Si-10 wt.% CNT composite.[31] (Reprinted from Srinivasa, R.; Bakshia Anup, K.; Keshri Arvind, A. A Comparison of Mechanicaland Wear Properties of Plasma Sprayed Carbon Nanotube Reinforced Aluminium Composites at Nano and Macro Scale. Mater. Sci. Eng. A528, 3375–3384. © 2011, with permission from Elsevier.)

increase in yield and ultimate tensile strength is 58 and 62%, respectively. For Cu-coated 1.0% CNT/Al composite, increment in yield and ultimate tensile strength is about 121 and 107%, respectively. CNT reinforcement to composite coatings prepared by Laha et al.[39] using thermal spraying methods has been shown to improve the hardness by 72%, elastic modulus by 78%, marginal improvement in tensile strength and 46% decrease in ductility with 10 wt.% CNT content. Anglaret et al.[40] reported increments in young's modulus, yield strength and tensile strength by 3, 27, and 18%, respectively. This enhancement in strength was explained to be due to the reinforcement effect of the CNTs, and in particular, to the formation of aluminum carbide phases due to the high process temperature. Stress-strain curve for Al-1.5 wt.% CNTs is shown in Figure 14.10.

CNT (1.0 wt.%) reinforced aluminum matrix composites were fabricated by isostatic pressing followed hot extrusion techniques by Deng et al.[41] Mechanical properties of composites were affected by nanotubes content. The tensile strength and Young's modulus reaches the maximum when CNTs content is 1.0 wt.%. The maxima of tensile strength and Young's modulus of the composite are 521.7 MPa and 102.2 GPa, respectively; which are enhanced 35.7 and 41.3%.

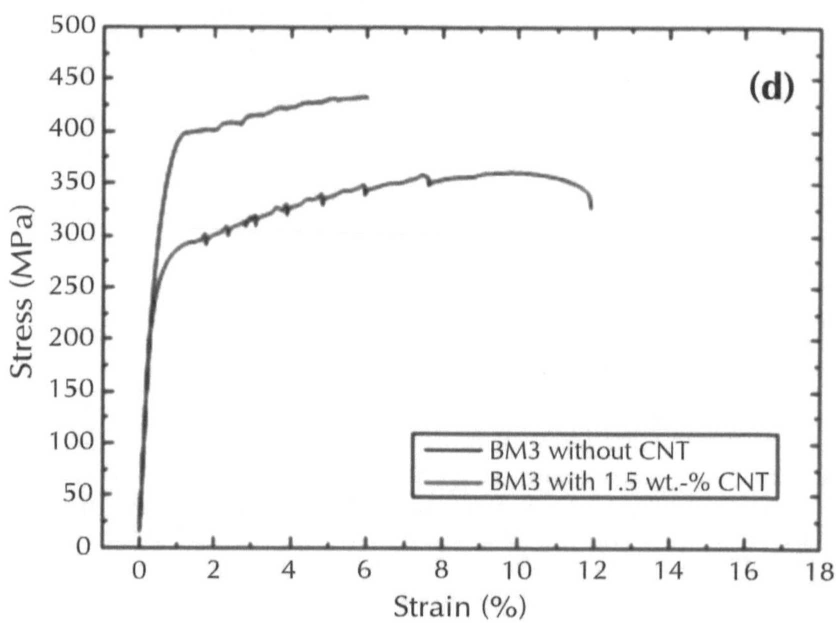


**FIGURE 14.10** Stress-strain curve for Al-1.5 wt.% CNT composite.[40] (Reprinted from Julien, S.; Blanka, L.; Nicole F.; Eric, A. Mechanical Reinforcement of High-Performance Aluminium Alloy AA5083 with Homogeneously Dispersed Multi-Walled Carbon Nanotubes. Carbon. 50, 2264–2272. © 2012, with permission from Elsevier.)

## 14.4 CONCLUSION

In conclusion, much advancement has been made over the last few years in CNT-reinforced aluminum metal matrix. The aluminum MMCs reinforced with nano particles shown significantly better properties than composites incorporated with micro particles. The mechanical alloying process has the important effect on uniform distribution of nanoparticles in metal matrix. We have seen that the conventional routes of fabrications such as PM, mechanical alloying and sintering, extrusion, hot pressing and electro- and electro-less deposition have been applied for synthesizing CNT-reinforced aluminum nanocomposites with success. In order to enhance the dispersion of CNTs in metal matrix, other noble methods like nano-scale dispersion and plasma spark sintering is employed. CNTs containing aluminum matrix composites are being developed and projected for various applications in distinct fields of engineering ranging from tribological coatings to functional applications, such as thermal management in electronic packaging or structural applications, such as load-bearing members and hydrogen storage and catalysis. Among all the processing techniques discussed above, PM is the most cost-effective and captive process. This technique has emerged as promising route for the fabrication of CNT-reinforced MMC and can be used to develop a new nano-composite material without sacrificing any functional properties.

## KEYWORDS

- **aluminum nanocomposites**
- **mechanical behavior**
- **carbon nanotubes**
- **metal matrix**
- **aluminum matrix**

## REFERENCES


1. Bakshi, S. R.; Lahiri, D.; Agarwal, A. Carbon Nanotube Reinforced Metal Matrix Composites – A Review. *Int. Mater. Rev.* **2010,** *55,* 41–64.
2. Kelly, A. *J. Mater. Sci.* **2006,** *41,* 905–912.
3. Rohatgi, P. *JOM.* **1991,** *43,* 10–15.
4. Santhosh Kumarand, S.; Somashekhar S., Hiremath. Microstructure and Mechanical Behavior of Nanoparticles Reinforced Metal Matrix Composites – A Review. *Appl. Mech. Mater.* **2014,** *939,* 592–594.
5. Tjong, S. C.; Ma, Z. Y. Microstructural and Mechanical Characteristics of In Situ Metal Matrix Composites. *Mater. Sci. Eng. R.* **2000,** *29,* 49–113.
6. Iijima, S. *Nature.* **1991,** *354,* 56–58.
7. Morsi, K., et al. Spark Plasma Extrusion (SPE) of Ball-Milled Aluminum and Carbon Nanotube Reinforced Aluminum Composite Powders. *Compos. Part A.* **2010,** *41,* 322–326.
8. Ci, L.; Zhenyu, R.; Jin-Phillipp, N. Y.; Rühle, M. Investigation of the Interfacial Reaction between Multi-Walled Carbon Nanotubes and Aluminium. *Acta Mater.* **2006,** *54* (20), 5367–5375.
9. Esawi, A. M. K.; El Borady, M. A. Carbon Nanotube-Reinforced Aluminium Strips. *Compos. Sci. Technol.* **2008,** *68* (2), 486–492.
10. Morsi, K.; Esawi, A. Effect of Mechanical Alloying Time and Carbon Nanotube (CNT) Content on the Evolution of Aluminum (Al)–CNT Composite Powders. *J. Mater. Sci.* **2007,** *42,* 4954–4959.
11. George, R.; Kashyap, K. T.; Rahul, R.; Yamadagni, S. Strengthening in Carbon Nanotubes/Aluminium (CNT/Al) Composites. *Scripta. Materialia.* **2005,** *53,* 1159–1163.
12. Kuzumaki, T.; Miyazawa, K.; Ichnose, H.; Ito, K. Processing of Carbon Nanotubes Aluminium Composite. *J. Mater. Res.* **1998,** *9,* 2445–2449.
13. Choi, H. J.; Kwon, G. B.; Lee, G. Y.; Bae, D. H. Reinforcement with Carbon Nanotubes in Aluminium Matrix Composites. *Scripta. Mater.* **2008,** *59,* 360–363.
14. Deng, C. F.; Zhang, X. X.; Wang, D. Z.; Lin, Q.; Li, A. B. Preparation and Characterization of Carbon Nanotubes/Aluminium Matrix Composites. *Mater. Lett.* **2007,** *61,* 1725–1728.
15. George, R.; Kashyap, K. T.; Rahul, R.; Yamdagni, S. Strengthening in Carbon Nanotube/Aluminium (CNT/Al) Composites. *Scripta. Mater.* **2005,** *53,* 1159–1163.
16. Pérez-Bustamante, R.; Gómez-Esparza, C. D.; Estrada-Guel, I.; Miki-Yoshida, M.; Licea-Jiménez, L.; Pérez-García, S. A. Microstructural and Mechanical Characterization of Al–MWCNT Composites Produced by Mechanical Milling. *Mater. Sci. Eng. A.* **2009,** *502,* 159–613.
17. Poirier, D.; Gauvin, R.; Drew, R. A. L. Structural Characterization of a Mechanically Milled Carbon Nanotube/Aluminium Mixture. *Compos. Part A.* **2009,** *40,* 1482–1489.
18. Tucho, W. M.; Mauroy, H.; Walmsley, J. C.; Deledda, S.; Holmestad, R.; Hauback, B. C. The Effects of Ball Milling Intensity on Morphology of Multiwall Carbon Nanotubes. *Scripta. Mater.* **2010,** *63,* 637–640.
19. Choi, H. J.; Shin, J. H.; Bae, D. H. The Effect of Milling Conditions on Microstructures and Mechanical Properties of Al/MWCNT Composites. *Compos. Part A.* **2012,** *43,* 1061–1072.

20. Liu, Z. Y.; Xu, S. J.; Xiao, B. L.; Xue, P.; Wang, W. G.; Ma, Z. Y. Effect of Ball-Milling Time on Mechanical Properties of Carbon Nanotubes Reinforced Aluminium Matrix Composites. *Compos. Part A.* **2012,** *43,* 2161–2168.
21. Hansang, K.; Mehdi, E.; Kenta, T.; Takamichi, M.; Akira, K. Combination of Hot Extrusion and Spark Plasma Sintering for Producing Carbon Nanotube Reinforced Aluminium Matrix Composites. *Carbon.* **2009,** *47,* 570–577.
22. Nicole, F.; Julien, S.; Blanka, L.; Eric, A. Influence of the Concentration and Nature of Carbon Nanotubes on the Mechanical Properties of AA5083 Aluminium Alloy Matrix Composites. *Carbon.* **2014,**
23. Jinzhi, L.; Ming-Jen, T. Mixing of Carbon Nanotubes (CNTs) and Aluminium Powder for Powder Metallurgy Use. *Powder Technol.* **2011,** *208,* 42–48.
24. Singhal, S. K.; Renu, P.; Mamta, J.; Rajiv, C.; Satish, T.; Mathur, R. B. Carbon Nanotubes: Amino Functionalization and its Application in the Fabrication of Al-Matrix Composites. *Powder Technol.* **2012,** *215–216,* 254–263.
25. Jeyasimman, D.; Sivaprasad, K.; Sivasankaran, S.; Narayanasamy, R. Fabrication and Consolidation Behaviour of Al 6061 Nanocomposite Powders Reinforced by Multi-Walled Carbon Nanotubes. *Powder Technol.* **2014,** *258,* 189–197.
26. Lin, J.; Zhiqiang, L.; Genlian, F.; Linlin, C.; Di, Z. The Use of Flake Powder Metallurgy to Produce Carbon Nanotube (CNT)/Aluminium Composites with a Homogenous CNT Distribution. *Carbon.* **2012,** *50,* 1993–1998.
27. Xudong, Y.; Chunsheng, S.; Chunnian, H.; Enzuo, L.; Jiajun, L.; Naiqin, Z. Synthesis of Uniformly Dispersed Carbon Nanotube Reinforcement in Al Powder for Preparing Reinforced Al Composites. *Compos. Part A.* **2011,** *42,* 1833–1839.
28. Cong, et al. Fabrication of Nano-Al Based Composites Reinforced by Single-Walled Carbon Nanotubes. *Carbon,* **2003,** *41,* CO1–851.
29. Liu, Z. Y.; Xiao, B. L.; Wang, W. G. Ma, Z. Y. Developing High-Performance Aluminium Matrix Composites with Directionally Aligned Carbon Nanotubes by Combining Friction Stir Processing and Subsequent Rolling. *Carbon.* **2013,** *62,* 35–42.
30. Kurita, H.; Kwon, H.; Estili, M.; Kawasaki, A. Multi-Walled Carbon Nanotube–Aluminium Matrix Composites Prepared by Combination of Hetero–Agglomeration Method, Spark Plasma Sintering and Hot Extrusion. *Mater. Trans.* **2011,** *52* (10), 1960–1965.
31. Srinivasa, R.; Bakshia Anup, K.; Keshri Arvind, A. A Comparison of Mechanical and Wear Properties of Plasma Sprayed Carbon Nanotube Reinforced Aluminium Composites at Nano and Macro Scale. *Mater. Sci. Eng.* **2011,** *A528*, 3375–3384.
32. Pérez-Bustamante, R.; Estrada-Guel, I.; Antúnez-Flores, W.; Miki-Yoshida, M.; Ferreira, P. J.; Martínez-Sánchez, R. Novel Al-Matrix Nanocomposites Reinforced with Multi-Walled Carbon Nanotubes. *J. Alloys Compound.* **2008,** *450,* 323–326.
33. Adnan Maqboola, M.; Asif Hussaina, F.; Ahmad, K.; Nabi, B.; Ali Hussain, Myong Ho, K. Mechanical Characterization of Copper Coated Carbon Nanotubes Reinforced Aluminium Matrix Composites. *Mater. Character.* **2013,** *86,* 39–48.
34. Laha, T.; Kuchibhatla, S.; Seal, S.; Li, W.; Agarwal, A. Interfacial Phenomena in Thermally Sprayed Multiwalled Carbon Nanotube Reinforced Aluminium Nanocomposite. *Acta Materialia.* **2007,** *55,* 1059–1066.

35. Javadi, A. H.; Sh. Mirdamadi, Faghihisani, M. A.; Shakhesi, S.; Soltani, R. Fabrication of Well-Dispersed, Multiwalled Carbon Nanotubes-Reinforced Aluminium Matrix Composites. *New Carbon Mater.* **2012,** *27* (3), 161–165.
36. Joanna L.; Mariusz, A.; Małgorzata, L.; Jolanta, Janczak-Rusch.; Krzysztof, J. Kurzydłowski, Evaluation of Thermal Stability of Ultrafine Grained Aluminium Matrixcomposites Reinforced with Carbon Nanotubes. *Compos. Sci. Technol.* **2011,** *71,* 1881–1885.
37. Balani, K.; Zhang, T.; Karakoti, A.; Li, W. Z.; Seal, S.; Agarwal, A. In Situ Carbon Nanotube Reinforcements in a Plasma-Sprayed Aluminium Oxide Nanocomposite Coating. *Acta Materialia.* **2008,** *56,* 571–579.
38. Christopher R. B.; Jaana-Kateriina, G.; LauriKollo, Hansang, K.; Marc, L. Hardness of Multi Wall Carbon Nanotubes Reinforced Aluminium Matrix Composites. *J. Alloys Compound.* **2014,** *585,* 362–367.
39. Laha, T.; Agarwal, A.; McKechnie, T.; Seal, S. *Mater. Sci. Eng. A.* **2004,** *A381,* 249–258.
40. Julien, S.; Blanka, L.; Nicole F.; Eric, A. Mechanical Reinforcement of High-Performance Aluminium Alloy AA5083 with Homogeneously Dispersed Multi-Walled Carbon Nanotubes. *Carbon.* **2012,** *50,* 2264–2272.
41. Deng, C. F.; Wang, D. Z.; Zhang, X. X.; Li. A. B. Processing and Properties of Carbon Nanotubes Reinforced Aluminium Composites. *Mater. Sci. Eng. A.* **2007,** *444,* 138–145.

# INDEX

## P



## Q



## R